The potential for weeds to affect the livelihoods of agricultural producers and to reduce endemic biodiversity is widely recognised. But what are the ecological attributes of weeds that confer this ability to interfere with human activities?

Until recently, the discipline of weed science has developed almost exclusively as an empirical subject with an emphasis on optimising the performance of herbicides. Roger Cousens and Martin Mortimer place weed management within an ecological context, in which the focus is on the manipulation of population size. The dynamics of abundance and spatial distribution are considered at both geographic and local scales. The basic processes of weed dispersal, reproduction and mortality are described, together with the factors that influence them. Management is shown to modify patterns of behaviour that are intrinsic to populations. This is done with the aid of simple models and an extensive review of the literature. Special attention is given to the evolution and management of resistance to herbicides.

This book therefore provides weed scientists with a conceptual framework. It also gives ecologists access to the extensive database on the population ecology of weeds.

Dynamics of weed populations

Dynamics of
weed populations

ROGER COUSENS
Department of Agriculture
Western Australia

and

MARTIN MORTIMER
Department of Environmental and Evolutionary Biology
University of Liverpool, UK

CAMBRIDGE
UNIVERSITY PRESS

Published by the Press Syndicate of the University of Cambridge
The Pitt Building, Trumpington Street, Cambridge CB2 1RP
40 West 20th Street, New York, NY 10011–4211, USA
10 Stamford Road, Oakleigh, Melbourne 3166, Australia

First published 1995

Printed in Great Britain at the University Press, Cambridge

A catalogue record for this book is available from the British Library

Library of Congress cataloguing in publication data
Cousens, Roger.
　　Dynamics of weed populations / Roger Cousens and Martin Mortimer.
　　　　p.　　cm.
　　Includes bibliographical references (p.　　) and index.
　　ISBN 0 521 49649 7. – ISBN 0 521 49969 0 (pbk.)
　　1. Weeds – Ecology.　2. Vegetation dynamics.　3. Weeds – Control.
I. Mortimer, Martin.　II. Title.
SB611.C67　1995
632′.58–dc20　95-949　CIP

ISBN 0 521 49649 7 hardback
ISBN 0 521 49969 0 paperback

SE

Contents

Preface

Traditionally, the task of increasing crop productivity has been seen as the role of crop breeders and agronomists, drawing respectively upon the sciences of plant genetics and plant physiology. From an agricultural science point of view, weeds of crops have been seen simply as a nuisance to be removed using the best available technology to hand. Much of weed science has therefore been devoted to the provision of 'tools' for weed removal. Only in relatively recent times has weed management been viewed as a problem to which ecological principles can be applied.

For historical reasons, then, the study of weeds has been divided between two groups of scientists. Agronomists and horticulturalists have seen the presence of weeds as a pragmatic problem to be solved (how can the weeds be killed?). Plant ecologists, on the other hand, have seen cropped land as somewhat unnatural, human-managed habitats and weeds as particular organisms that are able to exploit such habitats (i.e. as academic curiosities). Books considering weeds have followed this dichotomy, although there has been more emphasis on the practice of weed control than on weed ecology. Indeed, despite the fact that much of the temperate regions and an increasing area of the tropical regions of the world are farmed or managed in some way, it is surprising that plant ecologists have given agroecosystems relatively so little attention. As with other areas of pest control, the agroecosystem–natural ecosystem dichotomy has persisted to the detriment of weed science. However, with increasing concern over the preservation of biodiversity, many ecologists are becoming interested in alien invasions threatening more natural habitats ('environmental weeds'). In this book, we are concerned with the ecology of weeds and the understanding this brings to the management of weeds. Our aim is to consider ecological principles and the experience of agriculturalists, and to examine the management of weed populations and the implications of changes in management practices.

Currently, we believe there are many weed scientists (and agronomists) who would like to know more about ecological methods and theory, but who have previously considered plant ecology as an academic preserve and who have recoiled from the conceptual approaches used by ecologists. Although many weed scientists will have worked on aspects of weed ecology, they have perhaps not been aware of the broader significance of their work. On the other hand, there are many ecologists who are now interested in applied ecology and more specifically in agro-ecology, but who find the traditional 'production agriculture' view taken by most agriculturalists to be a well-worked but narrow approach. The weed science literature includes many demonstrations of the application of plant ecological research, particularly in the use of models. Equally, theoretical ecology provides a rigorous framework within which to place the extensive collection of observations on weeds which are derived from empirical experimentation. It is this interface that we have addressed in this book and we tread the precarious pathway of trying to interlink two historically separate areas of plant science. We hope therefore that both plant ecologists and weed science practitioners will find interests in common in the text. In tracing this pathway we have inevitably been selective and no doubt in consequence neither interest group will be fully satisfied!

The book is concerned with approaches and understanding. It is not a review of data on particular weeds in specific countries. The approaches are applicable to weeds in all countries. We try to draw on the best examples from a range of locations; wherever possible we try to be international. However, some countries have dominated research in certain areas, such as Australia in weed invasion studies, and the UK in modelling. We trust the more parochial will view this as a challenge and not as a fault with our coverage of the data. Inevitably, we will miss out some studies regarded as pre-eminent in some countries, but hopefully the alternative examples which we have chosen will adequately illustrate the principles.

Although our focus is clearly on the weeds of agricultural systems, the principles are applicable to any situation where plant populations have to be managed, such as plants of conservation value in 'natural' habitats. For weeds, habitats need to be managed so that the unwanted species decline, whereas for endangered species we need to ensure their continued existence or to cause them to increase. Although the aims will differ, the ecological principles will be the same. This book can therefore be considered as a book on the application of plant population dynamics in general, but with its examples drawn from studies of weeds.

There have been other books on plant populations. They have been written largely from the point of view of understanding the biology and ecology of populations or the implications for natural selection and evolution, and less from the point of view of population management. In such books, population *dynamics* is treated as just one of the inevitable outcomes of the ecological processes and is of no special significance. However, when we are interested in population management, the focus changes fundamentally: the dynamics is now central to our considerations. What makes populations increase; what makes them decrease; what will their ultimate levels be; how will those levels be reached? Our principal aim in this book is to establish a framework of interlocking components, which together help us to predict and to manage the sizes of populations.

Scientific names of species have been used throughout, to avoid confusion between common names in different countries. However, even this poses problems since synonyms may also be used in different countries. For example, *Brassica kaber* is used in North America to denote the species mostly called *Sinapis arvensis* elsewhere. Rather than make a taxonomic revision of all species, we have tended to use the name given in the cited references.

We are indebted to a number of people for reading and commenting on earlier drafts and parts of individual chapters, in particular Bruce Auld, Diane Benoit, Aik-Hock Cheam, Jonathan Dodd, Gurjeet Gill, Peter Gould, Linda Hall, Richard Medd, Peter Michael, Stephen Moss, Dane Panetta, Stephen Powles, Philip Putwain and Peter Thomson. Of course, we have not necessarily incorporated all of their suggestions, nor satisfactorily answered all of their criticisms! But the book has been very much improved through their efforts. We are also grateful to Lindsay Campbell for providing an e-mail 'gateway', and to Jane, Sue, Christopher, Ruth and Hugh for being so patient with us over the last few years.

RDC and AMM

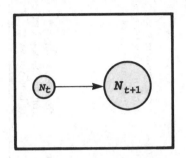

1

Weed population dynamics – the framework

Why is it that weeds have, over the centuries, consumed so much of the attention of farmers and, more recently, conservationists and scientists? A pragmatic answer to this question would probably involve the potential for weeds to affect the livelihoods of agricultural producers, or to reduce biodiversity. But what are the ecological attributes of weeds that confer this ability to interfere with our activities and to occupy our attention? The cosmopolitan occurrence of many weed species often bears witness to three common characteristics. Their often close association with human-managed habitats affords them a powerful *means of dispersal*. Their ability to increase rapidly in abundance after introduction into a habitat and potentially to dominate a plant community confers *colonising potential*. Thirdly, their ability to tolerate a wide range of habitat conditions and to reproduce ensures long term *regenerative capacity*. These, then, are some of the key features tending to ensure the success of weeds and to elevate them to human attention: the abilities to invade, to dominate and to persist.

No one definition of the term *weed* has been coined to universal satisfaction, but most definitions are anthropocentric. Weeds are commonly vascular plants which are natural impediments to human activities or health, or which are considered to cause unacceptable changes to natural plant communities. The subject of this book is weed population dynamics, namely the changes that occur in the abundance, distribution and genetic structure of populations of weed species. Changes in the abundance of agricultural weeds have been observed throughout history (van der Zweep, 1982) and, by pollen analysis, have been documented from pre-historic times. Such observations raise a number of key questions. For example, why does the abundance of one species change dramatically from one year to the next, while another species remains constant? Why is it that one species, in the course of a few years, escalates to be a major problem over a whole region, while another species declines to extinction?

Table 1.1. *The impact of weeds on loss in a range of land use systems. Illustrative sources of loss are given*

Category of loss	Arable & horticultural crops	Grasslands	Forestry	Amenity/nature reserves
Quantity Commodity yield reduction	Loss of crop yield through interference, weed parasitism	Herbage loss through interference	Yield loss in nursery & mature plantations	Loss of desirable species through long-term competitive exclusion/succession
	Alternate hosts to plant pathogens, viruses and insect pests	Alternate hosts to plant pathogens, viruses and insect pests	Alternate hosts to plant pathogens, viruses and insect pest	
		Loss of grazing land through invasion		
Production inefficiency	Clean seed bed preparation	Reduced herbage utilisation, palatability & digestibility leading to lowered efficiency of conversion to grazer biomass	Scrub clearance for new plantation	Interference with commercial activities
	Increased time and labour of harvesting	Physical damage to animals	Interference with logging operations	Time spent on weed control rather than other conservation tasks
	Crop damage in the application of weed control agents			
Quality Reduction in commodity value	Smaller size, poorer appearance & crop contamination	Poisoning of stock	Reduced appearance & vigour	Reduced amenity/conservation value; loss of biodiversity
		Faults in hides/fleeces		

Source: After Mortimer (1990).

Our aim is to analyse the factors that regulate weed populations in the context of their ecology and evolution in response to control measures. We will then examine the relevance of that information to weed management. The setting is mainly agricultural, but the approach is ecological rather than agronomic. We will be concerned as much with *retrospective analysis* (why have some plants become weeds?) as with *forecasting* (which plants are likely to become weeds and at what level of abundance; how can we make them decline in abundance?). In this introductory chapter we discuss the reasons for studying the population dynamics of weeds and our conceptual approach to the subject.

The impact of weeds

The hazards arising from the presence of weeds in any production system involving plants may be viewed at two levels – damage to absolute yield of the product and damage to the value of that product. These are the quantity and quality damage functions of the economist (Norton & Conway, 1977; Mortimer, 1984) that in some manner relate weed abundance to loss of financial income. A damage analysis may be applied to business activities as diverse as farming, running a railway or managing a nature reserve, although the actual yields are of a very different type! Table 1.1 simplistically describes some of the kinds of damage that weeds may cause, in four land-use systems (Mortimer, 1990). This type of analysis can most readily be made in agriculture and horticulture but can, with more difficulty, be extended to an amenity and conservation context. Typically, crop yield loss will bear a proportional relationship to the abundance of weeds and the duration and severity of interference* with the crop (Cousens, 1985a). The commodity price of a cereal crop may relate to the degree of grain contamination by weed seeds (Cousens *et al.*, 1985a). In grasslands, pasture productivity as measured by accumulated grazer biomass will reflect recent plant community composition and the extent to which the pasture has been improved in relation to the food-plant preferences of the grazer (Snaydon, 1980). Conversely, assessing the damage that is done to a nature reserve by invasion of 'undesirable' species is perhaps best done at a qualitative level, although the costs of weed control in terms of exclusion or eradication programmes are quantifiable as are the costs of quarantine procedures at a national or state level.

---

* Throughout, we will use the term 'interference' to refer collectively to the various negative interactions that may occur amongst individual plants. This use recognises the fact that the effects of different processes, such as competition and allelopathy, will not be distinguishable in descriptive studies at the plant or population levels.

A second dimension in considering the hazards due to the presence of weeds is the recurrence of damage over time in successive production seasons. The persistence of weeds from one cycle of production to the next will inevitably lead to similar causes of damage in each season. The anticipation of damage in the future (risk aversion) (Auld *et al.*, 1987) will promote prophylactic weed control measures which may be viewed as opportunity costs. The scale of these costs may be considerable. The cost of tillage for weed control in seed bed preparation in vegetable cropping in California has been shown to far exceed the losses incurred due to animal pests and plant diseases in unsprayed plots (Wall *et al.*, 1979).

If left uncontrolled, many weeds are capable of reducing crop yield by over 80%. Therefore, in world agriculture there are routine and intensive attempts to control most serious weeds. Even so, direct losses in production occur commonly. Analysis of the pre-harvest losses caused by weeds in major grain crops suggests that they are of a similar magnitude to those caused by fungal pathogens and usually exceed losses due to insect pests (Cramer, 1967). In 1965 it was estimated that, as a global average, loss in yield directly attributable to weeds was approximately 10%. More recent comparisons made in 1980 (Ahrens *et al.*, 1981) for wheat and rice suggest that despite technological developments in weed control, losses due to weeds remain unaltered. In the developed world at least, monetary expenditure on herbicides alone far outweighs the individual expenditure on pesticides and fungicides (Jetsum, 1988). In the USA in 1986, $3625 million was spent on herbicides, accounting for 56% of total pesticide sales (LeBaron & McFarland, 1990a); in 1992, 190 million kg of herbicides were used in cropping, compared with a combined 34 million kg of insecticides and fungicides. Whilst precise estimates of national expenditure on weed control practices are almost impossible to come by, it was suggested in 1982 that over £100 million was spent per annum in the UK on herbicides alone (Elliott, 1982). In Australia, A$2000 million was calculated in 1986 to be the total cost of weeds and their control, of which $137 million was attributed to the direct cost of herbicides (Combellack, 1989); an estimate for the USA in 1994 was $20 000 million (Bridges, 1994).

The development of a weed flora

Historically, weed control measures have been pursued to minimise the damage done by weeds. The weed control practices involved periodic habitat disturbance, typically the virtual destruction of above ground plant biomass, and crop rotation. These measures were often applied regardless of the size of the weed infestation and may have become part of a cultural heritage (as in the Norfolk rotation – Cooper, 1983). On an ecological time scale they have acted as a powerful force of interspecific selection of the weed flora (Mortimer, 1990).

Plants species may be *pre-adapted* to be weeds in the sense that a species possesses a suite of life history characteristics which enables rapid population growth in the particular habitat conditions created and maintained by human activity. Pre-adapted weeds have been defined as those species that (a) are resident in a natural plant community within dispersal distance of the crop (or other habitat) and (b) come to predominate within the crop as a consequence of a change in crop and weed management practices. The successful invasion of a crop by a species from natural habitat, wasteland or hedgerow therefore depends on the match of life history characteristics of the weed to the habitat 'template' provided by the cropping system.

Soil cultivation practices act as a selective force in the development of a weed flora, because species adapted to survive intermittent habitat disturbance are naturally suited to a cropping environment (Smith, 1970). Neolithic agriculture in temperate Europe suffered from *Sinapis arvensis* and *Fumaria officinalis*, in addition to *Plantago* spp., *Stellaria media* and *Taraxacum officinale*, which were present in Palaeolithic times. All these species exhibit traits which individually are likely to ensure population persistence – seed dormancy in *Plantago, S. arvensis* and *F. officinalis*, short generation time in *S. media* and the ability to regenerate from tap roots in *T. officinale*.

Interspecific selection will also have occurred in early agriculture at harvest time, since weed species which had escaped the attention of the farmer were likely to be harvested and retained with crop seed for subsequent sowing. By the Bronze age, *Chenopodium album, Polygonum convolvulus* and *Thlaspi arvense* were associated with cropping and *Urtica urens* is first recorded in the Iron Age. Many of these later appearances were almost certainly migrant species travelling as contaminants of the seed of cultivated plants such as wheat and flax during human movements in Europe. Agriculture also led to range expansion throughout the Roman empire, particularly by plants of southern European origin, including *Aegopodium podagraria, Agrostemma githago* and *Chrysanthemum segetum*. Similar events would no doubt have occurred in the early agricultural systems in other parts of the world, such as South America and south-east Asia.

The introduction of alien plants to a country is a well known source of weed species. Where intensive agriculture is practised throughout the world

Table 1.2. *Examples of alien introductions that have become established in the resident flora. Some species may have been introduced via another region, rather than directly from their place of origin*

Date of probable introduction	Species	Probables place of origin
In Britain		
1763	*Rhododendron ponticum*	Iberian peninsula
1842	*Elodea canadensis*	North America
1860	*Galinsoga parviflora*	South America
1871	*Matricaria matricarioides*	North-east Asia
1917	*Avena ludoviciana*	Central Asia
In Australia		
1833	*Arctotheca calendula*	South Africa
1839	*Opuntia stricta*	South-east USA/West Indies
1843	*Echium plantagineum*	Europe
1860	*Cryptostegia grandiflora*	Madagascar
1913	*Chondrilla juncea*	Europe
In USA		
1739	*Convolvulus arvensis*	Europe
1849	*Digitaria sanguinalis*	Europe
1870	*Salsola kali*	Asia
1884	*Eichhornia crassipes*	South America
1906	*Melaleuca quinquenervia*	Australia

it is common for species not endemic to the region to be present as weeds. Table 1.2 lists some of the non-endemic weeds that occur in three continents and their probable source of origin. Lists such as these, however, reflect only those species that are successful invaders. They may well represent a very small proportion of the total number of immigrant species entering a country. Retrospective explanations as to why certain alien species have succeeded raises similar questions as to why some species endemic to a region are weeds whilst others are not. Some species, such as *Opuntia stricta*, may be introduced into regions free of the predators which prevent them from becoming a nuisance in their native region. Other species may be introduced into a region in which specific habitats are unoccupied (e.g. *Spartina anglica* in British and Australian estuaries) or, as is often hypothesised, where pre-adaptation of the species may confer greater fitness than endemics.

Agricultural practices are constantly changing, and new technologies are likely to affect the success of different weeds by affecting habitat characteristics and opportunities for dispersal. The introduction of seed cleaning is

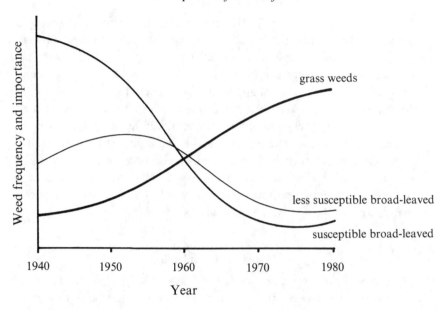

Fig. 1.1. Conjectured changes in the British arable weed flora over 40 years (after Fryer & Chancellor, 1970).

thought to have led to the virtual disappearance of *Agrostemma githago* in British cereal cropping (Salisbury, 1961). The introduction of reduced tillage systems and the reduction of crop rotations has allowed species with little dormancy, such as *Bromus sterilis*, to persist and become major problems (Froud-Williams, 1983).

Over the last 40 years, the continued and widespread use of herbicides has resulted in significant changes to the weed flora of arable land. Whilst botanical records are far from complete, Fig. 1.1. illustrates changes that are conjectured to have occurred on arable land in the UK. Broad-leaved species have declined in number and abundance whilst grass weeds have increased. It may be inferred from this that the introduction of selective herbicides specifically against broad-leaved weeds from the late 1940s has led to an increase in grass weeds, notably *Avena fatua* and *Alopecurus myosuroides*, possibly due to a relaxation in inter-weed interference. Whatever the reason, herbicides have markedly altered the composition of weed communities, but it is doubtful, however, whether herbicides have ever led to the eradication of weed species.

Mahn and Helmecke (1979) examined the effects of several herbicides at different dose rates on species composition of the weed communities within a cereal field over five years. The different herbicide treatments changed the density of individual weeds and the dominance relationships amongst species, but caused almost no change in the species present in the community. Similar conclusions can be drawn from the results of a much longer study of the effects of annual application of the herbicide 2,4-dichlorophenoxyacetic acid (2,4-D) on the weed community in a wheat crop (Hume, 1987). No weed species were eliminated over the 36 years of the experiment, and no new species were able to invade the community as a direct result of herbicide application (those species which did become established also became established in unsprayed control plots). The only changes in community structure resulting from treatment by herbicides were quantitative changes in the relative abundance of species.

The concept of a habitat

The behaviour of a weed population will depend, in part, on the nature of the environment experienced by individual plants. In this book we will use the term 'habitat' often and it is important to provide a precise definition from an ecological point of view. Increase in the size of a population will be achieved through reproduction of the individuals that survive to maturity and by gains from immigration. Survival may occur by persistence in a dormant state (as seeds in the soil or by underground perennating organs) or by escape from control as seedlings or plants (through chance or due to genotype, as in herbicide resistance). It is therefore the reproductive contribution of these survivors that is important in the growth of the population. We therefore define 'habitat' in an operational sense as *the sum of the factors at a point in space that may affect a plant's ability to survive and to contribute offspring to the next generation*. A 'favourable' habitat for a weed species is one that offers a high capacity for population growth. A 'poor' one offers little or no capacity for growth.

A habitat that is uniform in time and space will present a homogeneous environment for population growth. Naively, then, we might envisage a founder population in such a habitat to increase steadily in size to some maximum supportable density and for colonisation to occur at an ever expanding periphery. Such uniformity will rarely be encountered for several consecutive generations of population growth and we may naturally expect periods of unfavourable environment. Habitats in a temporally variable environment can be viewed as *predictable* if there is periodicity in the occurrence of favourable conditions. Clearly, habitats can be *seasonally* predictable, where a dominant factor determining growth is related to climatic conditions. Conversely, an *unpredictable* habitat is one in which there are erratic changes in favourability. An *ephemeral* habitat, however, is one in which brief periods of time that are favourable to growth are interspersed with long spells of unfavourable conditions. In relation to population dynamics, these periods of 'favourability' have to be judged in relation to the generation time of a weed.

Cropping history clearly has a dominant effect on habitat type. Continuous cereal cropping imparts a predictable habitat to an annual weed, where there is seasonal re-occurrence of the same type of crop and associated management conditions. Conversely, a repeating four year rotation of crops may be viewed as ephemeral habitat to an annual weed species with no persistent seed bank, whereas to a long lived perennial there may be an underlying predictability. Clearly, then, the judgement as to whether a habitat is 'good' or 'bad' has to be made in relation to the generation time of a weed and to its life history.

It is difficult, however, to attempt a simple match of life history characters with habitat type. In horticulture, short duration cropping and periodic soil disturbance will result in habitats likely to be unpredictable if not ephemeral. On the one hand, this will select for species reproducing by seed with short generation times, high fecundity and seed dormancy, and on the other for plants species with clonal growth forms and persistent perennating vegetative organs. Moreover, we must take into account all factors involved in the operational definition of habitat and these introduce different dimensions. In continuous arable cropping a crop protection policy deliberately avoiding reliance on one group of pesticide chemistries in an attempt to minimise the evolution of resistance (see Chapter 8) gives rise to a highly unpredictable habitat.

On a spatial scale, habitat types can be considered to be distributed uniformly or not. Non-uniform distributions constitute a mosaic of conditions for growth. The units of the mosaic can be further classified as isolated or contiguous in the context of a species' ability to disperse. Where favourable habitat units are isolated they may be beyond the maximum dispersal range of the species; where they are contiguous the species will be able to disperse amongst them. We consider the implications of this subsequently.

Studying weed populations

Weed populations need to be managed. As we have discussed, weeds may be invading, or threatening to invade, from another continent at one extreme or from an adjacent habitat at the other. The manager will need to keep the weeds out or to limit their rate of spread. Once they have invaded a location, the weeds may increase so as to reach unacceptable levels. The manager will need to regulate population levels (density) so as to minimise their impacts.

In order to make our weed management programmes as effective as possible, we need to understand the factors dictating the rates at which weeds spread, the rates at which they increase when they reach a location, the maximum extent to which they will increase and the ways in which these can be minimised or reversed. This knowledge may then help us to manage weeds efficiently, so as to minimise negative impacts.

In the following sections, we will provide the framework around which this book is built. We will divide our discussion along the dual purposes of weed management outlined above: to limit spread and to limit population levels. Central to the framework is an appreciation of the types of factors driving population change. At any given point in time, a population has a *state*, the set of attributes which can be used to describe it. These attributes include its spatial limits (range/boundaries), total population size (number of individuals), density at any point within its boundaries, genetic composition and phenotypic composition (such as the frequency distribution of plant sizes of which it is comprised). From the moment that a new population is founded, perhaps from a single individual, changes in the state of the population will occur. It is these changes with which we are concerned – the dynamics of the population. The causes of the changes ultimately will be either intrinsic to the population, driven by interactions amongst individuals in the population and therefore density-dependent, or extrinsic, governed by the species' environment (Table 1.3). The scale and type of response to either intrinsic or extrinsic factors will depend on the particular life history characters of the species and, to a varying extent, will be a reflection of both types of factor.

Understanding the spatial dynamics of populations

The concern with spatial dynamics, seen at the farm level in the use of hygiene measures when moving machinery from field to field, is mirrored at state, national and international levels in the imposition of quarantine regimes.

Most crop production systems introduce spatial heterogeneity by differentiating land into different habitat types by cropping management. From the air, farms are seen to be a patchwork of different crops, each of which is

Table 1.3. *Classes of factors determining the dynamics of plant populations*

Intrinsic to species (or ecotype)
Growth/mortality response to weather conditions
Growth/mortality response to edaphic conditions
Dispersal characteristics
Seed production characteristics

Intrinsic to population
Rate of increase in specified habitat
Upper limit to population density in that habitat

Extrinsic
Spatial distribution of suitable habitats
Variability in the habitat caused by weather
Variability in the habitat caused by management
Changes in the habitat caused by succession or other natural community
    processes
Presence and strength of dispersal vectors

---

managed differently and may contain a particular weed community
(Fig.1.2). Areas under cropping are delineated by boundaries, required for
instance by ownership, access, or topography. Within fields, some species
may be more abundant near field margins, others occur in moister parts,
and others may occupy areas reflecting former boundaries. Weed communities, therefore, are usually distributed unevenly across this landscape.

At this level the dynamics of the archipelago of patches of weeds is
integral to a strategy of weed control. A knowledge of the *dispersal rate*
amongst patches and to new areas (in colonisation) is important to
understanding spatial dynamics. The ability of a species to reach an area of
habitat will depend on its adaptations to dispersal, availability of dispersal
vectors (such as humans, livestock, wind, irrigation water) and the spatial
distribution of habitat types (scale of mosaic units).

Fig. 1.3 conceptualises this for a weed species which annually renews
from seed and is initially distributed across an area in a relatively homogeneous manner (state A). The species is abundant (high population
density) and is self-sustaining. Its persistence arises because of recruitment
of new individuals from seeds over generations. The imposition of weed
control practices may either result in state B or state C. In state B, the
consequence is a lowered population density over the entire area because
the habitat is less favourable. Nevertheless, the species remains widely
distributed but less abundant. In state C, the consequence of weed control is
fragmentation of the species into discrete sub-populations differing in

Fig. 1.2. Aerial photograph of the Gascoyne River valley of north-western Australia, illustrating the degree of complexity which an agroecosystem may contain, and the mosaic of habitats within which an invading weed must disperse. (Photograph courtesy S. Eyres, Department of Agriculture, Western Australia.)

density and in extent of isolation. Local populations may be self sustaining (C: 1 and 2) but others may persist (C: 3) only because of net recruitment of individuals from neighbouring sub-populations. Other sub-populations (C: 4) may become extinct because seed production is insufficient to maintain a viable population size within the local habitat type. Conversely, migration from existing populations may result in recolonisation and the establishment of new populations (C: 5).

Understanding the dynamics of population density

In an agricultural context, weeds may be controlled with the intention to maximise profit or to minimise risk of damage (Cousens, 1987). Weed management practices can therefore be viewed as aiming either to limit weed population density to acceptable levels or to eradicate weed species locally from the cropping environment. In the former case the goal is one of

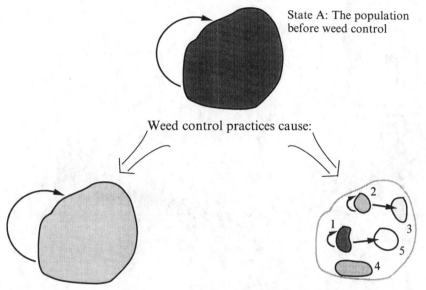

State A: The population before weed control

Weed control practices cause:

State B: Reduced growth rate, lowered density

State C: Fragmentation into sub-populations of varying density. Some of these are self-sustaining (1, 2), others are not (3, 4 and 5)

Fig. 1.3. A conceptual view of the effects of weed control practices on the spatial structure of a weed species. Populations with bold outline represent discrete populations; a faint outline indicates the former population limit. Relative densities are shown by degrees of shading and looping arrows indicate self-sustaining populations. Linear arrows indicate migration. See text for details.

halting population growth so that weed populations do not increase, whilst in the latter case the aim is to cause continuous population decline.

In consequence, the fundamental currency that will concern us is *rate of population growth* and we will be concerned with the ecological analysis of the factors which determine this rate. Such an approach requires a time scale on which to work and, as we shall see, this is in some cases conveniently determined by the length of the cropping season and in others by the generation time of the weed species itself.

All of the components of a weed management practice can be translated into their effects on the numbers and sizes of individuals within weed populations and thus on an individual's *likelihood of survival* and *reproductive success*. A census over a time period of the numbers of individuals surviving in the population and the number of recruits born into the next generation immediately enables estimates of population growth rate to be calculated.

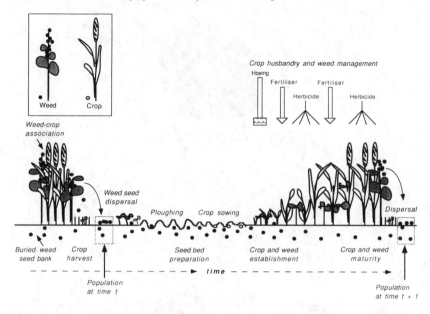

Fig. 1.4. Diagrammatic representation of the life-cycle of an annual crop, illustrating the temporal development of the above-ground vegetation and of the reserves of seeds in the soil. Two arbitrary census points are shown (as boxes), one generation apart, at which the weed population might be assessed.

Fig. 1.4 illustrates this approach in the context of a population of a hypothetical annual weed species in association with a cereal crop. At the time of crop harvest, seeds of the weed are disseminated to the ground surface and during (or at the end of) a fallow are incorporated into the soil where they augment a buried seed bank accumulated from previous generations. Episodic germination from this bank results in seedling recruitment both prior to and immediately after crop sowing. Thereafter, individuals of each species compete for limiting resources and may exhibit allelopathic interactions, the overall outcome of which is seen in the relative seed yield of each species at harvest. The diagram also notionally illustrates some of the important perturbations that this weed–crop association will experience. Losses to the weed population will occur between cropping cycles, not least due to stubble management and seed bed preparation – both seed and seedling population will suffer mortality from cultivation and spraying. After crop establishment, weeds may benefit from the input of fertiliser (as may the crop!) but be selected against by herbicide applications resulting in death of individuals or suppression of growth. At crop harvest, weed species may suffer loss due to the removal of seed in harvesting

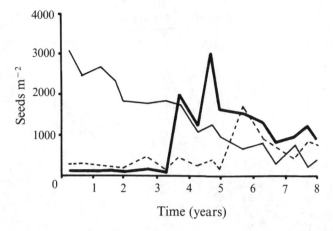

Fig. 1.5. Changes in the seed bank of an arable field (based on data in Dessaint *et al.*, 1990). Data are for experimental plots which were ploughed annually and received no herbicides. Plots were cropped with a rotation of oats + peas and winter wheat. *Sonchus asper* (bold line); *Amaranthus retroflexus* (thin line); *Alopecurus myosuroides* (dashed line).

machinery, yet on the other hand this process may promote dispersal of the weed. In this example, population growth rates from one generation to the next may be calculated from census points at any convenient stage in the life-cycle of the weed. Fig. 1.4 illustrates census times immediately post-dispersal, when only seeds are present. But if the entire population is to be considered at other census times, then plants above and below ground (seeds) must be enumerated.

Fig. 1.5 shows an example of the dynamics of seeds of just three species within the soil of an arable field. Populations were sampled over a period of eight years. One species, *Amaranthus retroflexus*, declined steadily; another, *Sonchus asper*, increased rapidly after three years, while *Alopecurus myosuroides* increased after five years. These, then, are examples of the dynamics of populations which we will explore: what, for instance, determines the rates of increase or decrease and the timing of the critical points at which these rates change?

Life-cycles and life histories

Central to the analysis of weed population dynamics is measurement of a state variable – the number of individuals per unit area. Since higher plants typically exhibit a range of life states throughout their growth and development, the definition of an 'individual' must encompass these states.

(a)

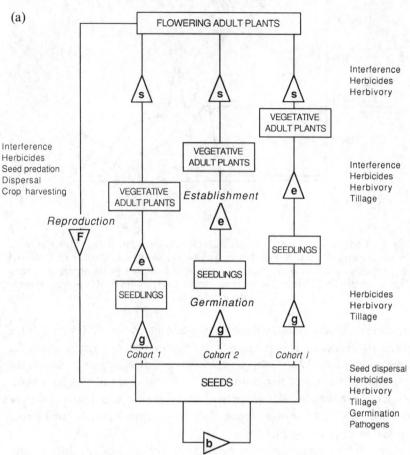

(b)

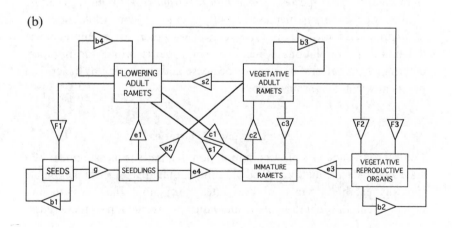

At its simplest an annual species (e.g. *Bromus tectorum*) will germinate from seed, establish an adult (photosynthesising) plant, flower and die, having set seeds which are disseminated to the ground. Alternatively, species with a clonal growth form (e.g. *Cyperus esculentus, Elymus repens*) may be perpetuated by fragmentation of plant parts (ramets) from which new individuals arise. The description of these differing life forms is typically handled by construction of a *life table* in which transitions between various life states are considered.

Two generalised flow charts are given in Fig. 1.6. In Fig. 1.6a, the life-cycle of a species that reproduces by seed is illustrated. Seeds germinate to give seedlings which then establish vegetative adult plants which flower, set seed and die. The possession of dormancy may result in episodic germination and staggered recruitment of individuals to the seedling and adult plant population. Thus, individuals in the adult plant population will differ in age, as will dormant seeds in the soil if seed longevity exceeds generation time of the plant. Populations of a mixture of ages are common in weed populations and whilst age range may be extensive in buried seed populations it is often restricted in adult plant populations because of weed control measures or because of the rapid development of a crop canopy. We may distinguish cohorts of plants (a cohort comprises individuals recruited into the growing fraction of the population at 'relatively' the same time) which track one another through the life-cycle. Individuals in each cohort may have different life histories (the lifetime experiences involved in growth, development and reproduction) according to the habitat they experience after birth. If individuals, regardless of age, die at roughly the same time (e.g. *Avena fatua*) and a species does not possess overlapping generations, then it is easy to compare the contributions to population growth made by different cohorts, since this will be the product of the size of the cohort, the chance of an individual surviving to set seed and the seed production per individual. In temperate winter wheat, for example, autumn germinating plants of *A. fatua* contribute more progeny to the succeeding generation than spring germinating ones.

In species such as *Senecio vulgaris, Carduus nutans* and *Poa annua* population structure may become more complex, since overlapping gener-

Fig. 1.6. Flow charts describing the life histories of (a) annual plants and (b) perennial plants (the latter after Lapham *et al.*, 1985). In (a), the main causes of loss of plants and reduction of seed production are shown in the margin. Letters in triangles indicate probabilities of germination (g), establishment (e), survival to maturity (s), survival in a 'bank' of individuals in the soil (b), other changes in plant state (c) and fecundity (f). See p. 130 for definition of 'ramet'.

ations of plants may occur. It has been argued that in plant populations plant size may be more important as a determinant of subsequent fate than plant age (e.g. Abrahamson, 1980), and a more precise description of a life-cycle will include adult plants of varying size. In such a scheme, individuals whilst remorselessly ageing may regress to smaller sizes. Recognition of an age/state or age/size classification allows more complex life tables to be constructed. Fig. 1.6b illustrates this for a species that may reproduce by seeds, by clonal growth and by vegetative propagules. This flow chart envisages recruitment of new individuals from seeds (as seedlings) or from reproductive propagules (as clonal individuals) and allows adult plants to 'generate' new clonal individuals through fragmentation of shoots. Such an approach to constructing life tables is applicable to all perennial weeds.

The above approach has been used to describe the life-cycles of a wide range of weeds, including *Poa annua* (Law, 1975), *Avena fatua* (Wilson *et al.*, 1984), *Euphorbia esula* (Maxwell *et al.*, 1988), *Senecio jacobaea* (Forbes, 1977), *Elymus repens* (Mortimer *et al.*, 1978), and *Cyperus esculentus* (Lapham *et al.*, 1985).

Overview

The previous sections illustrate in broad overview the ecological approach that we will be taking in this book. From the practitioner's point of view, the rationale for taking this approach is as follows.

Despite the high level of crop management and the array of effective options at the grower's disposal, weeds have remained a major problem. Even fields which have received herbicides annually for over 20 years may respond with a damaging weed flora if left unsprayed. In the last two decades, grass weeds such as *Lolium rigidum, Sorghum halepense* and *Avena fatua* have become increasing problems, demanding new herbicides or major changes in cropping to ensure continued productivity. Herbicide resistance is increasingly becoming a problem and in particular the phenomenon of cross resistance (where resistance has developed in response to one herbicide but the weeds are then resistant to other unrelated herbicides to which they have not been exposed). All of these observations illustrate the fact that, regardless of developments in weed control technologies, changes in weed abundance follow changes in farming practice. It is now widely accepted that programmes in which weed control is achieved almost exclusively by chemicals can be very unstable. There is also increasing public concern about the quantities of chemicals being used and their potential environmental effects. Indeed, in some countries, such as

Denmark, there is now legislation to reduce pesticide use; the use of particular herbicides has been banned in others. As a result there is a renewed emphasis on long term weed management and the integration of a range of methods of weed control. At the heart of this approach is the need to understand the dynamics of weed communities and their constituent populations.

Weed management is not, however, exclusively in the hands of the grower. Many countries have quarantine regulations or mandatory control ('Noxious Weed') legislation to prevent the importation and spread of known weeds. Such legislation is intended to prevent, or at least slow down, invasions by weeds likely to have an impact on agricultural production or on native plant communities. Hence, it is aimed at curtailing future population increases.

In addition to the aims of these legislative programmes, their very feasibility and costs are also dependent on the potential spread of the weeds under attention. Any economic justification for control measures must take into account the potential area that can be occupied, the rate at which the weed will spread and the abundance that it will eventually achieve. The likelihood of the programme succeeding will also depend on the effort to be invested and the frequency of the threat. Too little effort will make invasion by the species inevitable and money may be wasted. Clearly, administrators are interested in the degree of effort and expenditure to be invested. All of this hinges on understanding the population dynamics of weeds in both local and geographical contexts.

Since an epidemic of a weed is often initiated by long distance dispersal from another geographic region, spread on a geographic scale will be discussed first. In Chapter 2 we review the principal causes of weed dispersal and the directions in which weeds have spread around the world. Measurement of rate of spread at this scale is examined, along with the phases in the invasion process and how invasions should managed.

In the remainder of the chapters we transfer our attention to dynamics of weed populations at a local level, for example within a field. The size of a population depends on the balance between dispersal, 'births' and deaths of individuals. In Chapter 3 we examine dispersal processes and compare the distances by which propagules are moved by various vectors. In Chapter 4 the factors determining gains and losses at different stages in the life history of a weed are reviewed.

The dynamics resulting from these processes are then considered. We analyse the dynamics which arise as a result of intrinsic population processes in Chapter 5, assuming that immigration and emigration are balanced. In Chapter 6 we examine how extrinsic factors, such as land management, weather and interactions with other organisms, can modify the behaviour of populations. Dispersal, birth and death processes are then brought together to consider the dynamics of species within a field in Chapter 7.

In Chapter 8 the evolution and the management of herbicide resistance are explored, considering theoretical expectations and case studies. Finally, in Chapter 9 we discuss the extent of our present understanding of weed population dynamics and our ability to predict future changes in weed communities.

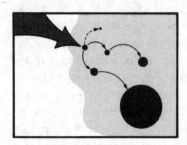

2

The dynamics of geographic range expansion

The spread of a plant species can be viewed at a number of levels – within a field, within a region, within a country or continent, and globally. These levels invoke very different considerations of weeds. For example, at the field level (c. 1–100 ha) the interest of farmers may be in the patchiness of weed infestations and their localised control, whereas at the country/ continental scale (10^6–10^{10} ha) the focus of administrators may be on quarantine and invasion. In this chapter we consider the spread of weeds on a geographic scale and the underlying issues relating to weed population dynamics. This is the scale appropriate to consideration of the dispersal of a non-endemic species and its subsequent invasion of a new region. It may also be relevant to formerly restricted endemic species which start to spread into other areas.

Groves (1986) argued that the process of invasion of an unoccupied region by a new species may be divided into three phases:

1. *Introduction.* As a result of dispersal, propagules (seeds or plant fragments) arrive at a site beyond their previous geographic range and establish populations of adult plants.
2. *Colonisation.* The plants in the founding population reproduce and increase in number to form a colony which is self-perpetuating.
3. *Naturalisation.* The species establishes new self-perpetuating populations, undergoes widespread dispersal and becomes incorporated within the resident flora.

It is only once the third stage is reached that the species is likely to be considered a nuisance and classed as a weed. Invasion and range expansion by a weed involves all three phases, which may have their own time scales and may not necessarily intergrade easily.

The main factors limiting species ranges have been argued to be barriers

21

to dispersal and availability of suitable habitat types (Krebs, 1972). Change of either of these two factors will allow range expansion. Removal of a barrier, by allowing dispersal, enables the entry of a species into areas of habitat previously unavailable to it. Range expansion (invasion) within those areas can then occur. Typical barriers to long distance dispersal include seas and oceans, mountain ranges, deserts and hostile climatic zones. For species poorly adapted to dispersal, even narrow regions of unsuitable habitat may provide effective dispersal barriers. An increase in the distribution of suitable habitat, such as through the spread of farming, may allow the concomitant spread of the species. The range of a species will be limited by the geographic extent of suitable habitat and the contiguity of favourable habitats in relation to dispersal traits of the weed (Opdam, 1990).

The agencies that disperse plants are central to the process of invasion and understanding their role is as important as defining habitat characteristics that may allow species persistence. It is clear that the successful immigration of alien plants into a geographic region and their subsequent spread as weeds has accounted for many of the important present-day weeds (see Table 1.2). Equally clear, however, is the fact that many immigrant species fail to undergo range expansion after migration to a new region. In 1919, Hayward & Druce (cited by Salisbury, 1961) recorded 348 alien plant species growing along the River Tweed, which was then the centre of the wool and cloth industry in the UK. Of these, 16% were of European and North Eastern origin, 4% Asian, 14% Australasian, 6% North American and 12% South American. Species with spiny or hooked seeds and fruits predominated, reflecting their mechanism of dispersal into the country, and ensuring their dispersal in the waste wool used widely in horticulture. Only four of the species, however, have become established in Britain (Salisbury, 1961). According to Crawley (1987), none of the 348 species now survives along the River Tweed.

A number of inter-related factors are therefore involved in the analysis of range expansion. Quite often the analysis will be retrospective, since introductions of plants are rarely experimental and are seldom closely monitored. To make sense of the available evidence, we pose a series of seven questions as a structure for this chapter:

1. Are there recognisable phases in the invasion process?
2. How can we measure rate of spread from the available data?
3. What governs rate of spread during an invasion?
4. What makes range expansion cease?
5. Have species spread in some directions more than in others?

6. Have particular modes of dispersal dominated spread between continents?
7. How should we manage invasions?

Are there recognisable phases in the invasion process?

In addressing this question, we will begin by discussing two examples of successful invasions. They illustrate that escape from natural enemies, resulting from the dispersal event, and availability of uncolonised habitat have yielded spectacular invasions. Their history has been documented, and they constitute two of the 'classic' cases of invasions.

Several species of cactus have successfully invaded Australia, the most economically important of these being the prickly pear *Opuntia stricta*. The part played by escape from natural enemies can be seen from the spectacular success of biological control. *O. stricta* was imported into Australia at least by 1839, possibly as a food plant for cochineal insects for dye production. It escaped and spread rapidly, so that by 1926 it had infested an area of 24 million hectares (Parsons & Cuthbertson, 1992). In at least half of this area grazing was totally impossible and farming ceased. A number of insect species was released to control this and other *Opuntia* spp., beginning in 1914. The most effective was the moth *Cactoblastis cactorum*, a predator of the cactus in South America but previously absent from Australia, which was introduced in 1926. By 1933 the moth had been so successful that more than 90% of the *O. stricta* infestation in Queensland had been destroyed and farming and grazing had resumed. Although still present, plants of *O. stricta* are infrequent and are held in check in most regions by *C. cactorum*. Other predators, such as the cochineal insect *Dactylopius opuntiae*, are currently providing control in areas where *C. cactorum* has been less effective. Presumably, had the weed originally been imported with its predators, it would not have spread as rapidly, nor would it have totally dominated habitats.

Spartina anglica, cord grass, provides a notable example of an immigrant colonising a previously unoccupied habitat. It is a rhizomatous, deeply rooting, perennial grass which has become widespread as a pioneer species in salt marshes in the United Kingdom. In salt marshes it is the only species that can colonise soft low-lying tidal mud flats experiencing up to six hours tidal immersion (Goodman *et al.*, 1969). It stabilises the mud sediments by means of extensive rhizome and root growth, whilst the filtering action of tillers and culms retains tidally moved sediments, thus gradually raising the level of the mud flats. It is a relative newcomer to the British flora, being first recorded at Lymington in Hampshire in 1892. This species arose through allopolyploidy. The hybridisation of *S. maritima* ($2n = 60$) and *S. alterniflora* ($2n = 62$) around 1870 produced the sterile hybrid S. × *townsendii* ($2n = 62$). Subsequent chromosome doubling to $2n = 120$ (Marchant, 1963) restored fertility and led to the species now known as *S. anglica,* a rhizomatous hemicryptophyte that propagates both by detached rhizome fragments and seeds. As a result of natural dispersal and introduced plantings it now occupies 24 000 hectares around the coast of north-west Europe (Long & Mason, 1983). The prolific spread of *S. anglica* has aroused considerable concern. In many locations its spread has been correlated with a decrease in the abundance of wading birds due to its occupation of feeding sites (Goss-Custard & Moser, 1988), and it has also threatened to colonise amenity beaches (Truscott, 1984).

Both of these examples are human-aided invasions and illustrate the speed with which range expansion may occur. Whilst they have occurred in recent history, charting their spread on a geographic scale and identifying phases in the invasion process remain impractical because of the lack of records.

Sometimes it would appear that there can be a distinct change in the rate of invasion by a weed. Species which establish local populations at the initial point of invasion may exhibit little increase in abundance for a considerable time, but then undergo sudden range expansion. *Mimosa pigra*, a native of Mexico, Central and South America, was introduced into Australia in the late nineteenth century, probably via Asia (Miller, 1988). For about 80 years it was restricted to the area around the city of Darwin. Although it became a problem in some locations, such as Darwin Botanic Gardens, it was restricted to wet habitats seldom visited by people. In 1952 it appeared 90 km to the south, and despite control efforts quickly spread. By 1987 it had formed 45 000 ha of impenetrable thickets in wetlands. Native communities are suffering from acute interference and the thickets provide refuges for feral animals. The wetlands in this region are of international significance and include the World Heritage listed Kakadu National Park. Several million dollars are now being spent to try to control *M. pigra* and to prevent its spread, using herbicides and biological control.

A two-part hypothesis is commonly proposed in explanation for this time-lag between colonisation and naturalisation. Firstly, populations may expand slowly at their periphery. The rate of areal spread will be governed by intrinsic demographic factors, in particular the finite rate of increase (see Chapter 5) and passive dispersal characteristics. Lack of suitable habitat and habitat contiguity relative to short dispersal distance act as a brake to range expansion. Secondly, spread is aided by further, external, dispersal agencies, which act to establish additional populations from which subsequent colonisation takes place. Salisbury (1961) and Baker (1965) proposed that in the lag phase the species must build up a sufficient 'infection pressure' before range extension can occur. It is not immediately obvious in biological terms what such an infection pressure is and the term is misleading. However, if long distance dispersal by external means is initially a rare event, then the probability of dispersal away from the source may be effectively zero until the population achieves a sufficiently high density. Thus, as the species slowly increases in abundance and spreads locally, the chance of successful long distance dispersal and new colonisation may increase until at some point a 'break-out' is achieved.

A much quoted example of the role of dual dispersal mechanisms is the spread of *Senecio squalidus* in the UK. The species was introduced into the Oxford Botanic Gardens in 1699. By 1799 it was recorded growing on walls around the city. It spread slowly at first, probably relying on the aerodynamic characteristics of its achenes and local air currents. When it reached the local railway line (Ridley, 1930), rapid spread throughout central England was then facilitated, partly as a result of airborne seeds being caught in eddies in train carriages for some distance before finally floating out of the windows (Druce, 1886). Similar observations have been made for *Chondrilla juncea* in Australia (Cuthbertson, 1967), although the movement of contaminated animals and other freight by rail was more likely to have been the primary cause of long distance dispersal. The spread of *Mimosa pigra* in northern Australia was probably accelerated by the use of heavy machinery to extend road systems, providing a means of long distance dispersal (Miller, 1988).

At the turn of the twentieth century in Britain *Chamaenerion angustifolium* spread rapidly on land designated for industry and railway development where fire was extensively used in clearance. This species, though native to Britain from Glacial times, was scarce at the turn of the century and its association with the habitat conditions that arise after firing is well known. Arguably, the development of roads and railways ensured habitat contiguity for this species and facilitated its spread (Salisbury, 1961).

How easy is it, then, to identify the phases of the invasion process in practice? One difficulty, as already stated, is that few invasions have been studied as they occur. Most are described retrospectively. Hence, although we may consider that a species remained in a small area and only spread outwards at a certain juncture, the evidence is only anecdotal and dependent on subjective powers of observation. Because we tend to classify, and may be looking to see a lag phase, we may be more likely to conclude that we

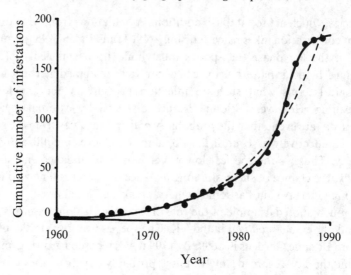

Fig. 2.1. Increase in the number of *Bromus tectorum* infestations in the province of Saskatchewan, Canada. The solid line is fitted by eye; the dashed line is the exponential curve fitted by Douglas *et al.* (1990).

have observed one. Indeed, this can be the case even if we have data on the area occupied by a species over time. It can be very difficult to determine whether spread is exponential (i.e. a constant proportional rate of increase) or two-phased. Douglas *et al.* (1990), for example, considered that there was a lag phase in the spread of *Bromus tectorum* in Saskatchewan, and yet their data are well described for most of the invasion by an exponential curve (Fig. 2.1), which would indicate no lag phase at all. It is often tempting to conclude that the early part of an exponential increase is a lag phase. Fig. 2.2 shows how similar the two types of spread can look, and how hard they would be to distinguish, even with statistical analysis. Yet they have such different interpretations! Is it possible that many of the cases claimed to show lag phases between colonisation and naturalisation are being incorrectly interpreted?

Bias in data may also be a problem causing misinterpretation of invasion dynamics. Awareness of a weed invasion will often change the effort expended in assessment of population size. In South Africa, *Nassella trichotoma* was known by farmers before 1930, yet the first herbarium specimen is for 1952. After a farmers' meeting there were suddenly a large number of reported occurrences (Wells, 1974). If the number of reports was to be plotted through time, the impression would be one of a lag phase, whereas the reality is that sampling intensity changed in a disjunct manner. A further example of this problem is shown in Fig. 2.3. *Chondrilla juncea*

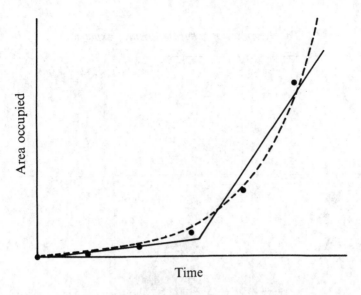

Fig. 2.2. Illustration of the difficulty in distinguishing exponential increase (dashed line) from a two-phase increase incorporating a lag phase (solid line). Hypothetical data are given as dots. Both lines give a 'good' description of the data.

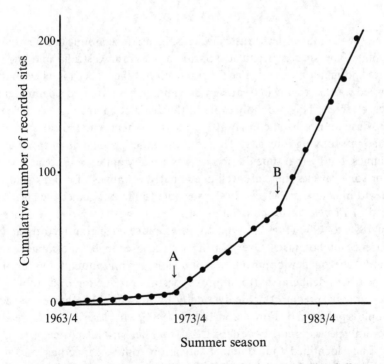

Fig. 2.3. Increase in the number of recorded infestations of *Chondrilla juncea* in Western Australia, showing one known change in recording effort (A) and another sharp disjunction for which no procedural event has yet been determined (B). Based on data from Dodd (1987).

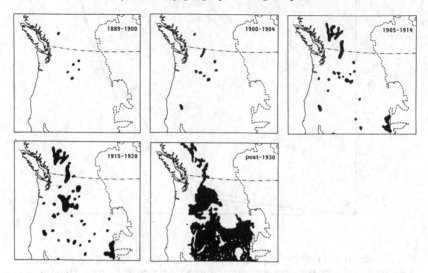

Fig. 2.4. Spread of *Bromus tectorum* in western North America (from Hengeveld, 1989, after Mack, 1981). The dotted line is the Canadian–USA border.

spread throughout eastern Australia, causing major economic problems in cropping. The species began to appear in Western Australia and each isolated population was targeted for eradication. When it became clear that the weed was increasing in frequency, the authorities mounted a campaign to search for it. This probably explains the disjunction around 1973. The figure has another disjunction in 1981, for which there is no indication of a change in sampling intensity. This could represent a change in invasion dynamics, but by that time the species was already quite widespread.

For some species there are well-documented examples of range expansion within countries. Mack (1981) was able to piece together the steady spread of *Bromus tectorum* from east to west in western North America from 1890 to 1930 (Fig. 2.4), mostly using plant collection records and experience of botanists. Over-grazing by cattle changed the rangeland habitats, offering new opportunities for invading exotic annuals. Dispersal was provided deliberately (to 'improve' pasture production and stability under grazing pressure) or inadvertently by humans. Another excellent example is provided by Forcella & Harvey (1988), who described the spread of several species in north-western USA from herbarium specimens (Fig. 2.5). They identified four dominant invasion routes, depending on the location of the initial invasion point. Invasions began (a) around Portland, Oregon, on the west coast and then spread eastwards, (b) in central

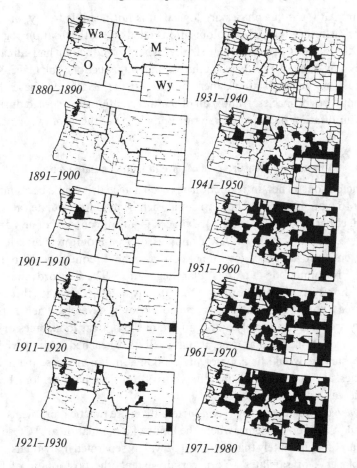

Fig. 2.5. Spread of *Bromus japonicus* in north western USA, compiled from herbarium records (from Forcella & Harvey, 1988). The outlines of the states of Washington (Wa), Oregon (O), Idaho (I), Montana (M) and Wyoming (Wy) are shown.

Montana, then spread further east, and finally westwards to the coast, (c) in eastern Washington, then spread southwards and finally east and west, and (d) in the east and west of the region, spreading in towards the centre. The predominance of dispersal along an east–west axis may be related to the railway system. However, rather than rail traffic directly causing dispersal, this probably reflects the direction of agricultural trade resulting in dispersal by contamination.

Even in these well-documented examples, however, it is almost imposs-ible to identify Groves' phases of invasions and to ascribe a duration to each of them. Although his phases are logical, it is not possible to say, on the basis of historical data, when colonisation ended and naturalisation began. From the date of first record we know when introduction had already occurred by, but not exactly when the original introduction took place. We can make no generalisations, therefore, about the lengths of phases for different types of species. Almost all of our information is on spread during the naturalisation phase, and even our knowledge of this is only coarse.

How can we measure rate of spread from available data?

In principle, once a new introduction has been detected, it would be a simple matter to map its spread. The map could then be used to calculate rates of spread, either in terms of distance advanced in any direction or of new area occupied. There are only a few examples of well-monitored invasions, mostly collected as part of government programmes to control weed spread under 'Proclaimed' or 'Noxious' Plant legislation. These records may be flawed, however, in that recording intensity may well have changed through time (see above). Interest in a species often only develops when it has already spread considerably and is an established threat to incomes or to native vegetation. By that time, it may well have reached a point where further surveys would be a major undertaking, and in any case the main priorities are then research into its control. Surveys have seldom been regarded as a research priority.

However, even if a survey is conducted, there are practical problems. In particular, how do we define where the limits of a population are? Close to the outer limit, it is likely that there will be widely separated individuals (the 'trail-blazers', or satellites). To a certain extent, the positioning of the boundary of the species' range therefore depends on the scales in which you are interested and at which you record. If the scale is coarse enough, such distinctions may cease to be important; on a small-scale map, a smooth line drawn around all known occurrences may be sufficiently accurate to provide the sort of information required. In the following discussion, we will ignore such practical problems and deal with the principles of measuring rates of spread. Given a set of data, how can we calculate a rate of spread? This depends primarily on the sort of data available.

Distances travelled from a source are perhaps most easy to monitor in a linear habitat, such as a river or a valley. For example, Gray (1960) reported that *Eichhornia crassipes* in the Sudan spread at a rate of 2.5 km per day down the Nile, and 1.4 km per day upstream. The rate of spread upstream was thought to result from plants being caught up in steamers, and was high in comparison with the spread downstream resulting from water flow. Terrestrial weed species, however, spread typically in two dimensions. Survey data for two-dimensional spread may be collected in a variety of ways.

Radial expansion

Radial expansion can be taken from maps showing the area occupied by the species over time, by drawing transects through them. This method was used by Lonsdale (1993a) to measure the rate of spread of *Mimosa pigra*. He determined that the rate of spread was 76 m per year, measured from aerial photographs. By experimentation, he showed that this rate could not be explained by wind dispersal alone. Other vectors, such as flooding, must be important.

Measurements of distance along actual transects through expanding populations are also available. Plummer & Keever (1963) recorded annually the outermost limits of *Heterotheca latifolia* along roads leading outwards from Athens, Georgia. They found that movement was about 9.5 km per year south-westwards, 11–13 km per year south-eastwards, and only 5 km per year north-westwards. They attributed the slower north-west rate of movement to prevailing wind direction, since the achenes have a pappus which, in updrafts, can carry some seeds for long distances. They also noted that roadside habitats are conducive to dispersal via a number of mechanisms.

The spread of *Galinsoga* spp. has received considerable attention in Europe. In the UK, it is believed that *G. parviflora* escaped from the botanic gardens at Kew, London. Lacey (1957) estimated that its early spread within London was at a rate of 1.6 km every 10 years. The early spread of *G. ciliata* in three locations averaged 1.6 km in two to five years. How fast did the species spread outwards from these initial foci to other parts of the country? We might examine records and look at the dates at which the species appeared along radii drawn outwards from London. It is apparent from the map of the UK given by Salisbury (1961), however, that this is no easy matter, since there are clearly new, distant records which appear well away from the main front and appear 'too soon' along the radii. This serves to illustrate an important point. Typically, plant invasions do not occur along a single front. Long distance dispersal initiates new outbreaks, which become the foci for shorter distance dispersal which then fills in the gaps between them. The picture may also be clouded by new introductions from overseas (Lacey, 1957).

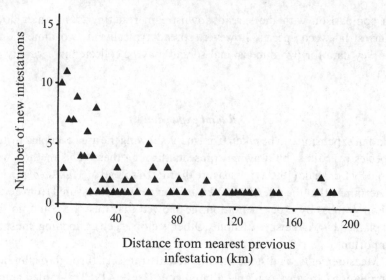

Fig. 2.6. Apparent dispersal distances for *Parthenium hysterophorus* in Queensland, Australia (redrawn from Auld *et al.*, 1987). Frequency of new infestations is given in relation to distance from the nearest previously recorded infestation.

Apparent dispersal distance

As discussed above, an invasion may commonly consist of sporadic, isolated outbreaks, followed by in-filling of the gaps. A given species, as a result of its capacity for dispersal, may be expected to display a characteristic range of distances moved. The frequency distribution of distances from new locations to the nearest previous infestations has been referred to as the 'spread pattern' (Auld *et al.*, 1987). An example of this is given in Fig. 2.6. for *Parthenium hysterophorus* in Queensland, Australia. Such frequency distributions can be summarised by regression modelling; Auld *et al.* (1982) proposed the equation

$$\log n = c - s \log d \qquad (2.1)$$

where *n* is frequency, *d* is distance from nearest previous location, and *c* and *s* are parameters. Auld *et al.* (1982) refer to *s* as the 'spread gradient'. The lower the value of *s*, the more the species will tend to spread by isolated outbreaks, rather than as an advancing front.

There are problems with interpreting these regression curves. When a species is just beginning to spread outwards from a single source, the relationship will represent the true frequency distribution of distances travelled. However, as more area becomes occupied, it becomes less certain that any new population will have come from its nearest neighbour. The distance to the nearest previous location will, of necessity, be shorter on average even though actual distances dispersed may not have changed; as a result, s will increase. The spread pattern will therefore tend to underestimate the distances actually dispersed as the number of foci increases. It is best described as an apparent dispersal distance.

Increase in area

For a newly introduced species, there may be either a lag phase, after which spread may be rapid, or a smoother exponential increase in occupancy (see p.27). Eventually, the species will be represented in all suitable habitats and spread will cease. Hence, we might expect the area occupied to increase in a sigmoidal fashion through time. The rate of increase, in terms of area added per unit time, will be given by the slope of the area versus time curve, dA/dt; the instantaneous rate of increase, in terms of proportional increase in area per unit time, will be $(dA/dt)/A$.

Consider the early phase of range expansion, before spread starts to become limited by site availability. If radial expansion away from a source is constant, we would expect (from simple algebra) the total area occupied to increase as a function of the square of time, and a plot of the square root of area against time will be linear. On the other hand, if area increases exponentially (implying that spread is not a simple constant radial expansion) a plot of log(area) against time will be linear. Lonsdale (1993a) showed for *Mimosa pigra* that the exponential model gave a better fit to his data.

The classic case study of the spread of *Bromus tectorum* in western North America by Mack (1981) has already been mentioned. As a result of his study, Mack was able to plot the area invaded against time (Fig. 2.7a). Transformation of the data to a logarithmic scale shows that the logarithm of the area occupied was approximately linearly related to time, between the years 1900 and 1930 (Fig. 2.7b). Linear regression of $\log_e A$ against time (t) gives $(dA/dt)/A = 0.12$. In other words, the area occupied was increasing at a rate of 12% per year over this period.

Most often, distribution data consist of the number of recorded locations, rather than area occupied. Medd (1987b) converted the locations of herbarium specimens of *Carduus nutans* to area occupied by calculating the number of 0.5° by 0.5° grid blocks in which specimens were collected (Fig. 2.8). Although an exponential curve gave a reasonable fit to raw data,

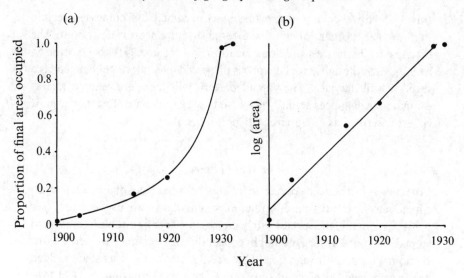

Fig. 2.7. Increase in the area occupied by *Bromus tectorum* in western North America (redrawn from Mack, 1981): (a) on an untransformed scale; (b) on a logarithmic scale.

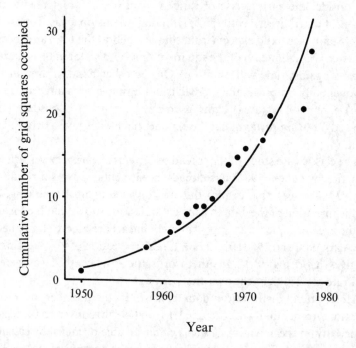

Fig. 2.8. Number of grid squares (0.5° latitude by 0.5° longitude) in which herbarium specimens of *Carduus nutans* were collected in Australia (redrawn from Medd, 1987b).

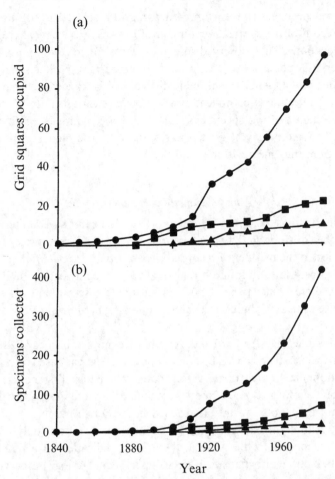

Fig. 2.9. Increase in *Echium* spp. in Australia, as determined from herbarium collections (redrawn from Forcella *et al.*, 1986): (a) number of grid squares occupied; (b) total number of specimens. *Echium italicum* (▲), *E. vulgare* (■) and *E. plantagineum* (●).

implying a constant proportional rate of increase in area occupied $((\mathrm{d}A/\mathrm{d}t)/A = \text{constant})$, much of the data would fit a straight line, which would imply a constant absolute increase in area $(\mathrm{d}A/\mathrm{d}t = \text{constant})$ and a declining proportional increase $((\mathrm{d}A/\mathrm{d}t)/A$ declines as t increases). However, there must be caution about the accuracy of rates of spread calculated from herbarium data and using grid squares to estimate area occupied.

Forcella *et al.* (1986) compared the rates of spread of three species of *Echium* in Australia (Fig. 2.9). They compared the method of using herbarium specimens to record presence in grid blocks (1° by 1.5°) with a simple record of the cumulative number of herbarium specimens (regardless of location). They observed similar patterns for the two approaches. However, they point out that the number of herbarium specimens may not detect the point at which the ultimate distribution is reached, as a result of continued collecting. A complication arising from such data is that herbaria may not retain multiple specimens from within a geographic region; these geographic regions may be greater in size than the grid blocks and two or more regions may meet within a grid block.

Increase in site occupancy

It is probably most common for rate of spread to be measured in terms of the number of (loosely defined) locations occupied. For example, Lacey (1957) plotted the number of recorded sites of *Galinsoga parviflora* and *G. ciliata* against time and found that increase was roughly exponential. For *G. ciliata*, there seemed to be a change in rate of increase, perhaps as a result of increased site availability or dispersal following bombing in World War II.

Douglas *et al.* (1990) recorded the cumulative number of sites of *Bromus tectorum* in Saskatchewan against time (Fig. 2.1). This was based on returns of questionnaires by farmers, giving dates when they first noticed the species on their farm. Some farmers were unfamiliar with the species; many would not have noted it until it reached high enough numbers to cause concern. Although the data must be treated with caution (only 5% of farmers responded to the survey), there is an indication of a relatively abrupt reduction in expansion rate in later years. A logistic curve for population size against time, often postulated for other organisms and used in modelling, would not be appropriate to describe the spread of this species in terms of site occupancy because the observed increase is asymmetric. The increase in number of recorded *B. tectorum* locations in Douglas *et al.*'s study is exponential over most of the period.

Instead of the number of point locations, the number of comital units (counties or other administrative regions, such as vice-counties) occupied can be used as the basis of monitoring spread. Forcella (1985) made a study of weed records for 199 counties in north-western USA. For those species reaching their final limits within the time-frame of the study, a logistic equation gave an excellent fit to the number of counties occupied through time. In order to see whether there was any relationship between early rate of range expansion and final area occupied, all species which had not reached their upper limit were excluded.

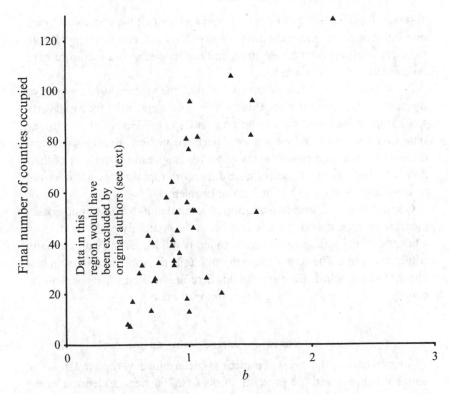

Fig. 2.10. Correlation between early rate of spread and the final number of counties occupied by various alien weeds in north-western USA. The data were re-analysed from Forcella (1985); rate of spread '*b*' is the slope of the regression of $\sqrt{n}$ against time (see text for details).

The equation

$$n = a + b \log t \qquad (2.2)$$

was then fitted to data for each of the remaining species, up to the point where they had occupied half of their final number of counties, where the *n* is the number of counties and *t* is the number of decades elapsed. Forcella was then able to plot the relationship between *b* and the area occupied. A re-analysis of Forcella's data, using instead the more appropriately shaped and better-fitting relationship

$$\sqrt{n} = a + bt \qquad (2.3)$$

is shown in Fig. 2.10 (although the conclusions to be reached are similar). The apparent correlation between rate of spread (*b*) and final area occupied should be treated with some caution, however, because of possible bias. The absence of species with low rates of spread and high ultimate areas occupied may be an anomaly, since the time-frame of the data set is not long enough for these to reach their upper limits and the species would therefore have been omitted from the analysis.

From this study, Forcella concluded that the species most in need of regional control measures are those which show a potential for rapid early spread, since these will tend to become widespread. He noted that by the time that several of the species were declared noxious (legislation introduced to ensure their control) they had already spread to over half of their final distribution; other species spreading more rapidly were not declared noxious and went on to become major problems.

One problem with using occupation of comital units (or grid squares) as a measure of spread is that their number will apparently reach saturation while the actual total ground area invaded is still increasing exponentially within the units. The larger the sampling unit, the sooner saturation will appear to be reached. Caution should therefore be used when considering county occupancy as a substitute for mapped area.

What governs rate of spread during an invasion?

As argued earlier, the range of a species is determined by the availability of suitable habitats and the presence of dispersal barriers. Extension of the range can take place when new, suitable habitats are created or when the barrier to dispersal is removed, such as provided by new trade routes. New habitats can be created when farming moves into a previously unfarmed area, for example, or when forests are logged. Once any dispersal barriers are removed, spread of a species is thus dependent on the rate of dispersal of the weed and the rate at which new habitats move ahead of it. In the case of early settlers, the rate of spread of habitat may have been limiting to invasive weeds; for most weed invasions in modern times it is likely that dispersal is more often limiting.

The potential rate of dispersal of weeds will depend on the relative importance of different modes of dispersal. On a geographic scale the efficiency of human-mediated dispersal is probably the most important. Forcella (1985) found that those species which spread fastest in north-western USA were those dispersed as contaminants of cereal and forage seed. Species adapted for wind dispersal did not show any tendency to move faster than those not so adapted. Distance dispersed by crop seed contamination will depend on trading behaviour, i.e. whether trade tends to be with neighbours or throughout a large region. Certification programmes for crop seed are likely to have reduced dispersal rate for many crop weeds. Hay can be traded over long distances in drought years and has been shown by Thomas *et al.* (1984) to be an important cause of dispersal.

An example of a range expansion in association with habitat expansion is given by the spread of some summer annual weeds into southern Canada. In recent years, cultivars of summer crops such as maize have been bred to mature in the shorter growing seasons in this region. As these crops have spread north-eastwards, so have weeds previously rare or absent, such as *Datura stramonium* (Weaver, 1985). It is likely that this crop provides a habitat more suited to the species than other crops grown previously in that region.

Decades earlier, as farming moved westwards (and later eastwards) from the coast in North America, farmland habitats would have moved in a similar fashion. Weed seeds would have inevitably been carried on clothes, in mud on footwear and by stock. Crop seeds taken from one area and sown in another would have inevitably been contaminated with weed seeds. Even today, despite increased hygiene, less weedy crops and seed cleaning, weed seeds will still be found in mud on boots, wool can be down-graded because of seeds caught in the fibres, and grain samples regularly contain weed seeds. A survey of cereal seed about to be sown in Utah in 1988 found that 31% of samples contained weed seeds; the average level of weed seeds in contaminated samples was 690 seeds kg^{-1} (Dewey & Whitesides, 1990). In Australia, the recent spread of *Solanum elaeagnifolium* is believed to have been largely the result of transportation of sheep from sales, where seeds from the farm of origin were held in the animal's digestive systems for long enough to be defaecated on the new property.

Many of the weeds found in crops throughout Europe are believed to have originated in the Mediterranean and western Asia, where farming first began millennia ago. Repeated waves of human migration would have resulted in dispersal of agricultural land use and of weeds from the south and east towards the north and west. Deforestation and cultivation created radically different conditions for plant growth and the endemic species of the areas opened to agriculture may well not have been the species best adapted to them. Salisbury (1961) noted that whereas there are many species of arable weeds in the UK of likely southern European origin, there are few of clearly northern European origin. However, pollen records indicate that the immediate response to Neolithic agriculture was an increase in those species that were already resident in the native flora at low abundance (Turner, 1970).

Why is it that the annual cropping habitat appears to be more suited to

Mediterranean species than to northern European species? It has been suggested that the former species are adapted to a higher frequency of disturbance (e.g. Groves, 1986), but the nature of that adaptation is rarely discussed. One component of the disturbance caused by cropping is physical movement of the soil, namely inversion. Presumably, species could adapt to this by becoming able to germinate from greater depths, by increased seed dormancy/longevity to survive burial, or the ability to survive the desiccation which results from the cutting of roots and the exposure of underground organs at the soil surface. There do not appear to have been any comparative studies of regional floras to test this. It is unlikely that natural Mediterranean habitats are more prone to soil inversion than temperate habitats. However, it is often argued that the long history of cultivation in the Mediterranean and western Asia has allowed their species longer to evolve adaptations to cropping.

The culture of annual crops favours annual species, by truncating the growing season. Weeds must be able to complete their life-cycle between the sowing of one crop and the preparation of land for the next. Poorly competitive species must be able to complete their life-cycles before the canopy closes, or between harvest (which removes much of the canopy) and land preparation for the next crop. In a Mediterranean environment, the climate (in particular seasonal drought) acts in a similar way to cultivation, selecting for species with a short life-cycle. By removing biomass, a seasonal drought also allows regular opportunities for seedling establishment, favouring species with rapid early development. Most temperate species, however, have evolved in perennial-based communities. Although an annual growth cycle may be enforced by cold winters, the vegetation is not removed regularly. In moving through Europe, therefore, southern weeds could have tracked the spread of annually truncated habitats, rather than necessarily genetically adapting to some new 'arable farming' habitat.

What makes range expansion cease?

Since spread begins because of the removal of dispersal barriers and/or by the creation of suitable new habitats, cessation of spread will be (a) due to the reimposition of a dispersal barrier or (b) by the species reaching the limits of suitable new habitat. If a species reaches the border of another country across which there is no trade, for example, the dispersal of the species will be impeded even though suitable habitat may be available over the border. The geographic limits of suitable habitats may be imposed by edaphic and climatic conditions acting directly on the weed, by the limits of another species (such as a crop in which the weed grows) or by management (see p.170). Although we know that some species are limited by soil type, such as *Alopecurus myosuroides* in Europe, most research has focussed on the effects of climate, in particular on predicting ultimate species ranges.

In assessing the economic implications of an invasion and the effort justified to control the spread of a species, it is essential to predict what the ultimate range will be. The larger the potential area, the greater the ultimate economic or environmental impact and the more justified is expenditure on control. Potential limits of occupation are only just beginning to be considered formally in making legislative weed control decisions.

There are several methods which can be used to infer ultimate geographic limits:

1. The present range in a region, in which the species is believed to be close to its maximum physiological limits, can be studied in relation to average weather data. By a process of trial and error, a subjective assessment can be obtained as to which lines of equal means (e.g. isotherms, isohyets, etc.) most closely match the known geographic limits. Stoller (1973), for example, found that the limits of *Cyperus esculentus* and *C. rotundus* in the USA roughly coincided with temperature minima in winter. Laboratory work confirmed that frost sensitivity may limit their ranges. To predict future limits in another country, these same isotherms could be plotted and assumed to represent the species' ultimate range.

   Panetta & Mitchell (1991) used the computer package BIOCLIM to analyse 11 climatic parameters at locations where *Homeria flaccida*, *Chondrilla juncea* and *Emex australis* had been recorded in Australia. This enabled them to describe the climatic profiles of the species. To examine the possibility of invasion of New Zealand by these species, and hence to see whether quarantine measures were justified, they matched the climates of locations in Australia with similar climates in New Zealand (see Fig. 2.11). They were able to conclude that *Chondrilla juncea* probably does not pose a major threat to New Zealand, since there were only small areas of the country of similar climate to the weed's range in Australia. *H. flaccida* was predicted to be climatically suited to a much larger area and an eradication campaign would perhaps be justifiable to keep the species out. Despite the presence of *E. australis* in New Zealand for over a century and the occurrence of large areas with suitable climates, this species has not become a problem. Hence, the use of climate to predict ultimate limits is no guarantee that a species will act as a major weed within that range or that the expense of a control campaign will be justified.

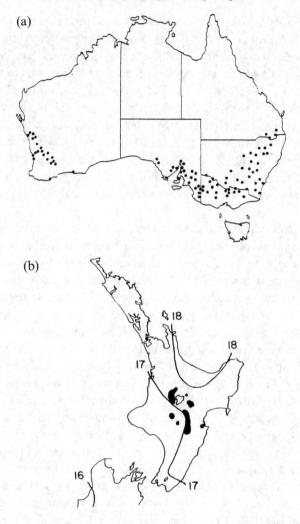

Fig. 2.11. Distribution of *Chondrilla juncea* in Australia (a), and the regions of the north island of New Zealand (b) predicted from homoclime analysis to be suitable for the species (from Panetta & Mitchell, 1991). The data points used in the homoclime analysis are shown on the map of Australia. Shaded areas on the map of New Zealand indicate climatically suitable locations; isotherms for mean January temperature are also shown. Maps are not to the same scale.

2. Rather than rely on geographic limits in another country/region, environmental data can be analysed from sites within the region currently being invaded. All locations with a similar environment to the occupied sites can be plotted, giving a conservative estimate of the total

occupiable area. The potential distribution of *Chondrilla juncea* in south-western Australia has been predicted in this way from recorded occurrences in eastern and western Australia (Panetta & Dodd, 1987) and used to recommend future control strategies.

3. Studies of the biology of species under controlled environment conditions can be used to infer likely limits to spread. Such an approach must be used with caution, since it is notoriously difficult to extrapolate from the laboratory to the field. However, Patterson *et al.* (1979) and Patterson (1990) have used studies of response to temperature to comment on the future importance of *Rottboellia exaltata* and *Panicum texanum* in the USA outside of their present distributions. Williams & Groves (1980) used studies of daylength and temperature responses of *Parthenium hysterophorus* to predict that it could potentially occupy a much larger area in Australia than at that time, in agreement with predictions of Doley (1977) based on gas exchange characteristics.

4. Growth simulation models can be combined with historical weather data to predict success of reproduction at any given location. Medd & Smith (1978) used such a method to determine whether *Carduus nutans* could successfully invade various sites throughout Australia. For each site a growth model was used to determine plant size at the onset of bolting, based on meteorological records. In turn, this was used to estimate potential seed production. Time to maturity was calculated from the number of 'growing degree days', to see whether seeds would ripen before the end of the growing season at a given site. Likely success at sites was determined by the scale of estimated seed production. From this, potentially invasible regions were predicted. The predictions agreed well with the known distribution at that time.

A problem with using climatic data to predict spread from present occurrences of a weed is that there is no guarantee that climate is currently the limiting factor. The species may still be spreading and current distributions may reflect chance historical events. Populations in different parts of the species range may have differentiated genetically, each with its own climatic preferences; the absence of one of the genotypes in a new country may lead to a poor prediction of ultimate spread. It may also be that the geographic range of a weed is determined by the climatic limits for growing particular crops, as appeared to be the case with *Datura stramonium* in Ontario, and not by the direct effects of climate on the weed. For example, the limits of many weeds in Australia coincide with those of the wheat-growing area, but it is unclear whether the weeds and wheat share climatic preferences or whether the weeds require this particular habitat

Table 2.1. *Origins of weeds in floras from two continents. Figures shown are approximate percentages of the total flora originating from each region of the globe. Such data should be treated with caution, since they may differ in the extent to which they include non-agricultural weeds or how extensive in distribution a weed has to be in order to be included in a national publication. For example, there are certainly at least some weeds of Australian origin found in the USA*

	USA	Australia
Europe, Mediterranean and W. Asia	52	33
Americas	42	28
Southern Africa	0	8
Australia, South and East Asia	0	24
Others, cosmopolitan and unknown	6	7

Source: Data are as given by Muenscher (1955) for the USA (mostly northern) and are extracted from Auld & Medd (1987) for Australia.

management system. If a crop (habitat type) spreads into a new area, will the crop reach its climatic limit first, or will the weed? If we are to predict ultimate species ranges, clearly we need to predict the ultimate range of the land management system, the climatic and the edaphic preferences of the weed genotypes present, and relate these to actual climate and edaphic environments.

Invasion direction: have species of weeds spread in some directions more than in others?

Where have most of the alien species arrived from? Species lists can be examined for different regions and for different habitats within those regions. The supposed origins of the species can be listed and summarised. Table 2.1. gives the origins of species in two floras in different continents. There are likely to be differences in the coverage between the source publications, especially in the types of species defined as weeds. Origins for many species are often poorly known. However, both floras share a high proportion of European weeds. The Australian flora has a high proportion of species from the Americas and a number from southern Africa. In contrast, the USA flora has few species from either Australia or southern Africa.

In part, these patterns of introductions can be understood from a consideration of early trade routes (Fig. 2.12). The Portuguese established a

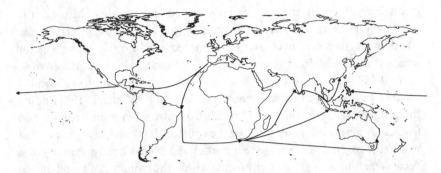

Fig. 2.12. Early European trade routes (outward legs only), reflecting possible dispersal routes for weeds. Some of these routes changed with time, such as that across the Indian Ocean to the East Indies.

route to Goa and beyond, via Brazil and the Cape of Good Hope. A route round Cape Horn, followed by many Pacific expeditions, included a stop in Brazil for supplies (Merrill, 1954). South American weeds, such as *Heliotropium indicum* and *Mimosa pudica* began to appear in south-east Asia from a very early date. The Spanish route to Asia was via Mexico. Ships sailed regularly from Acapulco to Manila for 250 years: Mexican weeds also made an early appearance in south-east Asia. It is logical to deduce that carriage along these trade routes was responsible for weed dispersal in both cases.

An important staging post on routes into and out of the Indian Ocean was the Cape of Good Hope. The nationalities able to use this facility have varied through time, according to conflicts between countries and the ownership of South Africa. The Dutch would have used it *en route* to their colonies. British ships servicing the penal colonies and new settlements in Australia usually stopped in South Africa. It is significant that several major weeds in Australia arrived from South Africa soon after that route was established. For example, the South African species *Arctotheca calendula*, which now infests farms throughout southern Australia, was first recorded there in 1833 (Burry & Kloot, 1982).

Modern trading routes offer a vast network of possible ways in which weeds may enter a country. It is in response to this that quarantine and 'noxious weed' control legislation has been established in many countries (see p.19), though it is significant that the number of new species that become naturalised does not appear to have declined as a result (Groves, 1986).

Once dispersed to a new area, population growth is essential for range expansion. Much European settlement was focussed in new temperate and mediterranean environments, similar to the climate of origin. Many of the weeds would therefore have been pre-adapted to the new locations (see Chapter 1). However, as Groves (1986) noted, the same would be true for weeds transported from parts of other continents with similar climates. Several Californian weeds, such as *Eremocarpus setigerus*, have invaded regions with mediterranean climates in Australia. Michael (1981) noted that, in Australia, the number of weeds originating from sub-tropical and tropical America increases towards the equator. It is hardly surprising that within the tropics most alien weeds are derived from other tropical regions. As was mentioned earlier, movements within the tropics can be related to trade routes, such as those from Brazil to southern Asia.

Trade between countries is usually in both outward and return directions. In some cases, however, it would seem that weeds establish primarily in only one direction. Whereas many weeds have moved outwards from Europe, relatively few have moved inwards from the settlements overseas (one example, however, is the entry of the progenitor of *Spartina anglica* (*S. maritima*) into Europe from the USA). There has been much speculation on why this should have been so. Three possible reasons are:

1. *Habitats* in the settled regions may have been in some way inherently more invasible than those in Europe, such that weeds from Europe were able to establish themselves. In contrast, species travelling in the opposite direction would not find habitats in Europe so easy to invade.
2. *Species* of European weeds may have co-evolved with agriculture over thousands of years. Agriculture is relatively new to some continents. European settlers, creating farm habitats in the new country, may have provided conditions in which only the European, farming-adapted, species could survive and reproduce. Species being transported to Europe would not be adapted to agriculture and would therefore not establish in farmland.
3. *Trade* may have been in both directions, but the nature of the trade may have been quite different. The commodities moving outwards might have included seed grain, feed and livestock destined for farmland, all of which could be contaminated by weed seeds. The commodities moving to Europe might have been predominantly for consumption or processing, and so any weed seeds in them would not have been dispersed to suitable habitats.

It is difficult to see why habitats in the newly settled regions should be more invasible than those in Europe. In most continents there are climatic regions similar to the Mediterranean, highly disturbed types of habitat, and communities of species well adapted to them. It seems hard to accept that these habitats are all in some way less 'filled' than in Europe. There are many relatively 'open' habitats in Europe (Crawley, 1987), particularly on farmland. Exclusion of exotic species from the overseas settlements through interference from other weeds would appear unlikely. A common difference, however, between the new and the original habitats of a species is a smaller number of natural enemies in the new location. Release from natural enemies may allow a considerable increase in the ability of some species to survive, grow larger and produce more seeds. When successful biological control agents are introduced, populations of the weed may be drastically reduced (see p.192), showing that dominance was not determined by abiotic or management components of the habitat.

There is some direct evidence that European and western Asian species have adapted to farmland, although evolution of weeds into farmland ecotypes is generally assumed rather than demonstrated. Froud-Williams & Ferris-Kaan (1991) described differences in plant morphology, seed size and rate of germination between *Galium aparine* growing in hedgerows and in adjacent fields. It is quite possible that the weeds of cropping are mostly species whose natural environment is seasonally disturbed, and they have simply taken advantage of the expansion of that habitat type through farming. However, native species which have evolved in seasonally disturbed habitats in other continents do not appear to invade their cropping systems so readily. For example, Amor (1984) found that only four species in Victorian cereal fields were natives of Australia. Although each of the floras in Table 2.1 has a high proportion of species from its own region, at least in the case of Australia a high proportion of these species is found in pastures, rangelands and relatively natural habitats. In California, Baker (1962) found that the proportion of native weeds increased with the degree of similarity between the human-modified habitat and the habitat in which those species occur naturally. Clearly, the more that habitats have been modified from their natural state, the more easily they are invaded by aliens.

Have particular modes of dispersal dominated spread between continents?

Although adaptations of species to dispersal can be identified, such as winged or plumed seeds for wind dispersal, burrs for dispersal by animals, etc., these are primarily concerned with transport over relatively short distances. Dispersal between continents, across oceanic dispersal barriers, must rely largely on the intervention of man. It is to be expected, therefore, that if we examine the alien weed flora of a region we would not find any

Table 2.2. *The most serious weeds in the world, as assessed by Holm et al. (1977)*

1. *Cyperus rotundus*	P	M
2. *Cynodon dactylon*	P	M
3. *Echinochloa crus-galli*	A	M
4. *Echinochloa colona*	A	M
5. *Eleusine indica*	A	M
6. *Sorghum halepense*	P	M
7. *Imperata cylindrica*	P	M
8. *Eichhornia crassipes*	P	D, Aq
9. *Portulaca oleracea*	A	D
10. *Chenopodium album*	A	D
11. *Digitaria sanguinalis*	A	M
12. *Convolvulus arvensis*	P	D
13. *Avena fatua* and relatives	A	M
14. *Amaranthus hybridus*	A	D
15. *Amaranthus spinosus*	A	D
16. *Cyperus esculentus*	P	M
17. *Paspalum conjugatum*	P	M
18. *Rottboellia cochinchinensis*	A	M

Notes: A – annual; Aq – aquatic;
D – dicotyledon; M – monocotyledon;
P – perennial.

particular dispersal adaptation to dominate. The Asteraceae, many of them wind-dispersed, often contribute a large proportion of a weed flora, but so too do largely non-adapted families, such as the Fabaceae, Brassicaceae and Poaceae (e.g. Michael, 1981).

Holm *et al.* (1977) listed what they regarded as 'the world's worst weeds' (Table 2.2). This status was accorded by virtue of the number of countries to which the weeds have dispersed, their abundance in those countries and their resulting economic impacts. Of these weeds, the great majority have no obvious adaptations for long distance dispersal. A few are probably adapted for dispersal by animals, but only one or two are clearly adapted for wind dispersal. The ability to spread vegetatively has a high frequency (some of the species rarely set seed), but this is likely to be related to ability to spread and dominate within a location, and to withstand control measures.

Instead of classifying the dispersal characteristics of weeds, and trying to generalise about geographic dispersal from these, we could consider the ways in which people have transported them to their new locations. Although ships have provided the primary means of locomotion across the sea, how did the species come to be on board, then unloaded and distributed? The following (based on Kloot, 1987) are just some of the varied reasons:

1. Ships, either empty or with light cargoes, once took on solid ballast to avoid capsizing. This may have been sand, gravel or soil. At another port, ballast would be off-loaded in favour of more cargo. Any weed seeds, or other propagules, could thus be taken from and distributed around ports. Many of these species may have been adapted to maritime habitats and, although establishing, may not have become widely distributed. Other species may have been carried inland on the wheels of carts or by draught animals. Although there were some studies of the floras of ballast heaps in the nineteenth century, the importance of this method of entry must remain largely speculation.

2. Animals were transported for food and for export. Hay, inevitably contaminated with weed seeds, would have been taken on at one port and discarded with faeces at the next. Some of this material would have adhered to animal's coats and would have been dispersed over the importing country with them. Animals may also have been given a respite from cramped ship conditions and allowed to graze near ports, thus both collecting and depositing weed seeds. It is easy to demonstrate that seeds are carried, for example in sheep fleeces, but it is difficult to assert with confidence that this was the cause of particular weed introductions (one exception may be the documented introduction of *Xanthium spinosum* into Australia on the tails of horses from Chile).

3. Hay and straw have been imported for reasons other than animal shipment. In times of drought, hay may be traded as a commodity in its own right. *Emex australis* was introduced from South Africa to South Australia in this way around 1840 (Kloot, 1987). Straw packaging was often used for shipment of fragile goods and would have often been discarded along with any weed seeds it contained. One of the methods of entry of *Imperata cylindrica* into the southern USA was as packaging for horticultural plants from Japan (Tabor, 1952). *Andropogon virginicum* is supposed to have been used for packaging whiskey and may have been dispersed in this way.

4. Shipments of crop and pasture seeds were, and are still, probably the most important mode of weed importation. The weeds will also be distributed widely and planted in an environment similar to that in which they were harvested. Early settlers would probably have brought contaminated crop seed with them. With the advent of stringent purity requirements for importation, coupled with efficient seed cleaning

methods, the number of weed seeds imported has declined. However, weeds can still enter a country as contaminants of seed, as illustrated by some notable recent introductions. *Nassella trichotoma*, an unpalatable grass which has taken over large areas of pasture in the southern hemisphere, was knowingly imported into the USA in several shipments of fescue (*Festuca arundinacea*) from Argentina in 1988 (Westbrooks & Cross, 1993). This has necessitated an expensive eradication campaign. *Parthenium hysterophorus*, a weed causing allergic reactions such as dermatitis, was introduced into Australia from Texas, probably in buffel grass (*Cenchrus ciliaris*) seed in 1960 (Michael, 1981). It now infests large areas of Queensland and the New South Wales government has instigated a major programme to try to prevent its establishment from across the state boundary.

5. International trade in fibres, such as wool and cotton for textile production, has been a major source of species introductions. Exotic species have often been found in the areas around textile mills in Europe (see p.22).

6. Species have been introduced throughout the world as ornamentals and have become problem weeds. We will give just three examples here. *Eichhornia crassipes* was displayed at an exposition in New Orleans in 1884. From there it was distributed by enthusiasts throughout southern USA. By the turn of the century it was reported to be clogging up waterways and interfering with shipping (Parsons & Cuthbertson, 1992). *Echium plantagineum* was grown as a garden plant in Australia and escaped to become one of the country's major weeds (Kloot, 1982); large sums of money are now being spent on its biological and chemical control. *Lantana camara*, originally from tropical America, was introduced into many countries as an ornamental and is now a major threat to tropical and sub-tropical forests, shading out native species and requiring intensive control measures. Some of the currently weedy forms of *L. camara* were first bred in Europe for their ornamental characters before being distributed further around the world. Kloot (1986) estimated that 359 ornamentals are now naturalised in South Australia alone. This is 40% of the total number of naturalised species and represents perhaps at least 10% of the number of ornamentals imported.

7. Early colonists used a wide variety of plants for medicinal and culinary purposes. Some of these escaped to become weeds. For example, chicory, salsify, watercress, fennel, blackberry, mullein and horehound were grown deliberately in Australia before they became weeds (Kloot, 1987).

8. A number of present-day weeds were first introduced as crops. Salisbury

(1961) speculated that the cosmopolitan weed *Chenopodium album* may have originally been cultivated in Neolithic Europe, as may have *Avena fatua*. Proso millet (*Panicum miliaceum*) was formerly grown as a grain crop in North America but is now a serious weed in maize production (Bough *et al.*, 1986). *Acacia melanoxylon, A. pycnantha* and *A. mearnsii* were introduced into South Africa from Australia for timber and for tanning, but have now spread into native vegetation (van den Berg, 1977).

9. Weeds have 'escaped' from other uses which may not be regarded as crops. Many species were imported for hedges by the British during colonial expansion. Particular problems have thus resulted from *Ulex europaeus* in New Zealand and *Hakea sericea* in South Africa. *Chrysanthemoides monilifera* and *Acacia* spp. (including A. *longifolia*, A. *cyclops* and A. *saligna*) have been transported in opposite directions between South Africa and Australia for sand dune stabilisation. Both now pose threats to native vegetation, the former around the south-eastern coasts of Australia, the latter in Cape Province.

Although we know for certain how and why some species have been introduced to a country, we have little data on most species within any weed flora. It is likely that some weeds were introduced for a number of reasons on several different occasions. We are therefore unable to categorise any flora accurately in terms of modes of introduction. Hence, we cannot conclude with any confidence whether particular modes of dispersal dominated spread between continents. We are, however, able to say for Australia, perhaps the most studied country in terms of weed invasions, that deliberate introductions, which have then 'escaped' to become weeds, have been extremely numerous.

Models of range expansion: implications for the management of invasions

One of the main reasons for measuring rates of spread is to be able to *predict* the behaviour of a species through time. Using models, we can simulate the invasion process, predict the likely economic implications of an invasion and compare the effectiveness of different control strategies. The philosophy of modelling and stages in the modelling process will be dealt with in more detail in Chapter 5. In this section, we will describe some of the models of the invasion process and the uses to which they have been put.

The simplest model is obtained by assuming that a species spreads outwards along a front at a constant rate in all directions. If the distance advanced each year is r, and assuming that the spread starts from a single point focus, the area A occupied after t years will be

$$A = \pi\,(rt)^2 \tag{2.4}$$

for which the rate of increase in area is

$$dA/dt = 2\,\pi\,r^2 t \tag{2.5}$$

and the instantaneous proportional rate of increase is

$$(dA/dt)/A = 2/t \tag{2.6}$$

It is clear from equation 2.6 that the percentage increase in area would decrease with time.

Auld *et al.* (1978/9) modelled the spread from several small sources (foci) in comparison with spread from one single source of similar total area. Suppose that spread outwards is a constant distance r per unit time, regardless of source size or density. Consider the increase in area of a single source, initially $\pi\,D^2$ in area, in comparison with four sources, each initially of area $\pi\,d^2$, such that $\pi\,D^2 = 4\,\pi\,d^2$ (i.e. $D = 2d$). One time unit later, the single source will cover an area $\pi\,(D+r)^2$ and each small source, assuming no overlap, will have expanded to cover $\pi\,(d+r)^2$. The sum of the small areas is then $4\,\pi\,(D/2+r)^2$, or $\pi\,(D+2r)^2$, which is larger than the area spread from the single source. In general, the area of the small foci will be $\pi\,(D+rt\sqrt{n})^2$, where n is the number of (equal) small foci and t is the time elapsed (Mack, 1985).

Moody & Mack (1988) used this model to compare the consequences of controlling the main area occupied by an invader with control of isolated satellite populations. They assumed that initially the total area of the satellites was very much smaller than the main focus. In the absence of control, the area occupied after time t would be

$$A = \pi\,(D+rt)^2 + n\,\pi\,(d+rt)^2 \tag{2.7}$$

If an outer ring around the edge of the main focus is controlled in the first year, equivalent to a proportion α of the initial area, then at time t the area occupied will be

$$A = \pi\,(D\sqrt{(1-\alpha)}+rt)^2 + n\,\pi\,(d+rt)^2 \tag{2.8}$$

If the satellites are all eradicated in the first year, then the area occupied will simply be that of the main focus, i.e. $\pi\,(D+rt)^2$. Clearly, as was predicted by the original model of spread, the area occupied by the species if satellites were eradicated will increase more slowly than where only the main focus was given a single control around its margin. The relative amounts by which the control measures will set back population expansion will depend on the size of α.

However, a control programme is unlikely to succeed if it is applied in only a single year. Also, satellites will be continually initiated. Moody & Mack (1988) expanded their model to account for this. They (a) assumed that new foci were only detected and controlled when they reach a critical (detectable) size; (b) allowed small and large satellites to expand at different rates, and (c) allowed the rate of satellite initiation to increase in a logistic relationship with total area occupied by the species. Annual control measures were simulated by either controlling the outer limits of the main focus, or by controlling a proportion of the satellites which exceed the critical size. They concluded that control of satellite populations was important in reducing rate of spread, and criticised programmes which concentrate on the main invasion front while leaving small satellites to increase.

In a similar simulation exercise, Auld & Coote (1980, 1981) included increase in population density within the area occupied. For any unit of area (they assumed each unit was a farm), level of infestation (P) increased according to the exponential model

$$P_t = P_0 \left[(1+c)(1-s)\right]^t \tag{2.9}$$

where c is the proportional increase and s is the proportion dispersed away from the unit. P was restricted to a maximum of 100. (Note that in the original and several other publications, P_0 was incorrectly given as P_1.) They assumed that dispersal was at random in a ring around each farm, and that no farm was successfully invaded until the dispersal into it reached a threshold level. They found that, overall, the number of farms occupied increased linearly with time, and that area increased faster from a number of scattered farms than from the same number of adjacent farms.

Menz *et al.* (1980/1) used this model to compare the costs of four strategies for controlling the spread of a localised weed infestation. One option was to contain the spreading population by maintaining a buffer zone around it equivalent to the annual rate of spread. This was the cheapest option if the rate of spread was low, but the most expensive if rate of spread was high. For high rates of spread the cheapest option was to try for complete eradication. Clearly, however, the costs will depend on how far the weed has spread before control measures are introduced.

As pointed out on p.33, the few data we have suggest that area occupied does not increase linearly with the square root of the area occupied, except perhaps in the case of data from censuses based on comital units. An exponential relationship is perhaps more realistic,

$$A = ae^{bt} \quad \text{or} \quad \log A = \log a + bt \tag{2.10}$$

$$dA/dt = abe^{bt} \qquad (2.11)$$

$$(dA/dt)/A = b \qquad (2.12)$$

The percentage increase in area in this case will remain constant over time. Various economists have modelled the economics of 'noxious plant' control, enforced by government legislation, using this model. The results are particularly sensitive to the rate of spread and the way in which 'externalities' are included (J. Roberts & C. A. Tisdell, pers. commun.). M. Hourigan (unpublished, 1985) analysed research and control measures for *Acroptilon repens* in Victoria, Australia. She found that the programme was justified if the rate of spread exceeded 4.6% per year for a low herbicide dose and 12.6% per year for double that dose.

Modelling of weed spread on a geographic scale can, therefore, be useful. Because of the serious limitations of data, the value of modelling does not lie so much in the quantitative prediction of spread, but in the establishment of qualitative principles and in the formulation of control strategies. We will return to considerations of spatial dynamics at other scales in Chapter 7.

Conclusions

It is clear that there is at present a paucity of detailed information on weed invasions. Retrospective studies of a small number of cases have been compiled, but there are almost always problems with their interpretation and rarely is it possible to calculate a precise rate of spread. With the present information, it is difficult to tell whether or not there are distinct phases to invasions, even though it is widely believed that an early lag phase is common. Because of the protracted timespan of invasions it is no easy matter to correct the lack of data by observing new introductions as they occur! In any case, monitoring is unfashionable and there are arguably more important things to do with scarce research money. Invasions will continue to occur, even in countries with quarantine restriction, and they will need to be managed. In this regard, an important point appears to be that invasions seldom consist of a single moving front. Isolated outbreaks, or 'satellite' populations, often appear well away from previous known occurrences. It is important in the management of invasions to find and control these outbreaks.

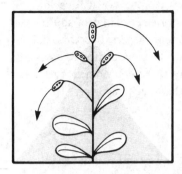

3

Dispersal within and between populations

The ability of propagules of a species to disperse within areas of suitable habitat and to reach new habitable areas across unfavourable habitat will be determined by:

1. the frequency and 'strength' of dispersal vectors,
2. adaptations to particular vectors of spread,
3. physical characteristics of the habitat.

For example, dispersal by wind will depend on whether propagules have appendages to increase their surface area, the mean wind strength and direction, and the roughness and topography of the ground or vegetation. Human-aided dispersal via harvesting machinery and tillage implements may be affected by the size of the weed seeds, the specific design of machine and the uniformity of the habitat across which it is being used.

Even after a species has spread throughout a site, dispersal may still be important to long term persistence. It may be, for example, that a species may only continue to occur in some habitats because of continued immigration from neighbouring, more suitable habitats. We also know that many habitat characteristics, especially in cropping systems, are unlikely to be stable, given that weather and land management will vary in time. An ability to disperse widely may be a life history feature essential to the maintenance of population size in such unpredictable habitats. Far from being incidental, immigration and emigration (the population processes resulting from dispersal) are therefore crucial aspects of plant population dynamics and often in the past have been ignored.

In Chapter 2, we discussed the role and importance of humans in long distance dispersal of weed species. For dispersal *within* a particular location, people will again be one of the possible vectors. Since the influence of human land management practices will vary in degree in different habitats, we might expect that short distance dispersal will vary with the type and intensity of management. In this chapter we will review our understanding of the processes involved in the dispersal of propagules from a parent plant, our knowledge of the different dispersal vectors and the dispersal distances which result. We will then be in a position to assess the extent to which dispersal distance is likely to be determined by human influence or by meteorological or inherent biological factors.

Patterns of dispersal

Not all of the propagules from a parent plant or stand of plants will travel the same distance. A population of propagules will result in a 'population' of dispersal distances. Individual seeds differ in their height of release, the timing of their liberation from the parent, their aerodynamic properties, the weather conditions at the time of release and the environment into which they descend. All of these factors are subject to variation. Dispersal distance is thus not a single quantity, but a frequency distribution of a range of values. To complicate matters, dispersal distances may vary in different directions, depending on the agencies causing dispersal. For example, prevailing winds may tend to blow seeds in particular directions; weeds growing beside a footpath may spread primarily along the path, in part due to adhesion of seeds to footwear.

We can characterise the spatial pattern of propagule dispersal by (a) counting the number of seeds landing (or number of seedlings emerging, if we are interested in 'effective' dispersal, i.e. dispersal success) in concentric circles about the seed source, (b) counting seeds in contiguous quadrats along transects in different directions away from the source, or (c) trapping seeds at discrete distances from the source. Using these data, we can represent the distances and directions dispersed in several ways. If dispersal varies in different directions, it could be described by a three-dimensional surface showing dispersal of propagules to a given distance in every possible direction. However, it is more usual in practice to consider either only one axis outwards from the source (implicitly assuming an absence of direc- tional trends), a single transect passing through the source, or dispersal in different compass directions (without reference to distance). In this section, we will consider five ways of analysing dispersal data from such studies.

Frequency distributions

Fig. 3.1. shows two hypothetical patterns of dispersal along a single axis away from a propagule source. A number of features of these curves are of

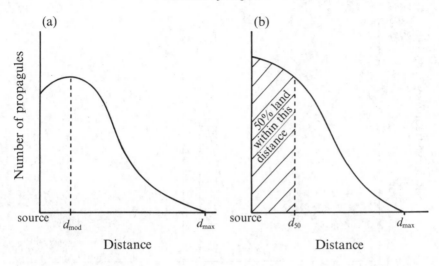

Fig. 3.1. Two hypothetical types of dispersal curve, showing the number of propagules dispersing to a given distance along a transect away from a source. Curve (a) shows a clear mode at a distance d_{mod} from the source, whereas (b) shows dispersal declining monotonically with distance. The maximum dispersal distance d_{max} is shown, along with d_{50}, the distance within which 50% of propagules land (in b only).

biological interest. The modal distance moved (d_{mod}) reflects the degree to which propagules have a tendency to disperse away from the parent plant. The maximum dispersal distance (d_{max}) gives a measure of how far the furthest propagules disperse. A measure of overall population performance, reflecting the width of the dispersal curve, can be obtained from the distance within which a given percentage of propagules disperses (e.g. d_{50} or d_{80}). The values of d_{mod} and d_{50} may well differ, depending on the exact shape of the frequency distribution.

The modal distance and the width of the distribution may be most important parameters to rate of spread within a suitable habitat. However, the maximum dispersal distance, d_{max} will govern ability to cross unsuitable areas and to establish satellite populations. Unless dispersal is very restricted (as in the case of heavy, non-wind-dispersed seeds), it is likely that d_{max} will usually be poorly estimated. This is because if the distribution curve has a long 'tail', the sampling effort involved in finding the 'few' at the tail will be considerable. In wind-dispersed or other highly adapted species, where a few seeds may be carried extremely long distances especially in extreme weather episodes, the upper limit to the frequency distribution may approach zero *asymptotically* and d_{max} may be effectively infinity. However, in order to compare the dispersal of different species and the effects of

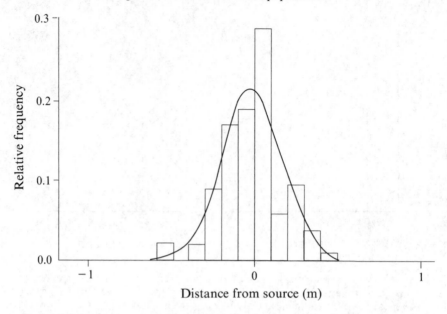

Fig. 3.2. A normal density function fitted to relative frequencies of seeds recovered in contiguous quadrats along a transect through a source. Data are for *Bromus interruptus* (redrawn from Howard *et al.*, 1991).

different dispersal agengies, in the remainder of the chapter we will define d_{max} operationally as the maximum distance to which dispersal is recorded in a typical study of dispersal from individual plants or stands of plants.

Data expressed as proportions of seeds landing in concentric circles or in contiguous quadrats along a single radius can be analysed by fitting suitable probability distributions to the relative frequencies. For example, the dispersal curve may resemble half of a normal distribution. Estimates of various dispersal parameters can be obtained by the appropriate statistical techniques from the parameters of the fitted distributions. Published tables of the area under a normal curve can be used to calculate the distance within which a given proportion of propagules disperse. Such distributions implicitly assume d_{max} = infinity, though the dispersal probability may be extremely small at only a short distance from the origin.

If dispersal is measured along a diagonal through the seed source, then a complete distribution may be fitted, rather than the half distribution outlined above. Fig. 3.2 shows results from a study by Howard *et al.* (1991), where plants of *Bromus interruptus* were sown closely together in a row in the field. Seeds were recorded in quadrats along transects at right angles to

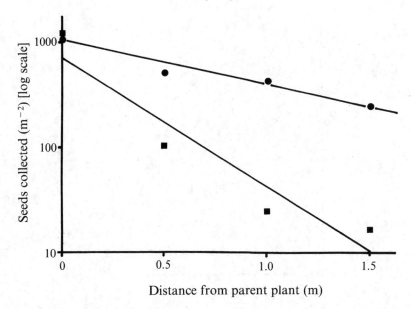

Fig. 3.3. Example of a regression model fitted to data on density of seeds trapped at a range of distances from a seed source. Data are for *Linaria vulgaris* (after Nadeau & King, 1991): ■, 1988; ●, 1989.

the row. The data show a reasonable fit to a normal distribution, which had a mode (d_{mod}) and a mean at -0.013 m. The modal distance was not significantly different from zero, suggesting that along this axis there was no tendency to move in a particular direction.

Regressions of density against distance

Instead of counting propagules in contiguous areas, dispersal curves may be sampled at discrete intervals, such as by using traps. It is possible to convert such data to relative frequencies for given distance categories and to fit a probability function. However, it is more common to use regression to relate trap counts to distance from a source. Linear regression of transformed variables has been used to describe such data. For example, Nadeau & King (1991) fitted the equation

$$\log n = a - bd \qquad (3.1)$$

to their data, where n was density of *Linaria vulgaris* seeds trapped at a distance d (Fig. 3.3).

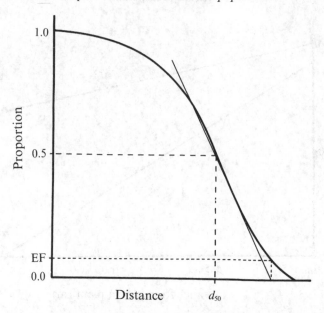

Fig. 3.4. Inverse cumulative frequency of seeds travelling to at least a given distance (after Smith & Kok, 1984). The figure shows the 'probable flight range' ($=d_{50}$) and the method of calculation of the 'escape fraction' (EF).

Auld (1988) used the regression equation

$$\log n = c - s \log d \qquad (3.2)$$

where d was the distance to which seeds dispersed and successfully produced seedlings; s may be referred to as the 'spread gradient' (see also p.32). Lower values of s reflect greater dispersal. The values obtained were 1.95 for *Carduus tenuiflorus*, 2.81 for *Onopordum acanthium* and 2.88 for *Avena fatua*, suggesting that of these species *C. tenuiflorus* has the greatest propensity for dispersal.

Cumulative frequency distributions

Rather than record the number of propagules landing at a given distance, microbiologists have considered the number of spores *remaining airborne* at a given distance. This, then, is the cumulative frequency of dispersal distance if we record inwards towards the source. This approach was used by Smith & Kok (1984) to examine the dispersal of artificially released *Carduus nutans* seeds. Parameters of interest can be estimated graphically. Fig. 3.4 shows that the d_{50} or d_{80} can be estimated by drawing lines directly from the appropriate percentage on the y-axis. Microbiologists have called the d_{50} the 'probable flight range'; d_{99} has been defined as the 'dispersal limit' (Gregory, 1973). However, the latter definition is only arbitrary and accepts that 1% of propagules will disperse beyond that distance. Another definition, again with major logical difficulties, is the 'escape fraction' (Smith & Kok, 1984). This is calculated graphically by (a) extrapolating the linear part of the cumulative curve towards the x-axis, (b) drawing a perpendicular from this point of intersection back up to the curve, then (c) drawing a horizontal back to the y-axis (see Fig. 3.4). Thus, the escape fraction is determined by the shape of the dispersal curve within the region of maximum deposition, rather than by any property of the 'tail' of the distribution. It use and interpretation must therefore be questionable.

Maximum establishment distance

It has been discussed above that estimation of maximum dispersal distance from curves is difficult, due to the small numbers of propagules dispersing long distances. However, some idea of d_{max} can be obtained by directly observing the greatest distance travelled by any propagule which forms a viable plant. This is not a measure of absolute dispersal, but of *effective* dispersal, and will be affected by the suitability of the habitat for colonisation at the location in which propagules find themselves. However, there are also statistical problems since the greater the sample size of seeds released the larger will be the expected maximum dispersal distance observed.

Direct experimental investigations of spread within fields are rare. Few farmers or managers would be eager to allow the introduction of any weed so far absent from a field. Auld (1988), mentioned earlier, introduced plants of *Avena fatua*, *Carduus tenuiflorus* and *Onopordum acanthium* to two pastures (on a research station) in which they had not been recorded. One year after introduction, plants of the three species were found up to 5, 7 and 5 m respectively from their parent plants, but the abundance declined rapidly with distance. After two years, they had spread further, but not as far as might have been expected from observations in the first year.

Anecdotal information on maximum dispersal distance is available for some species, although this is often imprecise and is seldom documented. We will give only one example here. At the Butser Hill Ancient Farm Project in southern England, archaeologists are attempting to re-create bronze-age farming techniques. They are growing ancient crop varieties and trying out what are believed to have been the types of implements used. In one small field of approximately 900 m², *Thlaspi arvense* spread from a small patch of about 1 m^2 and was ubiquitous within 10 years (P. J. Richards, pers. commun.). The only implement used for cultivation was a spade (although the cause of spread may have been wind, human traffic or harvesting methods, rather than the cultivation). Assuming that the original patch was near the centre of the field, the diagonal distance to a corner would be 21 m. If rate of spread was constant over the 10 years, maximum distance spread each year would have been around 2 m per year. Similar calculations might be possible using farmers' experiences with invading weeds.

Direction of dispersal

Where there is a clear tendency to disperse in some directions more often than in others, such as with wind as a dispersal vector, there is a need to analyse directionality. Usually this can be done simply, by ignoring distances and by merely counting total numbers of propagules along particular compass directions or in particular arcs around the parent plant. Two examples of this are given in Fig. 3.5. In both cases dispersal is predominantly in a direction reflecting the prevailing winds.

Dispersal by different agencies

Botanists have long recognised a range of adaptations for dispersal. These have been catalogued extensively (see Ridley, 1930 for a monograph and van der Pijl, 1969). Some of these adaptations have already been discussed in relation to dispersal on a geographic scale. However, within a site other mechanisms may be important. Explosive dehiscence in certain species, for example, will disperse seeds very locally and will probably play no direct part in geographic spread.

When assessing the relative contributions of the different mechanisms towards overall dispersal, it will be important to recognise which species possess adaptations to each mechanism. This is inevitably difficult because traits which can be inferred to be of 'adaptive significance' may have co-evolved for other functions as well. Adaptations which aid wind dispersal will often result in efficient dispersal by flowing water; seeds with an impermeable seed coat (probably an adaptation to impart dormancy) may be effective at dispersing in water, and yet it is difficult to argue that they have been selected for the dispersal ability of their seeds. In this section we will review current knowledge about dispersal by various mechanisms.

(a)

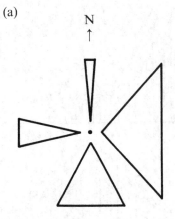

(b)

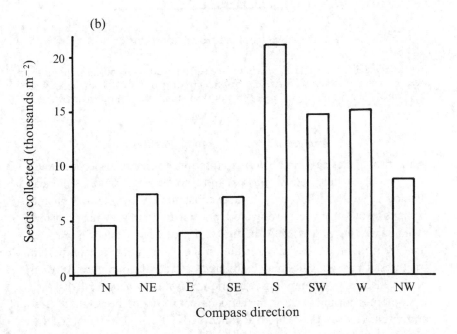

Fig. 3.5. Two examples of the presentation of data on direction of dispersal: (a) *Onopordum acanthium* (redrawn from Auld, 1988), where the area of each triangle represents the total number of progeny detected in each of four compass directions; (b) *Linaria vulgaris* (after Nadeau & King, 1991), where each histogram gives the mean density of seeds trapped at four distances along eight radii.

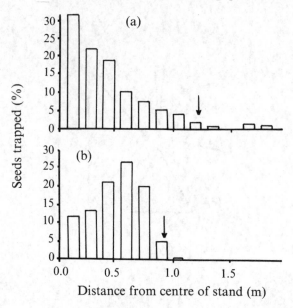

Fig. 3.6. Dispersal of *Panicum miliaceum* seeds from a 50 cm by 50 cm stand of the weed growing in either (a) maize or (b) beans (redrawn from O'Toole & Cavers, 1983). The maximum lateral spread of the plants is shown by an arrow.

Unaided dispersal (passive autochory)

Many propagules have no obvious adaptations to ensure dispersal. Examples would include many grasses and species with small, spherical, unadorned seeds. They may escape consumption by herbivores and, perhaps by maturing before any crop in which they are growing, they may avoid harvesting processes. At maturity, ripe seeds or fruits will fall from the parent plant, or they may remain on the plant until the stem or culm fractures. The propagules may land directly under the parent or, through being deflected by surrounding vegetation, may move a very small distance. If the plant is tall and sways in the wind, or it bows under the weight of seeds and perhaps lodges, the distance at which seeds fall off will be increased. It is to be expected, therefore, that such dispersal will be confined to a distance roughly equivalent to the plant's height. If the weed is growing in a tall crop, bending of the weed stems may be reduced and dispersal limited; if the weed is taller than other vegetation it may tend to fall further, spreading its seeds over a greater area.

O'Toole & Cavers (1983) collected seeds in traps placed within and outside of a 50 cm by 50 cm stand of *Panicum miliaceum*. This weed may

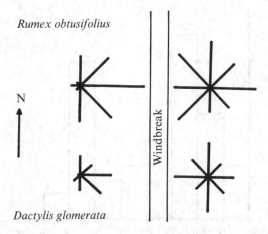

Fig. 3.7. Dispersal of seeds from individual plants of *Rumex obtusifolius* and *Dactylis glomerata* growing on either side of a windbreak (redrawn from Marshall & Butler, 1991). The length of line represents the number of seeds recorded in each direction.

reach a height of 1.5 m. In maize, the greatest concentration of seeds collected was at the centre of the stand (Fig. 3.6). In a shorter crop of white beans, the weed seeds were found in greatest numbers at the edge of the stand. Very few seeds were collected further from the stand edge than a distance equivalent to the height of the plants.

Howard (1991) recorded unaided seed dispersal by four species of *Bromus* away from a band of parent plants. She confirmed for these species that in monocultures at least 90% of seeds were deposited within a distance of 45–65 cm, equivalent to the approximate height of the grass inflorescences. In mixtures with winter wheat, dispersal was reduced such that 90% of seeds were recovered within 25–45 cm of the parents.

All but the heaviest seeds will be affected by gusts of winds as they fall, or by winds bending the parent and shaking seeds loose. Seed dispersal by species not obviously adapted for wind dispersal may still show a pattern affected somewhat by the wind (Marshall & Butler, 1991). *Dactylis glomerata* seeds were deposited mainly down-wind of the parent plants; behind a windbreak, the pattern of seed shed was more uniform (Fig. 3.7).

Explosive dehiscence (active autochory)

Some species possess morphological features to project their seeds away from the parents. In *Cardamine hirsuta*, for example, as maturing of the pod

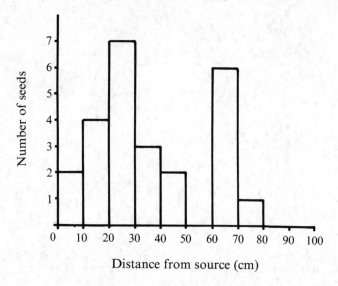

Fig. 3.8. Explosive dispersal by *Cardamine hirsuta* (based on data in Salisbury, 1961). Numbers of seeds were counted in concentric circles around the source.

occurs, the tissues of the two valves that separate the pod dry and contract to set up a mechanical tension. If a mature pod is touched, the valves tear away from the connections at their bases and coil back rapidly, imparting a projective force to the seeds and flinging them up into the air. In *Erodium* spp., a similar underlying process occurs as fruits dehydrate, but with a different morphological mechanism. The fruit divides into five sections, each acting like a sling-shot which sends a single seed into the air.

Salisbury (1961) observed the distances which *C. hirsuta* seeds were propelled in still air (probably from a single plant without surrounding vegetation). Despite the fact that this is not a tall plant (usually less than 25 cm), seeds were thrown up to 80 cm away (Fig. 3.8). The sample size was small (26 seeds), but more than 75% of seeds landed over 20 cm from the parent (d_{50} = 20–30 cm).

In a study by Stamp (1989), two species of *Erodium* (*E. botrys* and *E. brachycarpum*) were observed to propel seeds from isolated pods an average of 75 cm, whereas two species with smaller seeds (*E. moschatum* and *E. cicutarium*) threw them an average of only 54 cm from fruits at the same height. The pods were held at heights between 5 and 20 cm, but height had little effect on distance thrown. When other plants of the species were present around the parent, seeds were thrown an average of only 34 cm, illustrating the restricting effect of vegetation on dispersal.

Wind dispersal (anemonochory)

Many species can be recognised as having adaptations to assist in dispersal by the wind. In general, these adaptations increase the surface area of the propagule, hence decreasing terminal velocity by increasing aerodynamic drag. The pappus of members of the Asteraceae, for example, acts by slowing the rate of descent of the achenes, rather like a parachute; the slower the descent and the greater the release height, the further they are likely to be carried by the wind. In the case of fruits with wing-like structures (such as in the genus *Acer*), aerodynamic lift may also be generated as they fall.

Dispersal of particles and seeds by wind has been studied more than that by any other vector. Considerable attention has been given to predicting dispersal as a function of aerodynamic properties of propagules, wind speed and height of release. If it is assumed that there is no turbulence and that there is a steady wind speed at all heights, the distance travelled (d) can be calculated from

$$d = HU/V_s$$

where H is release height, U is wind speed and V_s is the terminal velocity of the propagule (Johnson *et al.*, 1981). For propagules with a terminal velocity of over 10 cm s^{-1}, this equation gives predictions equivalent to the value of d_{50} calculated when turbulence is assumed (McCartney, 1990). Most grass seeds, winged seeds and even many plumed seeds have a V_s of above 10 cm s^{-1} (McCartney, 1990). Myerscough & Whitehead (1966) quote values for pappus-bearing achenes of 14 cm s^{-1} for *Tussilago farfara* and 18 cm s^{-1} for *Epilobium montanum*. Salisbury (1961) found values of V_s of 35 cm s^{-1} for *Sonchus oleraceus* and 88 cm s^{-1} for *Galinsoga parviflora*.

For small seeds, where turbulence has a greater effect, such as for *Chamaenerion angustifolium* with V_s of only 6.5 cm s^{-1} (Myerscough & Whitehead, 1966), more complex models are required. It is also necessary to allow for the fact that wind speed will decline as a seed gets closer to the ground. Fig. 3.9 shows the predictions of such a model for three terminal velocities. It is predicted that d_{mod} will increase with release height but decrease with terminal velocity of the propagule. The value of d_{50} is predicted to be very much larger than d_{mod} (Table 3.1).

Terminal velocities are not easy to predict from the geometry of more complex dispersal adaptations. However, it is straightforward to obtain them empirically. Sheldon & Burrows (1973) compared the fall times of 18 species of Asteraceae in the laboratory. They found that both pappus

Table 3.1. *Predicted dispersal parameters (in metres) from a model which
includes turbulence. Seeds were assumed to be released from a height
of 1 m*

Wind Speed at 1 m (m s⁻¹)	Terminal velocity (V_s) (cm s⁻¹)							
	5		10		15		20	
	d_{50}	d_{mod}	d_{50}	d_{mod}	d_{50}	d_{mod}	d_{50}	d_{mod}
0.25	57	6	20	5	12	4	9	4
0.50	213	7	57	6	30	5	20	5
0.75	> 500	7	120	6	57	6	36	5

Source: From McCartney (1990).

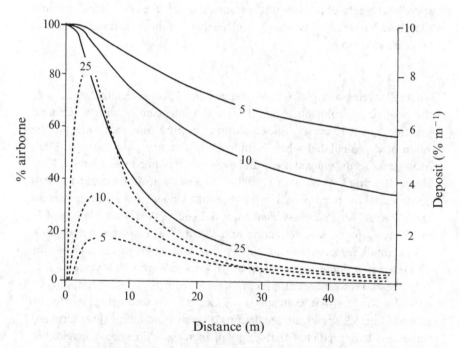

Fig. 3.9. Model predictions of the dispersal of particles by wind (redrawn from
McCartney, 1990). Predictions are given for particles of three different terminal
velocities (5, 10 and 25 cm s⁻¹); wind speed at release height was 30 cm s⁻¹; release
height was 1 m. Solid lines show the proportion of particles still airborne at that
distance; dashed lines show the fraction deposited per metre.

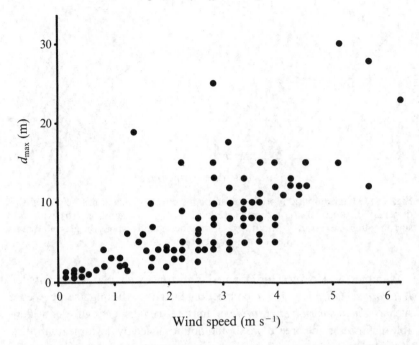

Fig. 3.10. Maximum dispersal distances of *Heterotheca latifolia* measured in the field (after Plummer & Keever, 1963).

geometry and the relative sizes of pappus and achene determined the time taken to reach the ground. A light achene with a relatively large pappus will take the longest to fall. The distance moved vertically will also depend on wind speed and the height at which dispersal units are presented. Sheldon and Burrows calculated that even in horizontal wind speeds of 4.4 m s^{-1}, the greatest dispersal distance would have been 11 m; most species were predicted to be blown less than 2 m from an isolated plant. The relationship between d_{max} and wind speed for *Heterotheca latifolia* obtained in the field by Plummer & Keever (1963) is shown in Fig. 3.10.

When achenes of *Carduus nutans* were artificially released from a height of 1.5 m in the open, wind speed increased dispersal (Smith & Kok, 1984). The value of d_{50} increased from about 15 m to 25 m from a wind speed of 0.76 m s^{-1} to 5.62 m s^{-1}. The 'escape fraction' (EF) remained very low (less than 3%), except for the slowest wind speed, with an EF of about 15%. The latter anomaly was explained in terms of surface characteristics of the experimental site (a car park). A black surface on a hot day at low wind speeds may create considerable turbulence, perhaps carrying a few achenes high into the air.

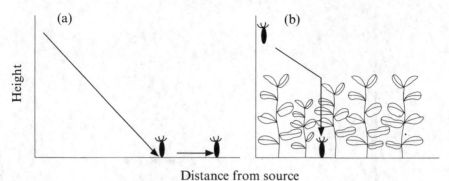

Fig. 3.11. Diagrammatic representation of the effect of vegetation height on the dispersal distance of an airborne seed: (a) without vegetation; (b) with tall surrounding vegetation. Without vegetation, once a seed reaches the ground it may then continue to be blown along.

Plummer & Keever (1963) noted only one seed per five plants of *Heterotheca latifolia* rising in convection currents. Although some weeds, such as *Chamaenerion angustifolium*, may be seen dispersing for considerable distances in the breeze, we should not be misled by the large quantities of thistle-down which we see airborne. Thistle seeds are often heavy and drop quickly; seldom does floating thistle-down ever have seeds still attached (Ridley, 1930). Although Sheldon & Burrows (1973) recorded a terminal velocity of 22 cm s^{-1} for *Cirsium arvense*, V_s for *Carduus tenuiflorus* was 79 cm s^{-1}, illustrating how much dispersal ability is likely to vary within the thistles.

Tall vegetation will restrict dispersal by interrupting the descent path of propagules (Fig. 3.11). We would therefore expect dispersal within a dense stand of uniform height to be less than that from isolated plants or from the edge of a population (Sheldon & Burrows, 1973), provided that the surrounding vegetation is lower. Plummer & Keever (1963) mapped the distribution of dyed pappus-bearing achenes after dispersal from *H. latifolia* plants. Highest densities of seeds were recovered within the stand in which they were produced; even outside of the stand, dispersal decreased rapidly with distance. No achenes were found outside of a range of 15 m from the edge of the stand, and most were found within 2 m. Michaux (1989) calculated that 58% of achenes produced within a 3 m by 3 m stand of *Cirsium vulgare* fell to the ground within the stand.

Even for isolated plants, data indicate that wind-assisted dispersal can be limited. Michaux (1989) trapped achenes of *Cirsium vulgare* in water-filled containers at various distances from a single plant of 1 m in height. It was estimated that 91% of the achenes fell within a diameter of 1.5 m. Even for *Senecio jacobaea*, where some achenes were carried up to 36 m, Poole & Cairns (1940) found that 60% landed around the base of the parent plants.

Despite the data reviewed above, showing that many seeds are deposited close to parent plants, it would seem from the limited data that dispersal curves with a mode away from the seed source are more common for wind dispersal than for other vectors. Salisbury (1961) presented data showing greatest frequency of dispersal by *Verbascum thapsus* at about 3.7 m from the parent. The size of the study plant was not reported by Salisbury, but individuals can exceed 1.5 m in height. In the study of *Cirsium vulgare* by Michaux (1989) quoted above, the greatest density of achenes was collected at 0.33 m from the parent in three out of four directions.

In the dispersal mechanism of *Papaver* spp., seeds are contained in a capsule on the end of a stem. The exit holes are above the seeds. To escape, the seeds either require sufficient energy to achieve a critical velocity to carry them out, or the stem needs to bend over to an angle at which they can fall out. The seeds lying closest to the bottom of the capsule will require the greatest velocity for escape or the greatest extent of stem bending. Oscillations of the stem are caused by the wind, and hence seed liberation will tend to occur only on windy days. The maximum conditions for seed liberation will occur at the end of the oscillations, at which point the seeds will have a velocity away from the parent. Hence, seeds will tend to be dispersed away from the parent, resulting in a 'halo' around the parent. In laboratory simulations of wind gusts, Salisbury (1942) recorded dispersal distances for three species of *Papaver*, ranging in height from 26 to 58 cm. Modal dispersal distances were between 50 and 100 cm; although the curves were not smooth, it appeared that d_{mod} was shorter for *P. hybridum* than for *P. argemone* or *P. dubium*. The values of d_{50} were similar to those of d_{mod}, i.e. 50–75 cm for *P. hybridum*, and 75–100 cm for the other two species. *P. hybridum* has shallower, more ovate capsules than the other two species, from which it is likely to be easier for seeds to be liberated. It also has shorter stems, which would result in shorter projection distances.

Although we tend to think of dispersal by wind in terms of seeds moving in the air, wind dispersal can be important along the soil or water surface. Tumble-weeds, such as *Salsola kali* and *Kochia scoparia*, are dispersed by wind as whole plants and may drop their seeds as they travel. Some grass panicles, such as *Chloris* spp., *Nassella trichotoma* and *Agrostis avenacea* in Australia, will also be blown considerable distances along the ground and may pile up along fences. Maximum dispersal distances by such plants may be in the order of kilometres. Most seeds landing on the water will at least temporarily float and can be blown by the wind to the bank, or to some distance away across rivers or lakes before they sink.

Dispersal by animals

Seeds may be carried on the outside of animals (ectozoochory), such as in mud on feet or caught up in fleeces, or on the inside (endozoochory), through ingestion. Quantitative studies of the distances moved by either of these means are rare.

It is not difficult to identify species adapted to ectozoochory, since they typically have hooked appendages. These are familiar to us, since they tend to cling to our clothing. Examples are *Bidens pilosa*, *Galium aparine*, and *Arctium* spp. Adherence to people, especially in heavily frequented areas such as along footpaths and in parks, may be an important long distance dispersal mechanism for weeds of urban areas. Many comments were made in the past about the collection of seeds in trouser turn-ups. These are, for the present, out of fashion – but not so socks! In an agricultural context, many seeds are found entwined in the fleeces of sheep, but again the importance of this for dispersal within fields is unclear, since it is very difficult to dislodge them. It may be that hooked fruits are adaptations to ensure long-distance dispersal by at least a small proportion of seeds, rather than for dispersal of the majority. Most of the hooked seeds of *Galium aparine* produced in field margins, for example, probably fall to the ground beneath the parents and do not get dispersed on animals.

Many aquatic weeds have very small seeds. These have no hooks, but they may still be efficiently dispersed by animals. It has been suggested that the tiny seeds of *Cyperus difformis*, a weed of rice farming, may be dispersed via the feet and plumage of ducks feeding in paddy fields. Yet again, although this is feasible, there are no conclusive data on its importance relative to other mechanisms. Ducks move around considerably within and between paddies. However, each *C. difformis* plant may produce many thousands of seeds, which soon sink to the bottom; it is again likely that only a minority of the seeds are dispersed in plumage or on feet.

Another form of ectozoochory can be referred to as 'inefficient predation'. Some seed predators harvest seeds and move them away from the parent plants before consumption. Ants, for example, may forage for many metres around their nests and return with intact seeds. Having moved the seeds, some may then be missed and may later germinate at the new location.

Endozoochory may be more important than ectozoochory in local dispersal. Many species of animal, particularly the larger herbivores, may ingest seeds and then defaecate some of them intact. Salisbury (1961) recorded a large number of species which germinated successfully from the dung of cattle, horses, pigs and goats, and from bird droppings. It is to be expected that seed coat characteristics may have some effect on successful passage through animals, since they will afford protection while in the gut. Some seeds may have their dormancy broken by ingestion, since acids are known to promote germination in some species. St. John-Sweeting & Morris (1990) found that, after ingestion by horses, species with smaller seeds and with a greater proportion of 'hard' seeds (seeds with an impermeable seed coat) tended to be defaecated with less loss of viability than those with larger and softer seeds. About 1% of *Asphodelus fistulosus* seeds ingested were defaecated, and they were rendered completely non-viable. In contrast, 17% of *Malva parviflora* seeds were defaecated, with almost no loss of viability.

The distance moved by seeds in this manner will depend on the rate of through-put of the animal and the distance away from the point of consumption that the animal moves. Lacey *et al.* (1992) recovered viable seeds of *Euphorbia esula* up to 4 days after being ingested by sheep and goats. In studies of the consumption of *Solanum elaeagnifolium* berries and *Reseda lutea* seeds by sheep, J.W.Heap (pers. commun.) found that most seeds were defaecated after 2 to 4 days. Seed emission had virtually ceased by 10 days, but one seed of *S. elaeagnifolium* was defaecated after 31 days. Piggin (1978) found that peak through-put of *Echium plantagineum* by sheep was also after 2–3 days, but with almost no viable seeds after 3 days. For horses, St. John-Sweeting & Morris (1990) found maximum seed emission after 3–4 days, with none being passed after 13 days.

The distance moved by farm animals varies with farming system and even with breed of animal. Clearly, in intensive farming where fields may be small, animals may range over an entire field every day and could defaecate almost anywhere in the area. Cattle in strip-grazing systems with electric fences may tend to roam less than in other systems; browsing cattle seldom stay still for long. Sheep in large fields tend to favour certain areas in which to spend the night ('sheep camps'). In these places they tend to concentrate their excreta, and nutrient-rich areas are formed with characteristic weeds. It is also likely that ingested weed seeds are emitted at greater concentrations in sheep camps than elsewhere in the fields. Under rangeland conditions, with few barriers, movement of animals can be considerable.

Tribe (1949) found that Cheviot sheep in Scotland grazing a 0.4 ha pasture moved an average of 4.2 km per day. On a hillside in New Zealand,

Cheviot sheep moved 1.8 km per day, compared with 1.2 km by the longer legged Romney Marsh breed (Cresswell, 1960). In a lowland pasture, the Cheviot moved 2.3 km per day and the Romney Marsh 1.9 km per day. England (1954) recorded movement of 3.4 km per day on bare pasture and only 1.9 km on a pasture with a high level of feed. Again, behaviour varied with breed. In a review of cattle movement, Hancock (1953) quoted distances of 5.3 km per day in rangeland and from 1.8 to 2.8 km per day in paddocks. In rangeland, cattle may move very much greater distances than these when water and feed are far apart.

Wild animals, such as birds, may also be important seed dispersers. Seeds of *Rubus* spp., for example, will be ingested by birds, which then fly off and deposit the seeds in their faeces (Jordano, 1982). Defaecation usually takes place while birds are perched, perhaps explaining the preponderance in many places of *Rubus fruticosus* agg. along fence lines and hedgerows. Proctor (1968) found that two migratory birds could, at least in theory, move weed seeds considerable distances. The killdeer (*Charadrius vociferus*) could pass viable seeds of *Malva parviflora*, *Convolvulus arvensis* and *Abutilon theophrasti* up to 152, 144 and 77 hours after ingestion respectively (seeds were supplied to caged birds). The least sandpiper (*Erolia minutilla*) defaecated viable seeds of *M. parviflora* for up to 123 hours after ingestion. Although the distance which seeds are carried by birds could therefore be considerable, the home ranges of non-migratory species (or of migratory species when not actively migrating) may often be restricted to a few hundred metres.

In summary, dispersal by endozoochory is likely to be considerable and seeds will be spread rapidly across fields. Although no dispersal curves are available, it is likely that they would be relatively flat, with larger values of d_{50} than most other dispersal vectors. The efficiency of this dispersal vector, however, will depend on the proportion of seeds produced by a plant which are actually consumed. As with ectozoochory, it may be that many seeds favoured by passage through animals will fall to the ground and enter the seed bank without ever being ingested.

Dispersal by water (hydrochory)

Many species growing in or near water liberate their seeds on to the water surface. The seeds, fruits or other structures containing seeds may be buoyant (e.g. *Alisma plantago-aquatica* and *Sagittaria sagittifolia*, or they may sink and their seedlings rise to the surface (e.g. *Juncus* spp.: Ridley, 1930). If the water is moving, either within its usual banks or as flood water over wider areas, the propagules will be deposited away from their parents.

Often there may be an interaction with wind, blowing propagules towards the banks, where semi-aquatics can establish.

In irrigation systems, water may be pumped or siphoned into fields. It is probable that seeds or small plants (such as *Lemna* spp. and *Azolla* spp.) floating on the water surface will be spread into and throughout fields by this means. Particularly in the smaller irrigation channels, where water may move rapidly, sediments (containing sunken seeds and seedlings) may be agitated and carried into the fields. Within the fields, plants close to the inflow will have their seeds moved further along by the irrigation water. Again, there are few if any quantitative data to indicate the importance of these mechanisms. Kelley & Bruns (1975) collected seeds of 84 species in irrigation water. Many of these would not be classed as specifically adapted to water dispersal. However, irrigation water is an effective dispersal vector for them nonetheless. Wilson (1980) found only 37 species in another irrigation system.

Xanthium occidentale fruits ('burrs') can float for up to 30 days (Hocking & Liddle, 1986). Since this species often grows along the banks of watercourses, fruits falling into a river can be dispersed for considerable distances. They may also be spread in flood-waters. In Australia, the species was recorded as it spread 200 km downstream in one river in 5 years, and 50 km in 3 years along another (I.Miller, cited by Hocking & Liddle, 1986). However, there are no data on frequency distributions of distance dispersed from parent plants for this species.

Vegetative spread

Many plant species are capable of spreading vegetatively. Indeed, some species, such as *Cyperus esculentus*, are rarely if ever seen to reproduce by seed. Over the past few years the subject of clonal growth has received considerable ecological attention (Jackson *et al.*, 1985). It is worth mentioning two ecological 'strategies' by which species or biotypes may be compared (Lovett Doust, 1981). Those types which spread quickly along a main axis, producing new plants ('ramets') separated by some distance, such that they invade a large area but do not dominate it, have been termed 'guerrilla' strategists (e.g. *Ranunculus repens*). Those which spread steadily outwards along a front, occupying and consolidating as they go and hence engulfing the area into which they spread, may be called 'phalanx' strategists (e.g. bamboo). The behaviour of clones can vary with the type of vegetation into which they are expanding, nutrient availability and other factors.

Most studies of vegetative spread of weeds have concentrated on the

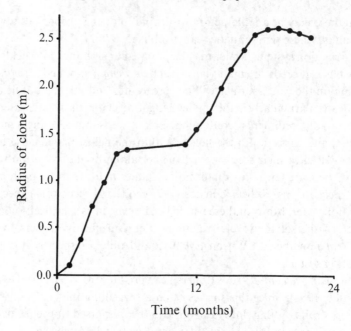

Fig. 3.12. Vegetative spread of *Cyperus esculentus* in the absence of interfering vegetation. The figure reflects two annual cycles of growth (redrawn after Lapham, 1985).

clone edge, rather than on dispersal within the clone. A typical experimental technique is to plant single vegetative propagules into an uninfested area and to measure lateral growth through time by marking the outline of the clone. Lapham (1985), for example, followed the radius of clones of *Cyperus esculentus* over two years (Fig. 3.12). He found that clones increased by an average of 1.3 m per year over two years, and that within a year the rate of expansion was slow until a temperature of about 20 °C was reached. The radius decreased slightly at the end of the growing season, when shoots at the periphery, presumably not sufficiently established, died back.

Horowitz (1973) made a similar study of *Sorghum halepense*. Over 2.5 years, clones spread outwards by an average of 3.4 m; variation between clones was large. The average area occupied after 2.5 years was 17 m². At the height of the growing season the rate of spread was 20 cm/month in one year and 10 cm/month in the next year; however, over an annual cycle the rate of growth was greater over the second year than over the first. The spread was not always even around the clone, depending on directions of growth of the main rhizomes. By and large, over the duration of the study,

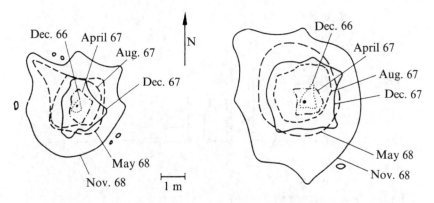

Fig. 3.13. Vegetative spread of two clones of *Sorghum halepense* (from Horowitz, 1973). Contours show the position of the clone edge at successive recording dates.

differences in direction cancelled out so that clones remained fairly circular (Fig. 3.13).

Amor & Harris (1975) found large variations in the rates of spread of clones of *Cirsium arvense* in an Australian pasture. Averages in three successive years were 1.48 m, 1.57 m and 0.80 m per annum. In reviewing other work on *C. arvense* spread, Chancellor (1970) quoted values of 1.25 m clonal spread from a first year seedling, 5 m from a second year plant, and up to 12.2 m from an established clone. Neither Amor & Harris nor Horowitz studied spread in the presence of other weeds or crops.

Spread by tillage

Once seeds have been deposited on the ground, machinery used to cultivate the soil can move them. The same will be true for rhizomes and other plant fragments capable of regeneration: farmers with *Elymus repens* in their fields may have to frequently unclog its rhizomes from seed drills. How far are seeds and rhizomes moved, and what are the effects of different implements?

Movement by cultivation is quite simple to record. Howard (1991) sowed seeds of rape seed (canola) in 0.23 m wide bands on the soil surface. These were then cultivated by either spring tines, a power harrow cultivator or an empty seed drill. The distribution of seedlings was mapped after emergence. It was found that most seeds had stayed close to their original positions after the passage of the drill (Fig. 3.14). For both the drill and the power harrow d_{mod} was close to zero. However, the power harrow had a much

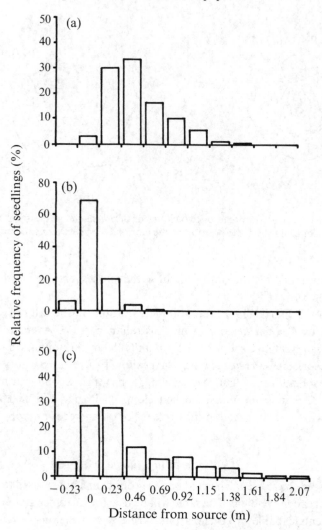

Fig. 3.14. Dispersal of seeds of canola, used as a weed seed surrogate, by cultivation implements (after Howard, 1991). Seeds were sown in a 23 cm band and cultivated from left to right by: (a) spring tine; (b) seed drill; (c) Roterra power harrow. Note the different scales on vertical axes.

longer 'tail' to its distribution, extending out to a d_{max} of about 2 m, compared to 0.7 m for the drill. In contrast to the other two implements, the value of d_{mod} for the spring tine cultivator was away from the origin, close to 0.4 m. Movement perpendicular to the direction of the implements was a matter of only a few centimetres.

There have been surprisingly few studies with which to compare these results. In describing the weed flora of a bog from which the peat had been removed and the area cropped, Curran & MacNaeidhe (1986) commented on the dynamics of *Polygonum lapathifolium*. Seeds were introduced into the field in mud from a ditch. Over a period of two years, the edge of the population moved about 4 m in a direction at 90° to the cultivations. Although this species can be dispersed by water, no flooding over the area was noted and that could not explain its spread. Gasquez & Darmency (1989) reported movement of triazine-resistant *Chenopodium album* perpendicular to the direction of cultivation at 1.6 m per year. They noted that spread was much greater in the direction of tillage, without giving an estimate. They also gave an equation relating area occupied and time, namely

$$A = 25.5 \ t^2$$

It is not possible from either study to determine the proportion of the perpendicular movement which is actually due to cultivation. It may be that much of the dispersal perpendicular to cultivation was due to seeds being shed by these fairly tall plants (see p.64).

Dispersal by combine harvester

Of any of the implements used by farmers, the one with the greatest potential for movement of seeds is the combine harvester. Although many weeds will mature and drop their seeds before the crop is harvested, some will have at least a proportion of their seeds held at a height at which they will be taken up into the combine. Larger seeds may be retained as contaminants in the grain and removed from the field. Smaller seeds will be ejected from the combine with the chaff or with the straw (Petzold, 1956), and may thus be redistributed about the field. The distance moved within a field will depend on the type of combine, the way it is set up (e.g. the size of its sieves for separating grain from stems and chaff), its speed and the quantity of material going through it. In some important respects, dispersal by combine harvester is similar to that by a grazing animal; a combine is an artificial grazer.

Results from studies of dispersal by combines have been somewhat variable. They no doubt reflect different methodologies, species and machinery. McCanny & Cavers (1988) laid out rectangular strips of cloth as seed 'traps' at regular intervals from a source stand of *Panicum miliaceum* in a maize crop. Combines were driven over the source and along a weed-free

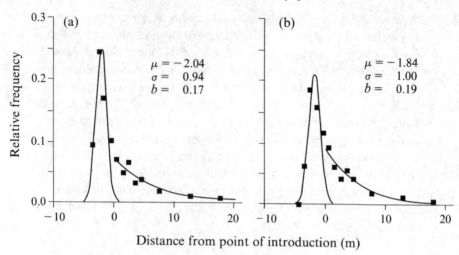

Fig. 3.15. Dispersal of seeds of *Bromus* spp. by a combine harvester: (a) *B. sterilis*; (b) *B. interruptus* (redrawn from Howard *et al.*, 1991). Normal distributions have been fitted to relative frequencies backwards from the source (mean = μ, standard deviation = σ); exponential regressions have been fitted to forward movement (parameter = b).

strip, moving across the traps as they went. They found that seeds were deposited fairly evenly along the 45 m strip, apart from a peak close to the source. There was little difference between the dispersal of two seed forms. About 2% of the weed seeds still on the parent plants at the time of harvest were carried down the entire length of the strip.

Howard *et al.* (1991) placed painted plants of *Bromus interruptus* into the path of a combine harvesting a winter barley crop. They also placed painted seeds of *B. interruptus* and *Bromus sterilis* directly into the header auger of the combine. Seeds were recovered from sheets of polythene dispensed by the combine as it moved. The majority of seeds were moved backwards from the point of uptake, with the modal distance being 3 m behind the source (Fig. 3.15). This backwards movement presumably resulted because the combine was long and the speed of plant material within it exceeded the forward speed of the machine. The number of seeds still being deposited in a forward direction was very low by a distance of 20 m from the source. Detailed analysis of these data suggested that the frequency distributions may be composed of one 'population' of seeds moving straight through the combine (giving a normal distribution deposited behind the source), and another 'population' which became lodged in sieves and ledges inside and outside the combine and released at a slow, steady rate (giving an exponential decline forwards of the source).

Table 3.2. *Typical values of two parameters for a range of dispersal vectors. Values are based on the very limited available data. Where dispersal is likely to be related to plant height, distances are given as multiples of height (*H*)*

Vector	d_{max}	d_{50}
Unaided	$1H$	$0.5H$
Explosive	$3H$	$0.5–1.0H$
Wind:		
'Gliders/parachuters'	100 m	$2H$
'Shakers'	$5H$	$2H$
'Tumblers'	several km	?
Water	> 100 m	?
Animals:		
Farm & birds	> 100 m	> 100 m
Ants	20 m (?)	5 m (?)
Cultivation	2–5 m	0.5 m
Harvesting	20–100 m	5–50 m

Combines differ in the pattern in which they deposit processed material; to aid incorporation of straw during subsequent cultivations, some may even be fitted with special choppers and spreaders. In a study by Ballaré *et al.* (1987a), three combines were compared in their dispersal of *Datura ferox* seeds when harvesting a soybean crop. Each deposited seeds in a characteristic width of strip, between either two, four or nine crop rows. Two of the combines deposited seeds up to 21 m forwards from the seed source; of these, one had its modal deposition 3.5–7.0 m from the source, the other had its peak in the first sampling unit, 0.0–3.5 m from the source (Fig. 3.16). No attempt was made to measure deposition backwards from the point of entry. The other machine moved seeds at least 98 m, with a modal distance of 35–42 m.

How important are the various vectors of dispersal?

Quantitative studies of weed dispersal have been few. Even for wind dispersal, which has been studied more than any other aspect, field data on populations of seeds are still uncommon. It is therefore difficult to make confident generalisations about the relative importance of different vectors. In Table 3.2 we attempt to compare dispersal parameters on the basis of available data. Although entries in the table are based on a certain amount of guess-work, they at least indicate the likely scales of dispersal distances. It is clear that for most vectors the distance moved by the majority of the

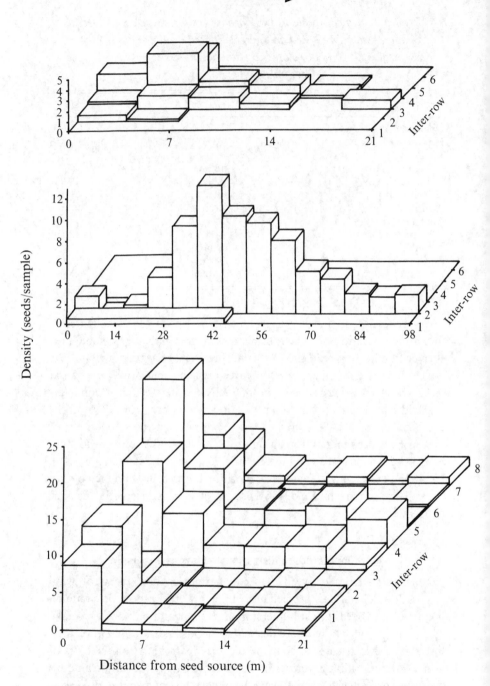

Fig. 3.16. Dispersal of *Datura ferox* seeds by three types of combine harvester (after Ballaré *et al.*, 1987a). Source of seeds was a narrow strip of plants perpendicular to the direction of the combine. Seeds were collected between the soybean rows.

population will be quite short. The best data, for wind dispersal, confirm that even where there are clear adaptations to a vector, the efficiency of the process is poor. Most wind-dispersed seeds do not move very far. It is likely that the same is true for fruits adapted to dispersal on the outside of animals, where the animals are grazing freely. As far as we know to date, within an arable farming system the biggest potential mover of seeds is the combine harvester. Within a pasture or rangeland it is probably the grazing animal; in aquatic systems it is probably water.

At the start of this chapter, we raised the hypothesis that if human activities are an important cause of dispersal, seeds may be moved furthest in the most heavily managed systems. At least for agricultural production systems, this is not supported by the data. Although combine harvesters may move seeds considerable distances, the same is true of grazing animals in less managed rangelands.

Within any agricultural system, there will be species behaving in different ways. In winter wheat crops, for example, *Avena fatua* matures at a similar time to the crop, is tall enough to be cut by a harvester and has large seeds. As a result, a high proportion (up to 75%) of seeds may be taken up into a combine, and hence may be dispersed widely. *Veronica persica*, on the other hand, is short and usually matures before the crop, dropping its seeds to the ground. Few seeds will be dispersed by the combine. Similarly, in pastures seeds of some species will be ingested by animals more often than others, and some will retain their viability better. Clearly, if we are to assess the importance of dispersal vectors to different species in various habitats, we need far more data on the fates of seeds. What proportion of seeds falls straight to the ground? What proportion is eaten by animals or 'ingested' by a combine harvester?

Few studies have attempted to separate out the influence of dispersal by different vectors within a population of weed seeds. One such example, however, is described by Howard *et al.* (1991). For an annual grass in an arable crop, the important vectors will be passive dispersal straight to the ground, cultivation and harvesting machinery. For *Bromus sterilis*, 60% of seeds fell to the ground under gravity, with a maximum dispersal of about 60 cm. Of harvested seeds, 34% were retained by the combine, resulting in only 27% of the seed population being dispersed by the combine. For these seeds, the modal distance which they were moved was 2 m backwards from their source, although some weed seeds moved up to 20 m forwards. Once the seeds reached the ground, some of them were lost (presumably by predation, etc.). Those remaining were moved a maximum of 1.8 m by a rotary cultivator (though less by other implements), with a modal distance

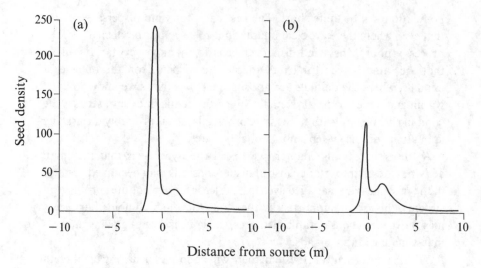

Fig. 3.17. Predicted seed dispersal resulting from a combination of natural dissemination, combine harvesting and cultivation: (a) *B. sterilis*; (b) *B. interruptus* (redrawn from Howard *et al.*, 1991). It has been assumed that all machinery travelled from left to right.

of roughly zero. A statistical model was used to predict the net effects of the dispersal vectors. Despite the potential of the combine harvester for dispersal, 65% of seeds would remain within 1.3 m of their source (Fig. 3.17). Maximum dispersal distance was predicted to be around 10 m.

Vegetative spread is perhaps less risky as an investment into occupying a local area than spread only by seed. Vegetative progeny may have an initial size advantage over seedlings for competing with other vegetation, since they may have access to a greater initial energy resource. Seeds must also land in a microsite where they can germinate and establish (such as a gap). In species with vegetative reproduction, successful invaders into a field can then spread with some certainty to invade the area immediately around them. We might then expect the overall distribution within the field to be more patchy than that of a non-vegetative species. A suitable study to confirm this might to analyse the distributions of species within the same genus, such as *Cirsium arvense* (vegetative and seed) and *Cirsium vulgare* (seed only), in fields containing both.

In Chapter 2 it was pointed out that the ability to spread vegetatively is a common trait in the list of the world's worst weeds of Holm *et al.* (1977). There may well be a degree of circularity in this observation, since the 'worst' weeds may be defined both by their wide geographic distribution and by their ability to dominate habitats. Dominance of plant communities seems to be common amongst vegetatively reproducing species. It was also observed in Chapter 2 that there is no obvious predominance of seed dispersal adaptations by species successful at dispersal between continents. Are weeds, however, better at short distance dispersal than non-weeds? The scarcity of dispersal data makes this a difficult question to answer. Within a family such as the Asteraceae there are weeds which are clearly good wind dispersers (such as *Taraxacum* spp.), while there are others with heavy seeds which probably do not move very far at all by wind (such as *Silybum marianum*). Within the Onagraceae there are several weeds with excellent dispersal abilities (based on observed escape fractions), such as *Chamaenerion angustifolium*; however, it is unclear whether those which have not become weeds are incapable of establishing in other than their native habitats or whether they have poorer dispersal abilities. Comparative studies of dispersal of successful weeds and non-weeds within the genus *Epilobium*, for example, might well prove informative.

Conclusions

Although dispersal of seeds by wind has been given a considerable amount of attention, both empirical and theoretical, dispersal by other vectors has seldom been studied in a quantitative manner. However, it is possible to conclude with some confidence that most seeds, even in wind-dispersed species, remain very close to the parent plant. The importance of different dispersal vectors will depend not only on the morphological adaptations of the species, but also on their phenological development in comparison with the timing of farm operations, in particular harvesting. This would make a fruitful area for future study.

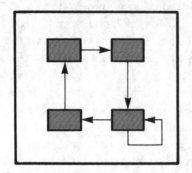

4

Processes involved in the regulation of population density

We argued in Chapter 1 that populations may fluctuate in size as a result of both intrinsic and extrinsic factors. There will be many such factors, the influence of each varying throughout the year and according to the developmental stage (seed, seedling, or adult) of the individuals within the population. In this chapter we will examine the causes of gains and losses of individuals to and from plant populations *within a generation*. This will enable an appreciation of the many factors and life history events which together determine the overall dynamics of population density over several generations.

The approach which we will take is phenological. Weed species in general display a range of life histories (see Chapter 1). For example, some species may germinate only in the spring and summer, then flower and die within the same year ('summer annuals'); others may germinate in the autumn and winter, then flower in the following spring or summer ('winter annuals'). It is clear that recruitment to the plant population from the seed population in the soil can be highly seasonal, as can seed production and adult plant mortality. Moreover, the phenology of a species predisposes plants and seeds to differing mortality risks due to seasonal climatic variation and due to crop husbandry.

Crop management practices are related inherently to seasonal events and may be major causes of weed mortality – indeed, the achievement of high levels of weed mortality is a major preoccupation of many farmers. For example, many weed seedlings emerging before crop sowing will be killed off by cultivations and seedbed preparation (Fig. 4.1). The timing of cultivations will determine when the weeds reach particular developmental stages, which can be critical to their chances of surviving a herbicide. Many herbicides will kill seedlings, but not mature plants. Herbicides are applied at times specified according to the growth stages of weed and crop, so as to

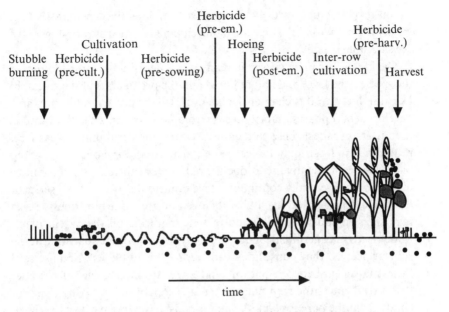

Fig. 4.1. Illustration of the development of a mixture of an annual weed and an annual crop, showing the main management events influencing losses from the weed population. For simplicity, the crop is shown as a cereal, the weed as a broad-leaved species (for key, see Fig.1.4). Herbicides may be applied at various times and an appropriate nomenclature has been developed by the agrochemical industry: abbreviations are pre-cult. = pre-cultivation, pre-em. = pre-emergence, post-em. = post-emergence, pre-harv. = pre-harvest.

maximise mortality of the weeds while minimising the damage to the crop.

Any attempt, therefore, to understand the processes leading to population fluctuations will need to consider developmental stages in relation to environmental events. At its simplest, the sequence of developmental stages in the life of an annual plant can be depicted as a cycle, from germination of seeds from the seed 'bank' in the soil, through to replenishment of the seed bank by reproduction – the plant *life-cycle*. The division of the life-cycle into a small number of simple categories (Fig. 1.6) enables us to focus on the gains and losses ('fluxes') at each stage, and to investigate the factors causing their variation. We can indicate on life-cycle diagrams the stages affected by key extrinsic environmental events, such as human-induced mortality, and by intrinsic population processes, namely density-dependence. Even when we turn to consider populations comprising overlapping generations and several cohorts, in which a range of developmental stages occur simultaneously (such as commonly for *Poa annua*), species with clonal growth (such as *Cyperus rotundus*), or long-lived perennials (e.g. *Prosopis glandulosa*), the division of development into arbitrary stages is of considerable benefit.

The life-cycle can be viewed in many ways, depending on the biology of the species, the level of detail required and the broad aims of the research. For simplicity, in this chapter the life-cycle will be divided into only three developmental phases, namely seeds (from here onwards the term seed is used in its broadest sense to denote a natural dispersal unit or diaspore, rather than in its strict botanical sense), plants and, for clonally spreading plant species, vegetative reproductive parts. Nevertheless, it is difficult to divide up an essentially continuous developmental process into a series of discrete categories and inevitably there is a degree of arbitrariness. Seeds may spend a considerable period of time maturing on the parent plant. Many seeds may germinate in the soil, only to die without appearing at the soil surface; do they constitute dead seeds or dead plants? Because seed losses may occur between maturity and entry into the soil, we will consider that a seed enters the seed phase when it is capable of surviving independently from the parent plant. A plant is considered to have been recruited from the seed bank at the point when a seedling (or shoot from a vegetative storage organ) is observed at the soil surface ('emergence'), i.e. the point at which it is no longer entirely dependent on seed reserves. Some of the many causes of gains to and losses from each of the three phases, seeds, plants and vegetative reproductive organs, will now be described in turn.

The seed phase

Gains in the number of seeds in a given area (or volume of soil) can result from immigration and emigration, as was addressed in the previous chapter. Here, however, we will consider a situation such as occurs at the centre of a homogeneous population, where migration in both directions will tend to cancel out. What then are the principal causes of gains and losses of seeds in a weed population?

Seed gains

The number of seeds produced by a plant in its lifetime (its *fecundity*) will depend on the size which it can achieve and the proportion of its resources that it invests in reproduction (its *reproductive effort*). Both plant size and reproductive effort are partly characteristic of a given species (and genotype), but they can also be strongly modified by factors intrinsic and extrinsic to populations. These modifying factors include time of seedling emergence, the proximity of neighbouring plants, herbivory, disease and factors affecting seed set and viability (such as pollinator availability and behaviour). Most is known about the effects of neighbours.

Species characteristics

Seed production will be determined by plant developmental and growth processes. The growth forms of plant species may be described as either determinate or indeterminate. When a terminal bud on a vegetative shoot is developmentally initiated as a floral meristem, continued vegetative bud production and internode extension will cease on the shoot. In consequence, shoot growth of that axis is precluded. Whilst cultivated plants may be selected for extreme determinism, with a single axis ending in an inflorescence (as in cultivated *Helianthus* species), most weeds show either indeterminate or ultimate determinate life-cycles.

Annual poppies in temperate regions (e.g. *Papaver rhoeas*) exhibit stems that terminate in a capsule bearing seeds. In the early life of an adult plant, a number of vegetative shoots are formed at the base of the plant which, once over-wintered, extend in height and produce a flower. In a similar manner, grass weeds such as *Bromus diandrus* and *Avena fatua* undergo a vegetative phase in which tillers are produced and, on receipt of appropriate environmental cues for flowering, the tillers produce inflorescences. In these instances, the weeds undergo an initial period of indeterminate growth in which vegetative plant size increases, but subsequently the growth of each shoot becomes determinate with the expression of an inflorescence. Fruiting then leads to senescence and death of the shoot. If all shoots of the plant display this characteristic then death of the whole plant ensues. Annual and ephemeral weed species characteristically show this behaviour. (Botanists recognise plants as *monocarpic* or *polycarpic* according to flowering behaviour. Monocarpic species flower once and die. Contrastingly, polycarpic species exhibit more than one flowering episode before death ensues.)

In contrast, the perennial grass weed *Elymus repens* possesses a potentially indefinite indeterminate clonal growth form. The species exhibits orthotropic (horizontal) shoots or rhizomes which extend the basal area of the plant by lateral branching. Episodically, rhizomes become plagiotropic (grow vertically) and produce an above-ground clump of tillers which seasonally flower and die. Rhizome growth is perennated, however, from lateral buds at tiller bases.

Senescence due to possession of a determinate growth form may occur in weed species, but often other external agencies may terminate the life of a plant. In *Senecio vulgaris*, for example, flowers are borne laterally on a spike and flowering continues with vertical extension. Growth and flowering is usually terminated by climatic change (winter). Similarly, in *Galium aparine*, a primary stem bears many lateral branches on which flowers are borne, and an individual plant can achieve considerable size by such lateral spread. This 'mix' of determinate and indeterminate growth poses difficulties in attempting to measure the fecundity of weed species.

It is not always easy to assess seed production. For species which flower over a short period, the entire population of new seeds can be assessed directly from a single count of the seeds attached to the parent plants, provided that dissemination has not begun. For less synchronous species, which are still producing new flowers as seeds from earlier flowers are dehiscing, we might assess numbers from counts of pod or fruit 'sites' on the stems, and multiply this by an estimate of the average number of seeds per pod/fruit. Such use of yield components has long been common in agriculture and horticulture (e.g. Ryle, 1966). Alternatively, we might set seed traps underneath plants and collect seeds as they fall (assuming dispersal is uniform and close to the parent), but this would then not account for losses of seeds while still on the plant. For these practical reasons, therefore, seed production per plant can be difficult to estimate and can become a protracted exercise.

Salisbury (1942) is a classic source of seed production data for many temperate species. His estimates of seed number for individual weed plants ranged from 39 for *Veronica hederifolia* to 23 138 for *Sonchus asper*. Examples of seed production by annual weeds are given in Table 4.1. Holm *et al.* (1977) gave values ranging from 150 seeds plant^{-1} for *Xanthium spinosum* to 500 000 seeds plant^{-1} for *Striga lutea*. Although some of the most fecund weeds (such as parasitic species) produce the smallest seeds, it is not possible to generalise that seed size and fecundity are correlated. Amongst the data given by Salisbury (1942) for temperate weeds, there is no statistical correlation between these two attributes.

For most species, Salisbury (1942) and Holm *et al.* (1977) unfortunately fail to give details of the environment in which plants were raised: were they growing as isolated plants, in crops or pastures, or in dense populations of their own species? It is probable that they were spaced plants. As will be seen below, the growing environment will have a considerable effect on seed production.

There is some evidence that the relationship between fecundity and plant size may be a relatively invariant characteristic of some species. The

Table 4.1. *Fecundity and individual seed weights of a range of temperate weeds*

Species	Seeds/plant	Seed weight (μg)
Cardamine hirsuta	640	178
Specularia hybrida	829	335
Anagallis arvensis	902	551
Senecio vulgaris	1127	160
Arenaria tenuifolia	1569	42
Sisymbrium thalianum	1650	31
Papaver hybridum	1674	158
Thlaspi arvense	1948	1245
Papaver argemone	1998	145
Linaria minor	2168	67
Erodium moschatum	5445	2220
Sonchus oleraceus	6136	420
Papaver dubium	13777	128
Papaver rhoeas	17070	138
Sonchus asper	23138	300

Source: As given by Salisbury (1942).

relationship between fecundity and plant size within a population can usually be described by a simple allometric equation:

$$s = c\,w^k \qquad (4.1)$$

where s is seed number per plant, w is plant weight and c and k are parameters (Firbank & Watkinson, 1986). Examples are given in Fig. 4.2. For some species, such as *Abutilon theophrasti*, there may be a minimum plant size below which there is no seed production (Pacala & Silander, 1987). In which case, a constant can be introduced into equation 4.1 to incorporate this size threshold, i.e.

$$s = c\,w^k - a \qquad (4.2)$$

where the lower biomass limit for seed production is $(a/c)^{1/k}$. In fact, for their data Pacala & Silander (1987) used a simple linear relationship, equivalent to equation 4.2 but with $k = 1$. Firbank & Watkinson (1986) reported that the value of k for *Agrostemma githago* was, indeed, very close to 1 ($s = 30.7w^{1.03}$, where w is in g dry weight).

When Watkinson (1981) grew *Agrostemma githago* with and without wheat he found no significant difference in the values of the parameters in equation 4.2. Cousens *et al.* (1988a) found little variation in the parameters

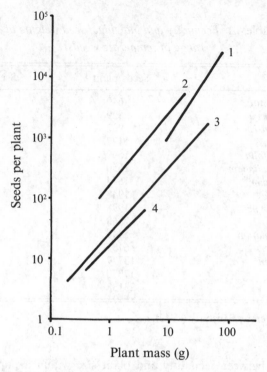

Fig. 4.2. Some regression lines for the relationship between seed production per plant and plant mass (after Watkinson & White, 1985): (1) *Rumex crispus* and *R. obtusifolius*, (2) *Plantago asiatica*, (3) *Bromus sterilis*, (4) *Setaria faberi*.

when a single stock of *Bromus sterilis* was grown in two fields at each of two geographic locations. It may be, therefore, that fecundity–plant size relationships are relatively constant for many species and a single relationship could be used at all sites. However, Rai & Tripathi (1983) found that for *Galinsoga parviflora* and *G. ciliata* fecundity–size parameters varied with soil moisture status. For obligate out-breeders dependent on insect pollinators, variability in fecundity at a given plant mass may be greater than for in-breeders, since pollination success may be more dependent on the weather and will be affected by the supply of pollinators. This remains to be tested rigorously.

Modifying factors

Even though fecundity–size relationships may often not vary as a result of environmental variables, plant fecundity can still vary considerably in response to site characteristics, soil moisture level, temperature, and growing season: a reduction or an increase in growth simply 'moves' a plant along the regression line. The variables eliciting a size response will depend on the physiology of the particular species. Many studies have been made of weed growth in pots, but relatively few have measured fecundity.

Zollinger & Kells (1991) found that a variation in soil moisture potential from 0 to -500 kPa caused a difference in *Sonchus arvensis* capitulum production of two orders of magnitude. Light intensity reduction from 1015 to 285 μE m^{-2}s^{-1} reduced capitulum production six-fold; there was little effect of a reduction in pH from 7.2 to 5.2. In a field study, Richardson *et al.* (1989) found that seed production by *Bromus tectorum* decreased with distance away from an irrigation source. They also showed that the timing of drought stress in pots could have a large effect on plant fecundity: plants were more sensitive to simulated droughts at anthesis than during seed fill. This observation is mirrored in the behaviour of small grain crops.

Many plants show fecundity responses to temperature during development, short periods at low temperature (causing 'vernalisation'), and daylength. For example, an increase in temperature from 15 to 25 °C caused an approximately five-fold decrease in *Avena fatua* seed production (Adkins *et al.*, 1987). Developmental responses to periods at 4 and 8 °C have been shown by some *Avena barbata* populations (Paterson *et al.*, 1976). In a study by Schuler (1986), most plants of *Avena* spp. ceased to flower at photoperiods below 10.5 hours, though they continued to grow vegetatively. It is to be expected, therefore, that seed production by species with determinate growth forms or where growth is truncated by a seasonal drought will vary with the time of year at which seedlings emerge. In general, the more rapidly plants mature, the less time will be spent increasing in size vegetatively and the lower will be the seed production. For example, in the USA, *Cenchrus longispinus* plants produced approximately 133 000 seeds per plant when sown in May, 49 000 when sown in June, 5000 from a July sowing, but only 40 from seeds sown in August (Boydston, 1990). *Tribulus terrestris* plants produced considerably fewer seeds when sown in August than when sown in May or June. In Australia, *Raphanus raphanistrum* seeds sown in May produced an average of 789 seeds plant^{-1}, whereas those sown in September produced only 7 seeds plant^{-1} (Cheam, 1986). For data collected from seven sowing dates, there was a positive correlation between fecundity and time to first flowering, indicating a negative response of seed production to rate of development.

Interference between plants will increase with the density and proximity of their neighbours. Plant size and seed production per weed plant will therefore be expected to decrease as the sowing density of a crop or

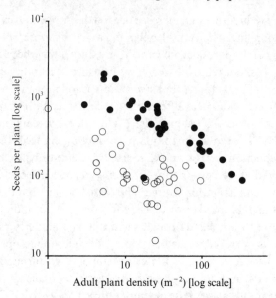

Fig. 4.3. Relationship between number of seeds per plant produced by *Avena fatua* and the surviving density of adult plants: ● in monoculture; ○ in a crop of winter wheat (redrawn after Begon & Mortimer, 1986, based on the data of Manlove, 1985).

beneficial pasture species is increased, and to decline as population density of the weed increases. For example, an *Echinochloa crus-galli* plant can produce over 100 000 seeds when growing in isolation (Norris, 1981), but only a few hundred when growing in a rice crop. An *Avena fatua* plant growing at low density in a cereal crop may produce 100 seeds or fewer, whereas an isolated plant which is able to tiller freely may produce 2000 or more seeds (Fig. 4.3). Plants emerging early on, within a crop or within a dense weed monoculture, will produce many more seeds per individual than plants emerging once the crop is well established (Peters, 1978; Mortimer, 1984). For example, plants of *Bromus rigidus* emerging between one and four weeks after the seeding of a wheat crop produced 57 seeds plant^{-1}, whereas those emerging between five and eight weeks after seeding produced only 3 seeds plant^{-1} (Cheam & Lee, 1991).

On the basis of our understanding of crop yield – crop density relationships, we might expect that as seed production per weed plant declines in response to weed density, seed production per unit area will reach a maximum, and may then decline at very high densities. Data summarised by Cavers & Benoit (1989) give maxima as high as 1 million seeds m^{-2} for *Amaranthus retroflexus*. However, a decline in seed production m^{-2} at high weed density has rarely been observed.

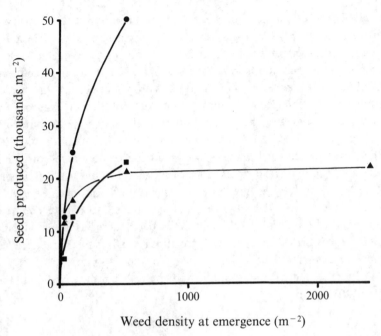

Fig. 4.4. Seed production by three temperate arable weeds: ● *Bromus sterilis*; ■ *Avena fatua*; ▲ *Galium aparine*. Curves are equation 4.5 fitted by non-linear regression. Parameter estimates for the equation are given in Table 4.2. (based on data from N. C. B. Peters, unpublished).

Fig. 4.4 compares seed production by three species of annual weeds, growing in wheat, in relation to weed density. *Galium aparine*, with ability of its individual plants to sprawl and to occupy a large area, is able to produce the greatest number of seeds per unit area at low density. However, its sprawling ability means that density-dependent effects start to appear at a very low density and it is the first of the species to reach its maximum seed production as density is increased. The two grasses, *Avena fatua* and *Bromus sterilis*, occupy a much smaller surface area and appear to approach their maxima at a much higher density. Cousens *et al.* (1985b) found that the maximum seed production per unit area by *Bromus sterilis* was greater in wheat than in barley and, not surprisingly, greater in a low density crop than a high density crop. Barley is generally found to interfere with weeds more than wheat, if both crops are sown on the same date (e.g. O'Donovan *et al.*, 1985).

Firbank & Watkinson (1986) used the equation

$$S = s_{max} N_w / [1 + a(N_w + \alpha N_c)]^b \tag{4.3}$$

to describe the relationship between seed production m^{-2} (S), density of *Agrostemma githago* (N_w) and density of a wheat crop (N_c). The parameters are: the seed production of an isolated plant (s_{max}), the relative effect of a crop plant in comparison to the effect of a weed plant (α), a coefficient describing the way that yield changes with plant density (a), and a parameter often referred to as 'the efficiency of utilisation of resources' (b). For values of b greater than 1 this equation allows for a maximum seed production, followed by a decline at very high densities. Watkinson (1980), in his development of the single species version of the model, originally stated that the biological significance of b is unclear, but that values of b greater than 1 '*reflect the fact that increasing density leads to a less efficient use of resources within a given area*'. It is difficult to equate empirical parameters with physiological quantities; indeed, there has been no formal proof that b *is* the efficiency of resource utilisation. It is safer to regard b as simply a parameter determining the shape of the response.

Within the density range studied in many experiments, the simpler asymptotic equation

$$S = s_{max}N_w/[1 + a(N_w + \alpha N_c)] \tag{4.4}$$

is often sufficient to describe seed production. If crop density is assumed constant, as in an additive interference experiment (Cousens, 1991), the equation

$$S = pN_w/(1 + qN_w) \tag{4.5}$$

can be used instead, where $p = s_{max}/(1 + a\alpha N_c)$ is the number of seeds per plant at very low weed density, $q = a/(1 + a\alpha N_c)$, and p/q is the number of seeds produced per unit area at very high weed density. As an example, Table 4.2 shows the parameter estimates obtained from fitting equation 4.5 to the data from Fig. 4.4. The lack of high density data means that the value of q is poorly estimated for the two grasses and the value of p/q involves considerable extrapolation. However, the parameter values indicate that although *Galium aparine* has 3.6 times the seed production per plant of *Avena fatua* at very low density, *A. fatua* has 1.3 times the maximum seed production per unit area of *G. aparine*. *Bromus sterilis*, with only 0.6 times the seed production per plant of *G. aparine* at low density, has 2.8 times the *G. aparine* maximum seed production per unit area.

Herbivory can reduce plant size directly by removing biomass, and hence reducing the resource supply for reproduction, and indirectly by reducing the ability of the plant to compete, again resulting in smaller plants with fewer resources for reproduction. An extreme example of insect herbivory

Table 4.2. *Parameters of equation 4.5 for three species in winter wheat in the UK*

	p (seeds/plant)	q	p/q (seeds/m^2)
Bromus sterilis	496	0.0081	61121
Avena fatua	234	0.0082	28399
Galium aparine	847	0.0387	21884

Source: Estimated from unpublished data of N.C.B.Peters.

is the complete defoliation of *Senecio jacobaea* occasionally caused by larvae of the moth *Tyria jacobaeae* in Europe; even if the plants recover, growth is likely to be seriously reduced, and so will seed production. Defoliating and leaf mining insects have been introduced in some countries to control invasive weeds (Harris, 1973). Other types of insect may reduce seed production by interfering with seed formation. For example, the gall-forming *Urophora* sp. reduced seed production of *Centaurea maculosa* at a site in Montana by 60% (Story, 1984). Another gall fly, *Rhopalomyia californica*, was collected on *Baccharis pilularis* in California and was introduced into Australia to control *Baccharis halimifolia* (originally from the east coast of North America). Early reports were that at heavily infested sites the gall fly was reducing seed production by 90% (McFadyen, 1985).

Plant diseases can reduce plant size and stop seeds from developing, in both cases leading to reduced fecundity. For example, Baker (1947) reported that plants of *Melandrium album* (= *Silene latifolia*) infected with the smut *Ustilago violacea* were shorter than uninfected plants and produced no seeds. This disease causes the abortion of ovules, and hence prevents reproduction. Paul & Ayres (1987a,b) found that infection of *Senecio vulgaris* by the rust *Puccinia lagenophorae* reduced plant size and ability to interfere with other plants, resulting in a decrease in fecundity of 46%. The ability of some pathogens to kill weeds has been used to great effect in the development of 'mycoherbicides', as well as in 'classical' biological control, and will be discussed later (see p.122).

Inefficient pollination is known to be a factor in the level of seed production in some plant species (Fenner, 1985). One of the few studies of the effects of pollinators on seed production by a weed is on *Sinapis arvensis* (Kunin, 1993). Plants were grown in a fan-shaped design, giving a range of *S. arvensis* densities, amongst backgrounds containing species with similar flowers, species with dissimilar flowers, or no flowers at all. Pollinators

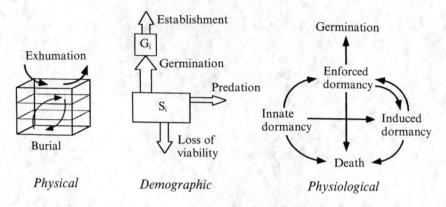

Fig. 4.5. Three sets of factors which determine the fate of individual seeds in the soil. See p. 104 for definitions of dormancy.

acted as floral specialists where only *S. arvensis* was available, as generalists where *S. arvensis* was growing amongst the similar-flowered *Brassica hirta*, and in a mixture of ways where dissimilar-flowered species were present. Visits to *S. arvensis* flowers declined with plant density in every treatment except where *B. hirta* was present. Seed-set was reduced in widely spaced plants in all cases.

Seed losses

Fig. 4.5 illustrates three sets of factors that interact to determine the fate of individuals in seed populations. In broad biotic terms, seeds can be lost from a seed bank by germination, death (loss of viability), or predation (in the widest sense). The magnitude and extent of each of these biotic factors will in part depend on the physical distribution of the seeds, in and on the soil. Moreover, physiological changes in the seeds and fluxes amongst different dormancy states may predispose seeds to particular fates.

Some seeds may remain on the parent for a considerable time, leaving them open to predation/removal by animals and machinery and long distance dispersal by wind. Others, such as many grass seeds, may dehisce very soon after maturity and fall straight to the ground. Species with dimorphic fruits, such as some thistles (Olivieri *et al.*, 1983) and the crucifer *Rapistrum rugosum*, may retain one type of seed on the dead or dying adult plant, while shedding another seed type early on. If cultivation follows shortly after dehiscence, burial will soon expose the seeds to a new range of mortality factors, such as soil pathogens, additional predators and germination (which, although it may result in birth of new plants, may be a cause of substantial loss from the seed phase). If cultivation does not occur or is delayed, many seeds will remain on the soil surface, where they may be more prone to predation from small mammals, death from fire (often used as a management tool) and germination followed by failure to establish. We can thus divide our discussion into pre-dissemination and post-dissemination losses; furthermore, it is useful for clarity to distinguish, within the post-dissemination category, between losses due to germination and losses by other means. After discussing the various individual causes of loss and the levels of reduction which results from each, we will consider the overall rate of loss of seeds from the seed bank resulting from their combined effects.

Losses while still on the parent

Most losses while on the parent plant can be attributed to either predation or, for weeds in crops, removal by harvesting machinery. The botanical descriptions of weed species in the flora of every continent often attest to the damage done to seeds whilst on the parent plant. It is, however, uncommon for the species of seed predator to be identified.

Many seed-eating birds are observed in cultivated fields, in pastures and in gardens. These may consume weed seeds, although the numbers of seeds lost in this way is uncertain. In Australia, crimson rosellas (*Platycercus elegans*) may feed on seeds of *Cerastium glomeratum* and other weeds growing in lawns (see Loyn & French, 1991); judging by the amount of time spent in this activity, seed removal may be considerable. In New Zealand, predation of seeds of the introduced *Carduus nutans* by the introduced goldfinch (*Carduelis carduelis*) was recorded to be 32% at one site (Kelly & McCallum, 1990). However, there are few other estimates of the proportion of weed seeds removed by birds.

In pastures, grazing animals (domesticated or wild) will consume vegetation containing weed seeds. The proportion of ingested seeds destroyed by animals will vary with the species and may be related to seed size (see also Chapter 3). In the study of seed passage through horses by St.John-Sweeting & Morris (1990), all seeds of *Asphodelus fistulosus* were destroyed, whereas some seeds of *Malva parviflora* and *Marrubium vulgare* survived intact. This was interpreted as being due to the lack of hard seed in *A. fistulosus*, resulting in germination and/or digestion during passage. Staniforth & Cavers (1977) found that whereas *Polygonum persicaria* and *P. lapathifolium* seeds passed intact through cottontail rabbits, the larger seeded *P. pensylvanicum* did not. The timing of grazing by farm animals to coincide with head emergence or anthesis in annual grasses is sometimes used in Australia to reduce future weed populations, particularly where those pastures will go into cereal cropping the following year. In addition, many herbivores are selective grazers and may leave forage (including seeds) of some species while eating others. Part of the selection may be related to plant morphology: for example, sheep can nibble on very short and prostrate species, whereas cattle can only remove taller species around which they can wrap their tongues. Hence, grazing behaviour may come to affect future community composition. Again, there are few quantitative studies of seed removal by grazing animals.

Predation of weed seeds may be deliberately enhanced in biological control programmes. Seed-eating insects have been introduced to control target weeds in a number of countries (Julien, 1992). The levels of predation achieved have varied (biocontrol is notoriously unpredictable in its success), but can reach as high as 90%. In one example, seed predation of *C. nutans* in New Zealand by the introduced beetle *Rhinocyllus conicus* was up to 49%, whereas at sites without this insect predation was very much lower (Kelly & McCallum, 1990). In their native Mediterranean Europe, in addition to bird predation, many thistle seeds may be eaten by insects. Sheppard *et al.* (1989) found that 0%, 63% and 90% of *Carduus nutans* seeds were eaten by insects, depending on whether the plants were summer annuals, biennials or winter annuals respectively.

Combine harvesters also act as 'predators' of seeds on the parent plants. The proportion of seeds removed from a field will depend on the proportion of seeds remaining on the parent at harvest time, which in part reflects the relative dates of maturity of the weed and the crop, and on the efficiency with which the combine separates weed seeds from grain. The more similar the sizes of the weed seeds and the grain, the less efficiently the weed seeds will be separated out and the more weed seeds will be 'predated'. Ballaré *et al.* (1987a) recorded losses of *Datura ferox* seeds from soybean harvesting of from 59 to 93%, depending on the type of machine. In this weed, mature seeds are held on the plant for a considerable time, and are mostly not dehisced before harvesting.

In *Avena fatua*, where a high proportion of seeds may be shed before winter wheat or spring barley crops are harvested, less than 2% of seeds may be removed from the field as contamination in the grain (Wilson, 1981). For a winter barley crop, which matures earlier than winter wheat and spring barley, fewer *A. fatua* seeds fall to the ground before harvest and as many as 17% of seeds may be removed with the cereal grain. In the latter case, only 23% of seeds had been shed before harvest, whereas in winter wheat the proportion could be as high as 75%. Moss (1983) found that

Alopecurus myosuroides began shedding seeds in early July and continued until late August. Winter wheat grain, harvested in late August, was contaminated little by *A. myosuroides* seeds, whereas winter barley harvested at the end of July was heavily contaminated.

It has been argued that removal of seeds by combine harvesters has been so efficient in the past that virtual extinction of certain species has resulted in some regions (Salisbury, 1961). Species such as *Agrostemma githago* and *Lolium temulentum* relied on being resown in crop seed saved from the previous year. Since the weed seeds were of similar dimensions to cereal grains, their removal by cleaning was not possible. However, with the advent of better seed-cleaning equipment and seed certification, these species have dramatically declined in Europe.

Losses in or on the soil (other than by germination)

Once seeds reach the soil surface, some of the same causes of mortality while on the parent may continue to act. There may be predation by birds, small mammals, earthworms and seed-eating insects (such as ants). In addition, decay by fungal attack may be important, especially in warm moist conditions, and in some situations fire causes considerable seed mortality. Losses which cannot be attributed to successful germination/emergence are often significant, but there seem to have been few quantitative studies of the factors involved. This may in part be due to the technical difficulties involved in designing the relevant experimental treatments to exclude potential agencies of loss. Laboratory studies have examined deterioration of seeds in storage (Bewley & Black, 1985), but it is unclear how applicable the results are to seeds in a soil environment.

Wilson (1981) recorded losses of up to 85% of *A. fatua* seeds in the autumn and early winter following their production. He reported that losses under cages (eliminating birds as predators) were as high as on uncaged plots. He suggested that losses were mainly by 'natural deterioration' due to environmental extremes at the soil surface, rather than by predation or microbial attack (although no formal attempt was made to determine causes of loss). Sarukhan (1974) attributed up to 54% of *Ranunculus repens* seed losses to predation (Fig. 4.6), citing voles as the likely predators on the basis of the type of damage seen and the presence of their runs. Seed decay was estimated to be up to 21% in the three *Ranunculus* species studied.

Reader (1993) placed plastic tube 'cages' over vegetation in old-fields to reduce seed predation. Emergence of seedlings of large-seeded species was increased, whereas the density of small-seeded species was unaffected by the

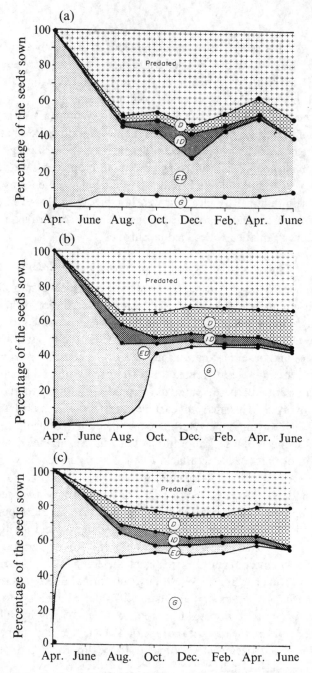

Fig. 4.6. Variation in the states of seeds in the soil over time (from Sarukhan, 1974): (a) *Ranunculus repens*; (b) *R. bulbosus*; (c) *R. acris*. Key: G, seeds recorded as germinating; ED, seeds in enforced dormancy; ID, seeds in induced dormancy; D, seeds which decayed. The remaining seeds, which were not recovered, were assumed to have been predated.

presence of cages. This suggested that seed predation was size-specific; it was considered that ants were the main predators. Predation by ants can be extremely high. In a pasture, Panetta (1988) recorded 62% removal of *Lolium rigidum* and 49% of *Chondrilla juncea* over a 24 hour period in summer; in autumn the equivalent losses were 33% and 26% respectively.

Earthworms are known to ingest the seeds of many plant species and either to move them intact to other parts of the soil profile or to render them inviable. The survival of seeds of 13 species passing through earthworms has been examined under laboratory conditions by McRill & Sagar (1973). For *Poa annua*, only 28% of seeds survived ingestion compared with 41% of *Poa trivialis*. The species most able to survive passage through the gut were *Sinapis alba*, *Lolium perenne* and *Trifolium striatum* (67% survival). It was also found that the percentage germination of seeds surviving ingestion was higher than in the seeds originally offered to the worms, suggesting that dormancy may be broken during passage.

The contribution by soil pathogens to seed mortality has not been well studied. Kiewnick (1964) found that surface sterilising *Avena fatua* seeds before sowing resulted in a fourfold increase in the numbers remaining viable after one year. Steam sterilisation of the soil had little effect on seed survival, indicating that the organisms responsible may have been associated with the seed and surrounding structures. Lonsdale (1993b) applied a fungicide to *Mimosa pigra* seeds; this reduced seed losses by 10–16%. Kirkpatrick & Bazzaz (1979) isolated over 25 species of fungi from surface-sterilised seeds of each of *Abutilon theophrasti*, *Datura stramonium*, *Ipomoea hederacea* and *Polygonum pensylvanicum*. Whereas most isolates reduced germination of *A. theophrasti*, only a few affected *I. hederacea*. Of the four species, *I. hederacea* and *A. theophrasti* had the greatest abilities to resist infection by fungi. It would seem that susceptibility to pathogenic fungi varies with species. Species certainly vary in the protection given to them by their seed coat, which is likely to impede penetration by fungi even after the caryopses have imbibed water. Hard seed coats, preventing imbibition, may also be extremely efficient in preventing fungal invasion. Study of seeds in the soil is not easy, and it is difficult to attribute exact causes of death. It may well be that infection by fungi may often be secondary, and not the primary cause of death.

Fire may be an important natural component of some ecosystems, such as arid rangelands. Weed populations in these systems may, sporadically, lose many seeds due to burning. Fire has also been deliberately used in some cropping systems, such as traditional 'slash and burn' systems, sugar cane and temperate cereal cropping. Crop straw and stubble have been burnt after harvest for a number of reasons, such as disease control and to avoid interference with sowing machinery. The practice has declined in recent years in many countries due to concerns over pollution and soil erosion. Modern machinery has also been developed to handle straw efficiently.

Fire may burn some weed seeds on the soil surface, and perhaps stimulate others to germinate through breakage of dormancy. The degree of loss from burning will depend mainly on the quantity of material burnt, its moisture content, its spatial distribution (whether evenly spread or in swaths) and wind speed, since these will determine the temperature of the burn and the area affected. Wilson & Cussans (1975) found that 68% of *A. fatua* seeds under spring barley straw swaths were killed by burning, giving a reduction over a whole field of 32%. They also recorded increased (up to 12% higher) autumn germination of seeds after an ensuing cultivation. Moss (1980a) also recorded losses of seeds under straw burning of 33 to 70% for *Alopecurus myosuroides*, with losses being highest on direct-drilled plots where seeds were closest to the surface. Even quite shallow burial is sufficient to protect seeds from burning, since temperature will decrease rapidly with depth.

Losses due to germination

Species vary in their degree and types of dormancy, and this will govern the rate of loss of seeds through germination. For example, Sarukhan (1974) recorded only 9% germination of *Ranunculus repens* in the year after production, with 32% remaining dormant in the soil. In *R. bulbosus* and *R. acris* there was 41% and 54% germination respectively, and in both cases there were almost no dormant seeds left in the seed bank. Depending on the species, the conditions experienced prior to burial (which may affect dormancy status) and the soil conditions after burial, dormancy may allow seeds to survive in the soil from a few months to many years. In general, seeds with only short term dormancy will germinate if buried too deeply and will be unable to emerge (presumably they rot away); those with long term dormancy mechanisms may be preserved by burial and will often only germinate when they are brought closer to the surface.

When seeds are shed and fall to the ground, they may be either *innately* dormant or non-dormant. Innate dormancy may simply diminish through time, at a rate set primarily by temperature – a process referred to as *after-ripening*. Particular sets of conditions, such as high temperatures or burial, may *induce* dormancy, whereas an abiotic trigger such as low temperature may break dormancy. In response to the environment, dormancy of seeds in the soil may cycle during the year (e.g. Baskin & Baskin, 1985). Even once these physiological forms of dormancy are broken, seeds may still not germinate, as a result of the absence of a factor required for growth, such as water, oxygen, certain temperatures or light: this is often termed *enforced* dormancy. It can therefore be appreciated that at any particular time the proportion of seeds in the soil capable of germinating and producing seedlings may vary considerably.

Many annual grass weeds, adapted to mediterranean climates, only have a very short period of innate dormancy, often just enough to enable them to avoid germination until reliable autumn rains and lower temperatures set in (Groves, 1986). After that time, many of these species will germinate no matter what conditions they are under, provided there is sufficient water for the seeds to imbibe. If they are close to the surface, they will have a high probability of successful emergence (though Froud-Williams, 1983, observed a high proportion of *Bromus sterilis* seeds which germinated at the soil surface but which failed to establish, probably due to poor seed–soil contact). If buried too deeply, they may germinate, fail to reach the surface, and die. Froud-Williams (1983) showed that all viable *Bromus sterilis* seeds will germinate when buried. Buried *B. sterilis* seeds were found to emerge from a maximum depth of 130 mm (Froud-Williams *et al.*, 1984). As a result, the species could be successfully controlled by tillage with a mouldboard plough in which good soil inversion (e.g. to 200 mm) is achieved, burying all seeds to a depth from which none could emerge.

In pasture, most seeds will remain close to the soil surface where they fall (although earthworms may move some seeds around within the soil profile). 'Hardseededness' (innate dormancy due to an impervious seed coat) in clovers breaks down over hot mediterranean summer conditions, probably due to large diurnal temperature changes at the soil surface (Quinlivan, 1971). This results in a pool of 'soft' seeds ready to germinate with the first rains of autumn. However, periods of rainfall over summer may initiate water imbibition and stimulate the early developmental phases of germination to begin. Since the rainfall over summer may be little and the moisture soon evaporates, seeds will often then dry out again. If the rainfall is sufficient, the seeds may pass a critical developmental stage beyond which the ensuing dehydration causes death of the embryo or seedling. It is likely that similar effects due to extreme wetting and drying cycles will cause mortality of weed seeds on the soil surface.

Seeds also remain close to the surface in 'minimum-tillage' cropping systems. Under more traditional tillage systems, however, such as mould-board ploughing, weed seeds will be incorporated into the soil to the depth of the ploughshare (Cousens & Moss, 1990). Thus, seeds may be buried to depths as great as 25 cm and then later be brought to the surface by the next cultivation. In most species, the majority of germination takes place within a few centimetres of the soil surface (see p.112). Chancellor (1964) found a correlation between average germination depth and seed size in 18 species of broad-leaved weeds. Some of the species, clearly requiring light, only germinated from within a few millimetres of the soil surface.

For seeds distributed throughout a soil profile, therefore, losses from germination will usually be greatest near the surface. Roberts & Feast (1972) showed that when seeds were mixed throughout the top 2.5 cm, seedling emergence and seed decline were greater than where seeds were spread throughout the top 15 cm. For *Chenopodium album*, for example, only 24% of seeds remained after five years when sown in the top 2.5 cm, whereas 66% were left when sown throughout the top 15 cm. Presumably, if seeds had been buried in a 2.5 cm band further down the soil profile, losses would have been less than in the 2.5 cm band at the surface. Germination may well be the reason that many researchers have found greater rates of seed decline close to the surface (such as Banting, 1974 and Miller & Nalewaja, 1990 for *Avena fatua*).

Cultivation, by bringing buried seeds close to the soil surface and enabling dormancy to be broken, causes some seeds to germinate and hence reduces the size of the seed bank. Roberts & Feast (1972) and Warnes & Andersen (1984) have shown that seed bank decline increases with frequency of cultivation. However, in general it is not feasible to use repeated cultivations to eradicate completely weed seeds from the soil. Unlike some annual grasses, many broad-leaved weeds have large seed banks and a high degree of dormancy; only a small proportion of seeds will germinate at a given time. In any case, frequent cultivations are likely to lead to increased soil erosion and are therefore not desirable.

The use of chemicals for stimulation of dormant seeds, forcing them to germinate and enabling them to be killed by conventional means (such as herbicides or cultivation), has been investigated over a number of years. Although stimulation has been obtained in the laboratory for many species, for example with gibberellic acid, potassium cyanide, hydrazine, butylate, sodium azide and ammonium nitrate (e.g. Hurtt & Taylorson, 1986) the results have been variable and have so far not been transferred successfully to the field, except in the case of the parasitic *Striga* spp.

The seeds of *Striga* are extremely small. Once they germinate, they must quickly attach themselves to the root of an acceptable host plant, or they will die. Chemicals, such as strigol, are exuded by the roots of some plant species, allowing the *Striga* seeds to detect the presence of a root system; in response to the presence of strigol they will germinate, attach to the host roots and grow into plants (Parker, 1983). Hence, seed mortality could be induced by application of strigol (or synthetic analogues) to the soil in the absence of a susceptible crop. The seeds would germinate but then die. The seeds must be in a receptive state, which is ensured by a warm, moist period for at least two weeks prior to application. However, strigol dissipates rapidly after application (Babiker *et al.*, 1988). It has been found that ethylene gas (or ethephon, which generates ethylene) will also act as a germination stimulant (Egley & Dale, 1970). When ethylene was applied between late April and late July to fields in the USA containing *Striga asiatica*, a 91% reduction in the seed bank was obtained (Eplee, 1975). Despite the efficacy of this treatment (three annual applications can give complete eradication), and its relatively low cost, less than 3% of the infested area in the USA is treated each year (Eplee, 1983). An alternative to chemical stimulation is a 'trap crop', a species such as cotton which releases germination stimulants but which is not an acceptable host (Wilson-Jones, 1952). The trap crop can be grown in rotation with susceptible crops to reduce the seed bank, as well as being a commodity in its own right.

Rates of loss of seeds

If the density of viable seeds remaining in the soil is recorded through time, the decline in numbers can be charted. The relative patterns of decline of a species under a range of conditions, or of various species under the same conditions, can then be compared quantitatively. However, there are a number of ways in which this can be done.

Many studies have shown that it is common for graphs of the logarithm of seed number plotted against time (in years) to be linear (Fig. 4.7). This implies that decline is exponential and occurs at a roughly constant rate. The equation for exponential decline is

$$N = N_0 e^{-bt} \tag{4.7}$$

or

$$\log_e N = \log_e N_0 - bt \tag{4.8}$$

where N is the number (or density) of seeds remaining, N_0 is the initial number of seeds, t is time, and the parameter b is a measure of the rate of loss of seeds on a logarithmic basis. The model implicitly assumes that seed number asymptotically approaches zero, i.e. that there is no maximum longevity. This may be a reasonable assumption within the duration of

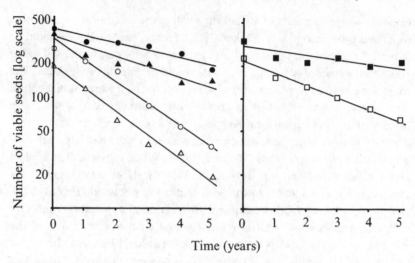

Fig. 4.7. Decline of seed banks of *Capsella bursa-pastoris* (●, ○), *Spergula arvensis* (▲, △), and *Papaver rhoeas* (■, □). Soil was either cultivated annually (open symbols) or undisturbed (closed symbols) (redrawn from Roberts & Feast, 1973).

most studies. When the exponential equation is differentiated, it can be seen that b is also the instantaneous per capita rate of loss, $(dN/dt)/N$.

A more popular measure is the proportional loss of seeds between two censuses. For the exponential model this can be calculated as

$$P = 1 - N_{t+1}/N_t = 1 - e^{-b} \qquad (4.9)$$

As the rate of loss becomes small, the values of b and $(1 - e^{-b})$, become similar. It appears possible from some papers that the distinction between these two measures of rate of decline may sometimes be confused.

Alternatively, rate of loss can be expressed as the time taken for the population to decline to some specified level. The 'half-life' is defined as the time taken for the population to halve. By substituting $N_0/2$ for N in equation 4.7, we obtain the half-life as

$$t_{1/2} = (\log_e 2)/b \qquad (4.10)$$

For the exponential model, all three measures of rate of loss (b, P and $t_{1/2}$) are independent of population density at any given time. The exponential model for seed decline is therefore both appropriate and convenient. Table 4.3 compares the three estimates of loss for *Abutilon theophrasti* (from regressions using data of Lueschen & Andersen, 1980). Since all three are related, they all show the same trend; loss was faster in systems with tillage.

Table 4.3. *Three measures of rate of loss of* Abutilon theophrasti *seeds.*
For definitions, see text. Raw data for the first three years of decline were
taken from Lueschen & Andersen (1980) and fitted to equation 4.7. (the
fourth year was excluded due to poor fit to the model). Treatments marked
with an asterisk include no cultivations

	b	P	$t_{1/2}$
1-plough fallow	0.78	0.54	0.89
2-plough fallow	0.96	0.62	0.72
*Chemical fallow	0.34	0.29	2.05
Continuous maize	0.67	0.49	1.03
Continuous oats	0.71	0.51	0.97
Maize/sorghum rotation	0.45	0.36	1.55
*Continuous alfalfa	0.22	0.20	3.12

There is therefore little to choose between the indices. Convention tends to dictate the use of the half-life. However, it is always necessary to ensure that data follow the exponential model before calculation of loss parameters.

A considerable body of comparative data on rates of loss for different species has been provided for weeds of horticultural fields. Roberts & Feast (1972) found that in cultivated soil losses by all species averaged 32% per year ($t_{1/2} = 2.5$ y), as compared to 12% per year ($t_{1/2} = 6$ y) in undisturbed soil. Under cultivation, the greatest loss rate was for *Veronica persica* (48% per year; $t_{1/2} = 1.1$ y) and the least was for *Fumaria officinalis* (20% per year; $t_{1/2} = 3.1$ y). In undisturbed soil, the greatest loss rate was 21% per year ($t_{1/2} = 2.9$ y) for *Vicia hirsuta* and the least was 6% per year ($t_{1/2} = 11.2$ y) for *Thlaspi arvense*. The greatest rates of loss recorded by Roberts (1962) in cultivated soil were 52% per year for *Capsella bursa-pastoris*, 50% per year for *Poa annua* and 49% per year for *Stellaria media*. Chancellor (1986) reported half-lives under pasture (following arable cropping) of over 20 years for *Fumaria officinalis* and *Aethusa cynapium*, 11 years for *Papaver rhoeas* and only 1.5 years for *Chrysanthemum segetum*.

It has been noted often that many annual grass weeds tend to have short seed longevity in the field. For example, Froud-Williams (1983) found that the seed bank of *Bromus sterilis* was exhausted by the spring after production, giving a half-life of very much less than one year. Chancellor (1986) quotes various studies of *Avena fatua* which have found half-lives of 1 year or less (but 3 years in one case). Martin & Felton (1990) found an average half-life of about 7 months for *A. fatua* seeds in a fallow. It is certainly of note that this species, which has been the object of so much research on dormancy mechanisms, can be relatively short lived in the soil!

Rather than estimating the time for the population to decline by 50%, some studies have examined the time taken for seed number to reach zero, i.e. the maximum longevity. For many species this therefore dictates extremely long experiments! The classic studies begun by Beal in 1879 and by Duvel in 1905 have been used to examine the maximum longevity of a range of species, some of them weeds (Priestley, 1986). In the Beal experiment, where seeds were buried in sand in upturned bottles and exhumed at intervals over 100 years, significant numbers of *Rumex crispus* and *Oenothera biennis* seeds survived to 80 years and *Verbascum blattaria* to 100 years. For the two grasses studied, *Bromus secalinus* failed to survive to the first (5 year) census, whereas *Setaria pumila* lasted 30 years. The study has been criticised for its artificial burial conditions, small sample sizes, the way that seed numbers were assessed (by counting germination after spreading the sand out in the glasshouse, hence dormant seeds were not accounted for) and the fact that the time of exhumation varied in some years (Bradbeer, 1988). However, the experiment remains a unique record of relative seed longevities.

Other records of maximum longevity come from assessment of floras of fields ploughed up after known times since previous cultivation. Some examples of these are given by Brenchley (1918). For example, she found that when fields were cultivated after 60 years there were significant numbers of *Polygonum aviculare* and *Atriplex patula* seeds remaining viable in the soil and able to germinate. At various periods in history, marginal land in Britain has been ploughed to increase food production in wartime; between these times, the land was under pasture. It is almost part of rural folklore that the crops in those fields were bright yellow due to abundant *Sinapis arvensis* germinating from the seed bank. Dorph-Peterson (1925, cited by Salisbury, 1961) found that 87% of *S. arvensis* seeds survived burial for 10 years, also indicating this species' potential longevity.

As an alternative to collecting repeated samples over time, the proportion of seeds surviving can be determined at just a single point in time. Since we would not, under similar environmental conditions, expect two decline curves to cross over (except owing to experimental or sampling error), repetition of sampling will be unnecessary if all that is required is a qualitative comparison of species or treatments. Roberts & Feast (1973) found a good correlation between a single census after 6 years and measures of loss rate based on a regression over several years.

Not all seed declines follow an exact exponential curve. In some data sets it is apparent that there is a much greater loss in the first year than thereafter. For example, Miller & Nalewaja (1990) found that the number of viable seeds remaining at two sites after 7 months was only 21% and 15% of the number originally buried (i.e. losses of 79% and 85%). Over the following 53 months, the number of remaining seeds declined further, to 11% and 13%, a reduction *over that period* of only 48% and 13%. A population of *Sinapis arvensis* studied by Thurston (1966) declined to half its original size within a year, but thereafter the decline was much slower. In both examples the faster initial decline may be because the populations of new seeds contained both non-dormant and innately dormant individuals (and perhaps enforced dormant seeds buried by cultivation). If the non-dormant fraction is the most abundant, there will be an initial rapid decline due to losses from germination, followed by more gradual decline due to degradation and slower germination as the remaining seeds are released from dormancy. For such reasons, in modelling weed population dynamics age of seeds in the seed bank is sometimes taken into account (e.g. Doyle *et al.*, 1986).

Although rates of seed decline have been compared extensively between species and between disturbed and undisturbed soil, there is little information on the effects of soil type or soil conditions. Salisbury (1961) stated that viability is retained longest under acid or waterlogged conditions. Kiewnick (1964) recorded losses of *A. fatua* 15% higher in a sandy soil than in a loamy soil. Lewis (1973) found a faster rate of decline in an acid peat than in a loam. Chepil (1946), however, found little difference in longevity between clay, loam or sandy soil. Clearly, more work is needed to compare rigorously rates of loss in different soil types and to explain the soil properties responsible for any observed variation.

The plant phase

Births (gains) from seeds

Gains to the plant phase, in species without vegetative reproduction, will be entirely from germination. The proportional loss of seeds from the seed bank through germination was discussed on p. 104. However, since many seeds may germinate but not successfully produce seedlings, losses from the seed bank and gains to the plant population may not be the same. As will be seen later, the subsequent fate of a plant will depend on when it (and the others in the population) germinates and the density of the population which results. In this section we will therefore discuss both the number of seedlings emerging and the timing of their emergence. In both instances we will pay special attention to the attempts that have been made at their prediction.

Proportion of the seed bank emerging within a year

The number of establishing seedlings will depend on the distribution of seeds in the soil, their requirements for germination and their ability to reach the surface from that depth. The requirements for breakage of dormancy and for germination have been reviewed extensively by other authors (e.g. Bradbeer, 1988) and will not be dealt with here.

In the absence of major soil disturbance, such as in a rangeland or pasture, seeds will accumulate near the soil surface. In contrast, cultivation may bury seeds to several centimetres. The effects of tillage on seed depth distributions have been reviewed by Cousens & Moss (1990). Not surprisingly, the greater the depth of cultivation, the deeper that seeds are distributed. A mouldboard plough will incorporate seeds more evenly than tined implements; over time the former will result in a relatively homogeneous depth distribution (Fay & Olson, 1978), whereas tines will caused a decreasing density of seeds with depth.

As discussed earlier in this chapter, although seeds can be spread throughout the soil profile, most germination will take place near the soil surface. Although *A. fatua* can emerge from as deep as 20 cm, most seedlings emerge from seeds in the top 10 cm (Holroyd, 1964). Moss (1985) found that most *Alopecurus myosuroides* seedlings emerged from seeds in the top 5 cm, with an average of 0.9 to 2.5 cm. Naylor (1972) found that 90% of *A. myosuroides* seedlings in a ploughed system were recruited from seeds in the top 2.5 cm of the soil. Moreover, 66% were from seeds shed the previous season. Moss (1980b) found that 80–90% of *A. myosuroides* seedlings were from new seeds in a direct drilled crop.

The patterns of emergence in relation to depth can be classified into two groups (Mohler, 1993), depending on whether or not an optimum below the soil surface is observed. The majority of studies have shown that germination declines monotonically with depth; for most of these the relationship

$$G = ae^{-bD} \tag{4.6}$$

can be fitted, where G is the proportion germinating, D is depth, a is the proportion germinating at the surface and b is a parameter describing rate of decline with depth. In Mohler's (1993) review, values of b between 0.03 and 3.6 were obtained, but with most values between 0.1 and 1.0. Some studies have shown an optimum depth, usually within 3 cm of the surface. Clearly, the ability to detect an optimum depth will depend on the number of depths observed and their positions. Most studies include only a small number of depths, usually four or five, covering the top 10 to 15 cm. Their accuracy in determining the optimum depth is therefore very poor. Since we would expect any optimum to be close to the surface, an experiment should include several shallow depths of burial and fewer deeper ones. Even with the best intentions, however, it is not easy to plant seeds at precise shallow depths.

Most reports of emergence in the literature, however, give estimates of total emergence and do not distinguish between depths of origin. Cavers & Benoit (1989), in reviewing the literature, found reports of from 1% to 12% of the viable seeds in the soil germinating in a year. The percentage is likely to vary with depth in the soil, and it may be that, say, although 10% of the *total* seed bank emerges, perhaps 90% of those at the optimum depth will emerge. If lack of cultivation leaves all seeds at the soil surface in conditions ideal for germination and there are few seeds deeper in the soil, percentage emergence may be very high. In a study of seed banks at eight locations in the Corn Belt of the USA, Forcella *et al.* (1992) found that emergence in a single year ranged from less than 1% of the seed bank in *Barbarea vulgaris* to 35% for *Setaria faberi*. Average emergence of grasses was higher (8.9%) than that for broad-leaved species.

Many studies of seed bank dynamics have been on grasses with relatively short-lived seeds, and the proportion emerging is likely to be higher than in more dormant species such as *Fumaria* spp. or *Papaver* spp. Various studies by B. J. Wilson (cited by Cousens *et al.*, 1986), suggest that emergence of old *Avena fatua* seeds in the soil will average about 10%, whereas only 3% of new seeds will produce seedlings in their first autumn if tillage is shallow, and only 0.2% will produce seedlings if the soil is ploughed. The figures may be higher for new seeds of *Avena ludoviciana*, where there is usually less innate dormancy. The data of S. R. Moss (summarised by Doyle *et al.*, 1986) give an average annual emergence of *Alopecurus myosuroides* of 20% of seeds in the top 2.5 cm. Under moist conditions, close to 100% of *Bromus sterilis* seeds will germinate, but their success in emerging will depend on how close they are to the soil surface. Most seedlings were able to emerge from 75 mm, but the maximum depth was 130 mm (Froud-Williams *et al.*, 1984).

There have been a small number of attempts to use empirical regressions to predict seedling densities emerging from seed banks. Naylor (1970) found a correlation between total number of *Alopecurus myosuroides* seedlings emerging in the field and numbers emerging from soil cores spread out in trays in the glasshouse. He proposed this as a bioassay for determining seed bank density, but did not go on to a full-scale test of his predictions. Others have found that the proportion of the seed bank emerging in a given year can be extremely variable, particularly in species with long-lived, highly dormant seeds (G. W. Cussans, pers. commun.). For such species, relationships between seed bank density and seedling emergence are likely to vary considerably between years. Emergence is also likely to vary between years as a result of weather and field management (cultivation in particular).

In a study of weed emergence in summer crops, Wilson *et al.* (1985) used regression to relate seedling numbers to seed bank density obtained from soil cores. In the first year of study, they found significant relationships for only four of the 13 species found in the seed bank, namely *Solanum rostratum*, *Chenopodium album*, *Portulaca oleracea* and *Helianthus annuus*. Two of these species were amongst the three most common in the seed bank. For a limited sampling frequency, it is to be expected that species represented poorly in the seed bank will not be predicted accurately. In the second year of their study, they found significant relationships again for *C. album* and *P. oleracea*, and for *Amaranthus retroflexus* and *Echinochloa crus-galli*. Three of these were among the four most abundant species in the seed bank. The uncertainty of detecting relationships, even for the most abundant species (*A. retroflexus*) suggests that the usefulness of regressions to predict seedling densities will be variable. Very high levels of sampling may be required to make the technique more reliable.

Forcella *et al.* (1992) recorded emergence at six sites and tried to relate percentage emergence to environmental variables. For *Chenopodium album*, emergence was linearly correlated with the number of 'growing day degrees' in a ten day period prior to the main emergence period. For *Amaranthus retroflexus* there was a curvilinear relationship with rainfall, again recorded in a ten day period shortly before expected emergence.

Seasonal timing of emergence

One technique for controlling weeds, used particularly before the advent of herbicides, is to wait until their emergence, cultivate the soil so as to kill them (perhaps going through this procedure several times) and then sow the crop. A knowledge of when emergence will occur will therefore be critical for predicting the response of a species to changes in crop management practices, especially those involving tillage.

We know from casual observations that, for example, if there is a hot, dry period (summer) there will be a flush of emergence of many species when rains come or temperature falls (autumn), and that after a cold spell (winter) emergence of some species will occur as temperatures warm up again (spring). We can categorise the species according to the season of their main periods of emergence, or as species capable of germinating all year round.

Detailed field observations of emergence are infrequent and are confined to a few classic studies. Roberts (1986) and Mortimer (1990) described emergence for a large number of species in the field. Species exhibited characteristic patterns (Fig. 4.8). Most species fell into three categories: (1) those emerging only in spring and early summer (summer annuals); (2) those emerging primarily in the autumn, but with some emergence through to spring (winter annuals); (3) those emerging virtually throughout the year, although often with slight peaks in spring and autumn. However, some species did not fall into any of these main categories. The durations of the emergence periods were largely independent of soil disturbance, although in some cases cultivation determined when the maximum emergence was, by bringing more seeds to the surface. The emergence period appeared to be determined mainly by temperature and rainfall events. Roberts & Lockett (1978) interpreted considerable differences in emergence period of *Veronica hederifolia* over 18 years as being due to differences in weather.

Murdoch (1983) buried *Avena fatua* seeds in winter at a depth of 25 mm and observed the timing of their emergence over two years. Seedling emergence appeared to coincide with minimum daily temperatures greater than 3 °C and maximum daily temperatures less than 17 °C. Peters (1991) also recorded emergence of *A. fatua*, in his case over three years. Emergence in winter and spring always occurred in warm periods following periods of mean daily temperatures below 4 °C. Autumn emergence was variable, but in one year it coincided with rainfall and a reduction in temperature to about 15 °C. More studies to relate emergence quantitatively to environmental parameters would enable better predictions of emergence periods and seedling numbers.

Predicting the timing of emergence within a year would appear difficult. Seeds may be in a range of dormancy states at any given time and those at different depths may be experiencing a range of environmental conditions capable of affecting germinability (see Zorner *et al.*, 1984, for *Avena fatua*). Mathematical models could be constructed to include these various processes and to predict the dynamics of emergence within a year from weather data. Perhaps because of the complexity required, there have been few attempts to do this. Data are available: laboratory studies of temperature and light responses, and of the factors required to break dormancy,

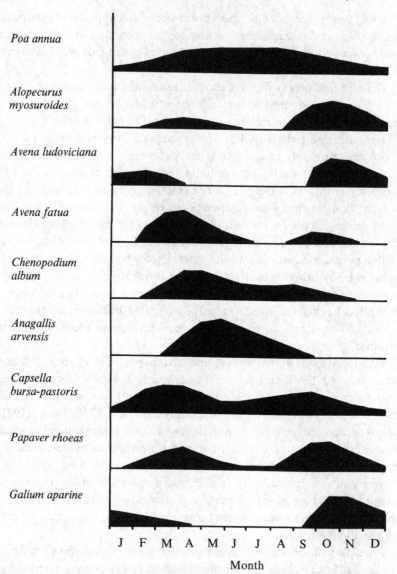

Fig. 4.8. Emergence periods of a range of temperate arable annual weeds in the UK (after Mortimer, 1990). The height of the shaded area indicates the relative frequency of emergence.

abound. Goloff & Bazzaz (1975) formulated a general model which predicted the start and subsequent rate of germination in relation to temperature and moisture based on monomolecular reaction kinetics. Other workers have focussed on the importance of temperature in particular.

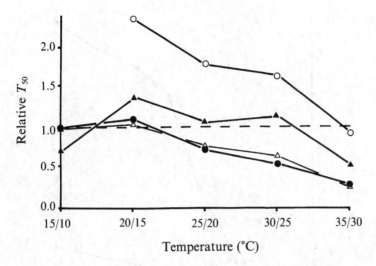

Fig. 4.9. Relationship between time to emergence and temperature (day/night, 12 hour photoperiod) for four weeds: ▲ *Chenopodium album*; ● *Amaranthus powellii*; △ *Setaria viridis*; ○ *Solanum ptycanthum* (redrawn from Weaver *et al.*, 1988). Relative T_{50} is the time taken from sowing to 50% emergence for the weed divided by the time taken by tomato; values greater than 1.0 indicate that the weed emerged after the crop. Seeds were sown at the soil surface.

Weaver *et al.* (1988) studied the effects of temperature on rate of emergence in pots. For tomato and four weeds (*Chenopodium album, Solanum ptycanthum, Setaria viridis* and *Amaranthus powellii*) they obtained base temperatures (i.e. the lower temperature limits for germination) and the number of day degrees taken from sowing to emergence. For various temperature conditions they then predicted the length of time for each to emerge (assuming that development of all species began at crop sowing). By expressing time to emergence of each weed relative to the crop, they were able to predict the sowing temperatures under which a tomato crop would emerge furthest ahead of its main weeds (Fig. 4.9), and would therefore experience the least interference. These predictions, however, remain untested in the field. Given the large volume of laboratory data on temperature responses of weed seeds, similar models should be possible for many species (Cousens & Peters, 1993).

Benech Arnold *et al.* (1990) established the requirements of *Sorghum halepense* seeds for germination and their rates of germination in relation to temperature in the laboratory. They divided the seed population into highly dormant, dormant and non-dormant seeds and derived simple rules for their behaviour. In their model, the probability of passing from one

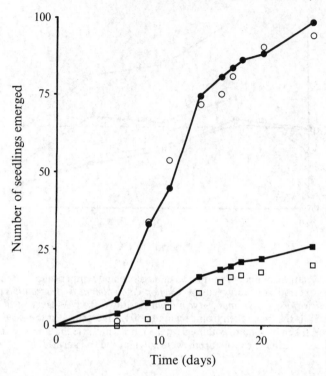

Fig. 4.10. Cumulative number of *Sorghum halepense* seedlings emerging from bare soil (●, ○) and from shaded soil (■, □). Simulations from a model are shown as solid symbols; observed means are shown by open symbols (after Benech Arnold *et al.*, 1990).

category to another was made to be a function of the number of elapsed diurnal cycles of a given temperature amplitude, and for which a certain maximum temperature was exceeded. Rate of germination of non-dormant seeds was assumed to be a function of the number of day degrees above a base of 8.5 °C since they entered that category. These rules were used to simulate numbers of seedlings and times of emergence on the basis of records of soil temperatures. Comparison with observed emergence showed good agreement with the model (Fig. 4.10).

For many species, the range of temperatures within which germination will occur varies during the year. The 'window' of possible temperatures opens and closes. For example, *Spergula arvensis* seeds exhumed from the seed bank in winter will not germinate at any temperature when incubated in water in petri dishes. In summer, they will germinate between 5 °C and 30 °C (Karssen & Bouwmeester, 1992). If the dynamics of the window is known (or can be predicted), germination will be expected when field temperatures fall within the window. Karssen & Bouwmeester(1992) modelled the germination window as an empirical function of the duration of cold temperatures (below a characteristic species base temperature), cumulative mean temperature, and mean temperature of the preceding period. They found that they were able to predict the emergence period with reasonable accuracy. Similar results were obtained for *Polygonum persicaria*, *Chenopodium album* and *Sisymbrium officinale*.

Deaths (losses) of plants

The level of mortality resulting from extrinsic (density-independent) factors, from intrinsic (density-dependent) population processes, and from interactions of the two, will depend on the susceptibility of the species to each factor, the timing of seedling emergence, the climate which the species encounters and the management regime to which it is subjected. Although there has been considerable discussion of whether density-dependent regulation of populations occurs in complex 'natural' (perennial-based) communities (Keddy, 1989), it is unlikely in any habitat that only one or the other group of factors will act in isolation.

It is often difficult to record mortality directly: techniques for identifying plants, such as mapping or tagging, are difficult to use within dense plant stands. It is much easier to record the *number of survivors* in fixed quadrats and to deduce mortality from two census dates (provided of course that no new recruitment occurs). Although it is then impossible to ascribe a cause of death to a particular individual, it is usually possible by close examination of the resulting 'survivorship curve' to correlate synchronous mortality episodes to particular weather or management events, or to periods where intense competition is likely.

In this section we will discuss our limited knowledge of the factors causing mortality. We will also discuss briefly the use of survivorship curves to classify temporal patterns of mortality.

Density-independent factors

In many of the habitats in which weeds grow, active measures will be taken to control them. In addition, there may be other aspects of land management which will directly or indirectly affect weeds. Hence, cultivation and herbicides are among the most important causes of density-independent mortality. In addition, weather contributes to mortality in both disturbed and 'natural' habitats.

Cultivations are used by farmers to prepare a loose seed bed for easy crop seedling emergence, but also to kill the weeds which will potentially interfere with the crop. In traditional temperate cereal cropping, this may entail a deep cultivation using a mouldboard or disc plough, followed by one or more shallower 'scarifications' using tines, discs or a harrow. The first cultivation will kill most of the existing vegetation, whereas the later scarifications kill newly emerged seedlings or any remaining established plants. The probability of death as a result of cultivation is likely to be primarily a function of the efficiency of soil inversion, soil properties, and the weather during and after cultivation. If carried out effectively, each cultivation is capable of killing up to 100% of plants emerged at that time. There are, however, few data on levels of weed seedling mortality actually achieved by farmers.

Mortality from herbicides, if they are applied at an early age or to very sensitive plants, may also be independent of density. Herbicide applications can be categorised according to the developmental stage of the crop or weed at which they are applied. Examples are pre-sowing, pre-emergence, post-emergence and pre-harvest herbicides. Mortality will depend particularly on the species of weed, since many herbicides can selectively kill particular target plants while leaving others (such as the crop and its relatives) virtually unharmed. Although chemical company advertising may lead users to expect 100% control of specified weeds every time, weed control is seldom perfect. Late emerging plants may escape the chemical; some seedlings may be protected by crop plants; some may by chance be missed by the spray jets. Weather conditions before and after spraying, particularly rainfall and temperature, will also affect chemical performance (Kudsk & Kristensen, 1992).

Many experiments with herbicides do not monitor weed mortality at all, merely score control according to a visual scale, or (in the case of grass weeds) assess numbers of inflorescences at maturity. Data on herbicide-induced mortality are surprisingly uncommon considering the vast amount of weed control research. However, we will present three examples here, illustrating how mortality can vary between sites, chemicals, application times and dose rates.

For broad-leaved weed control in cereals, many farmers use a single broad spectrum herbicide; some of these chemicals also kill certain grasses. Table 4.4 shows the results of three experiments on two such herbicides, the chemically related isoproturon and chlorotoluron. It can be seen that mortality was not 100% for most weeds; mortality varied between the herbicides and between sites for chlorotoluron. *Veronica persica* was killed

Table 4.4. *Herbicide-induced mortality (%) of weed floras in winter wheat in the UK*

Species	Herbicide		
	Isoproturon Site 1	Chlorotoluron Site 2	Chlorotoluron Site 3
Veronica persica	0	67	74
Aphanes arvensis	84	97	—
Poa annua	78	0	0
Chrysanthemum segetum	100	—	—
Stellaria media	—	93	99
Lamium amplexicaule	—	77	—
Legousia hybrida	—	99	—
Veronica hederifolia	—	—	0

Source: From Wilson & Cussans, 1978.

by chlorotoluron but not isoproturon; *V. hederifolia* was not killed by chlorotoluron.

Often weed problems occur for which there is no recommended herbicide; a range of chemicals and rates are then tested to find the best herbicide (often giving far from perfect control) and the minimum application rate. *Delphinium barbeyi* can cause poisoning of cattle in mountain rangelands in the USA and Canada. One of the chemicals tested for possible control was picloram (Fig. 4.11). It can be seen that mortality increased asymptotically up to a maximum efficacy. As is often found, variability amongst sites and years was greatest at intermediate application rates.

A balance must be found in timing herbicide applications so that the damage to a crop is minimised while an acceptable level of weed control is obtained. Considerable attention during herbicide evaluation is paid to the way that weed control varies with weed growth stage. An example is given in Table 4.5. It can be seen that for the four broad-leaved weeds, mortality from the herbicide ethametsulfuron methyl was never 100%. For all species, mortality declined when applied at the third time of application in comparison with the second time of application. This can be interpreted as the plants becoming in some way more tolerant of the chemical as they develop.

Biological control, in which pests or diseases are used to control weeds, is a major cause of mortality in some species. The example of *Opuntia* control with *Cactoblastis* was described in Chapter 1. That form of biological control, where an exotic agent is released, its populations build up of their

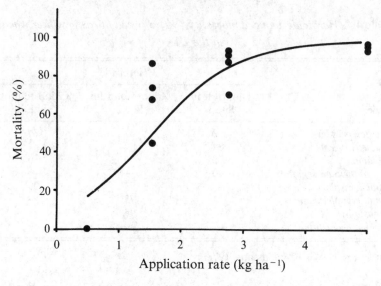

Fig. 4.11. Mortality of *Delphinium barbeyi* resulting from a herbicide (redrawn from Ralphs *et al.*, 1990). Picloram was applied in two years at each of two sites. The regression line is a logistic equation, fitted by the original authors and not constrained to pass through the origin.

own accord and the weed populations then decline, is termed inoculative, or 'classical' biocontrol. Since the abundance of the agent, and hence the mortality caused, will depend on the availability of its food supply (the weed), mortality is density-dependent. Inoculative biocontrol will therefore be discussed in the next section. Here we will discuss another technique, referred to as inundative biocontrol, where an agent is multiplied artificially and then applied to overwhelm the weed population.

Observations of diseased weeds led to the idea that endemic pathogens could be harnessed to the advantage of farmers. By collecting the pathogens and bulking them up in laboratory cultures, they can be applied as foliar sprays ('mycoherbicides', or more generally 'bioherbicides'). The first commercial release of a mycoherbicide was *Phytophthora palmivora*, registered in 1981 for the control of *Morrenia odorata* in citrus orchards in Florida. This was followed in 1982 by *Colletotrichum gloeosporoides* for controlling *Aeschynomene virginica* in rice and soybeans in southern USA. These pathogens are very successful in killing their hosts, commonly resulting in 95–100% mortality (Templeton *et al.*, 1979). Without repeated applications, however, the weed and pathogen would return to their former equilibrium, with the weeds at unacceptable levels; the method relies on

Table 4.5. *Herbicide-induced mortality (%) in relation to stage of weed development. All weeds were sprayed when* Sinapis arvensis *was at the cotyledon, 4–6 leaf or 6–8 leaf stages. Results are for ethametsulfuron methyl applied in Canada at a rate of 10 g ha^{-1} in the autumn*

	Growth stage of *Sinapis arvensis*		
Species	Cotyledon	4–6 leaf	6–8 leaf
Sinapis arvensis	86	96	61
Capsella bursa-pastoris	87	96	84
Chenopodium album	35	48	27
Amaranthus retroflexus	48	61	36

Source: From Buchanan *et al.*, 1990.

artificially increasing the pathogen population so as to overwhelm the target weed population. Despite major international research programmes, there have been few further mycoherbicides released commercially (Watson, 1989). At least some of the failures have been due to problems in developing formulations for commercial application and environmental limitations to pathogen development.

Extremes of weather, either hot/dry or cold, are often causes of plant mortality. Cold, particularly below freezing point, can be a major cause of mortality in both summer and winter weeds. Many summer annuals have poor frost sensitivity; mature and immature plants are likely to be killed by the onset of frosts (indeed, it is their inability to reach maturity before frosts which may prevent their spread into cooler climates). Seedlings of such species emerging very late in the season are unlikely to survive the winter. For example, Debaeke (1988) found that winter mortality of *Anagallis arvensis* in northern France was 100%. In the harsh winters of Canada and northern USA, mature plants and autumn-germinating seedlings of *Sinapis arvensis* (*Brassica kaber*) are killed by the cold; the population regenerates each spring from the seed bank.

Amongst more tolerant species, survival may depend on the severity of the weather and the growth stage of the plants. Ability to survive over winter has been shown to increase with plant size in *Capsella bursa-pastoris*, a species which can germinate throughout the year (Debaeke, 1988). In Europe, cold winters may kill many autumn-germinated *Avena fatua*, whereas in mild winters a large proportion of them will survive. Debaeke (1988) recorded winter mortality of *Stellaria media* of up to 50%, *Viola arvensis* up to 30%, *Galium aparine* up to 70% and *Papaver rhoeas* up to 60%. Mortality of *Capsella bursa-pastoris* ranged from 30 to 100%.

In many studies of weeds in winter crops, plant numbers are counted in autumn and then again in spring. The difference between the two censuses represents the balance of mortality and new emergence, and caution must be used when interpreting winter mortality from such data. The only reliable way of determining mortality is to tag plants and follow their fates.

Hot weather and lack of rainfall in summer are also likely to increase plant mortality. It has been mentioned previously that in the hot, dry summers of mediterranean climates, short periods of unseasonal rainfall ('false breaks') may cause seeds to germinate; insufficient moisture from lack of further rain will then result in their death. Unusually dry summers or droughts may cause the premature death of many species, before they can set seed. Unfortunately, there are few data on summer mortality of weeds; all we can say is that if the weather is harsh enough, mortality can reach 100%! It may be difficult, in a dense crop, to determine whether plants dying in the summer are being killed directly by drought or by competition for water and other resources.

Because of the timing of seedling emergence, cold-induced mortality will occur late on in the life of summer annuals, but early on in the life of winter annuals. For summer annuals, therefore, density-dependent mortality due to interference will already have occurred before plants are frosted, whereas for winter annuals density-dependent mortality will occur after most cold-induced (density-independent) mortality. This has implications for the way in which population models are constructed (see Chapter 5).

Density-dependent factors

Where plants grow in close proximity, they will interfere with each other via processes such as competition and allelopathy (Harper, 1977). In response to density, plants may show phenotypic adjustment, by changing their morphology or physiology in response to reduced resource supply, or they may die. In a population of plants developing from seedlings, the first responses may be reduction in growth; only later, and if interference becomes intense enough, will density-dependent mortality become important. Also, if interference is intensified by the initial population density being high, mortality will begin earlier than in lower density populations.

Density-dependent mortality is often referred to as 'self-thinning'. The inter-relationship between plant mortality and plasticity of plant size has been studied intensively in monocultures (see reviews by White, 1980, and Lonsdale, 1990). It has been found that as plant density decreases during

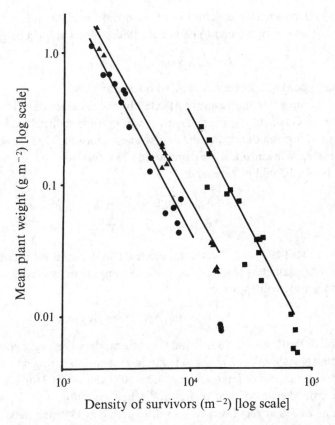

Fig. 4.12. Self-thinning lines for *Cichorium endivum* (●), *Agrostemma githago* (▲) and *Festuca pratensis* (■) (after Lonsdale & Watkinson, 1983). Regressions were fitted by principal components analysis; points well below the lines were omitted from the analysis, since they had not begun to decline in density.

self-thinning and plant weight increases during growth, the trajectory of $\log_{10}$(mean plant weight) in relation to $\log_{10}$(density) approaches and then follows a line which often has a gradient of close to -1.5 (Fig. 4.12). However, there has been little work involving weeds or weed–crop mixtures.

Rather than consider the time-course of interference in detail, for many purposes we may wish only to relate density of mature plants (N_p) to density of seedlings (N_s). This then measures overall mortality during the plant phase. The equation

$$N_p = N_s(1 + mN_s)^{-1} \qquad (4.11)$$

provides a reasonable description of monoculture data, where m is an arbitrary 'shape' parameter (Yoda *et al.*, 1963). Mortality will be given by

$$1 - N_p/N_s = mN_s/(1 + mN_s) \qquad (4.12)$$

Hence, as density approaches zero, so does mortality.

Self-thinning will, however, be affected by the presence of other species (Harper & Gajic, 1961), since resources will become reduced by a greater amount at a given density of the first species. Lonsdale (1981, quoted by Firbank & Watkinson, 1985) proposed that for two species mixtures equation 4.12 could be extended:

$$N_{p1} = N_{s1}[1 + m_1(N_{s1} + \gamma N_{s2})]^{-1} \qquad (4.13)$$

$$N_{p2} = N_{s2}[1 + m_2(N_{s2} + \delta N_{s1})]^{-1} \qquad (4.14)$$

where subscripts 1 and 2 refer to species 1 and 2 and m, γ and δ are regression parameters. As the density of one component species approaches zero, mortality will approach

$$1 - N_{p1}/N_{s1} = m_1\gamma N_{s2}/(1 + m_1\gamma N_{s2}) \qquad (4.15)$$

Firbank & Watkinson (1985) found that this model gave a very good fit to data on *Agrostemma githago* growing in mixtures with spring wheat. They obtained parameter estimates of $m_1 = 4.2 \times 10^{-4}$ and $\gamma = 0.62$ for *A. githago*. For a crop density of 300 plants m^{-2}, their equation would predict a mortality of 7% at very low weed densities, rising to 33% at a weed density of 1000 plants m^{-2}. It is likely that the model could be extended in the same way to mixtures of more than two species.

Mortality from herbicides is usually assumed to be independent of plant density. However, at least in theory herbicide-induced mortality could be density-dependent. A herbicide acting on weeds as they germinate or shortly after emergence will induce a mortality rate which depends on the efficiency with which the herbicide reaches the target and on environmental factors. At that time, because plants are still dependent on seed reserves, plant size would not have been affected by density and all plants would have an equal chance of death. However, if a herbicide is sprayed *after* plant size has responded to interference between weeds, its effects may then depend directly on density. At low density, each weed will experience a similar probability of death, since they are of similar sizes and herbicide interception and uptake will be similar. Also, at very high density where there is intense interference, each plant may be small and unbranched, again resulting in a low variance of size and hence of herbicide uptake. However, at intermediate densities variance in plant size may be enhanced by interference, resulting in a wider range of herbicide interception amongst plants. The establishment of a distinct plant size hierarchy may mean that smaller individuals are shaded by larger ones, and as a result the smaller ones may receive little if any chemical. Hence, plant mortality due to herbicides may be least at intermediate densities (see Chapter 6).

Another cause of mortality likely to be density-dependent is inoculative biological control. In isolated countries, such as Australia, many weeds will successfully invade without being accompanied by their natural enemies. By identifying these enemies in the region of origin, organisms can be imported and released to help control the weed in the new location. As of 1984, there had been 499 releases of exotic invertebrates and fungi for biocontrol, in 70 countries and directed towards 101 weed species (Julien *et al.*, 1984). Of these, 108 releases were made in Australia, 80 in North America and 71 in Hawaii. Because of the nature of the particular host–agent interactions, only some of these are likely to result directly in weed mortality: some may reduce fecundity or decrease the number of seeds released (as discussed earlier in this chapter).

When first introduced, the biocontrol agents may have an almost unlimited food supply and weed mortality will be set by the rate at which the agent can multiply. As they reach their greatest abundance, weed mortality may well reach levels close to 100%. Then, as the weed becomes rare, the agent will become limited by food supply, its own populations will decline and mortality of the weed will fall. Few data have been collected on annual mortality of weeds affected by inoculative biocontrol. A pragmatic explanation for this is that if the agent virtually wipes out the weed, why bother to assess weed mortality? Similarly, if weed mortality is only slight, it is more important to look for new agents than to assess clear failures. There are numerous estimates of density before and after the introduction of agents (Julien, 1992), but the resultant mortality will be spread over several years. Although the final effect of a successful agent may be close to a 100% reduction in the weed, mortality in any one year may have been considerably below that level; after the agent has had its greatest effect, the residual population may continue to undergo mortality from the agent, but without any change in population density.

In one example, Cofrancesco *et al.* (1984) studied the impact of releases of the weevil *Neochetina eichhorniae* on the aquatic weed *Eichhornia crassipes* in Louisiana. At one site they found that weed density declined from 117 to 41 plants m^{-2} over a period of five months, a mortality of 65%. This shows the dramatic effect that the weevil can have on *E. crassipes*; however, at other sites and in other countries the agent may fail to establish and may cause no mortality (Wright & Stegeman, 1990). In a study of biological control of *Hypericum perforatum* in South Africa, plants were sprayed regularly with insecticides to prevent attack by the gall midge *Zeuxidiplosis giardi* (Gordon *et al.*, 1984). Even in sprayed plots, mortality of seedlings (probably weather-induced) was about 80%; survival of plants open to attack by the midge was reduced by 15%.

Interactions between factors

Any density-independent factor causing mortality before density begins to affect growth will affect the intensity of later interference, by reducing density. Hence, frost acting independently of density will reduce the probability of density-dependent mortality in spring and summer. Although we tend to label factors regulating density as either density-independent or density-dependent, the two do not act in isolation. For example, grazing by some animals may be indiscriminate, removing a fixed amount of biomass. It may cause some mortality, either through removal of all meristems or by uprooting plants. It may thus act as a density-independent factor. However, by removal of leaf tissue and reduction of plant size, it may retard the onset of density-dependent mortality by reducing interference. It may therefore retard or prevent competitive exclusion and thereby maintain a higher community diversity. Dirzo & Harper (1980) found that when *Capsella bursa-pastoris* was grazed by a slug, the rate of mortality was actually reduced. They interpreted this as being due to the retardation of self-thinning as a result of defoliation.

As another example, herbicides applied at a time when their effects will be independent of density may affect the intensity of later density-dependent processes. It can be seen from equation 4.12 that, by reducing density, herbicides will reduce mortality from interference. However, it is also likely that plants escaping a herbicide, and those recovering from a sub-lethal dose, will be smaller than those growing where herbicide has not been used. As a result, they will interfere less strongly as individuals with each other and with a crop. Weaver (1991) showed that, at an equivalent density, plants recovering from or escaping metribuzin had far less effect on a soybean crop than plants in unsprayed plots. We would therefore expect that, for weed mortality, m_1 would be lower and γ would be higher after herbicide treatment. Overall, we might perhaps expect less mortality due to interference at a given density following herbicide use.

Temporal patterns of plant mortality

When the number of surviving plants in a population is plotted against time, the trajectory is often referred to as a 'survivorship curve' and its slope is the rate of decline. Deevey (1947) classified survivorship curves for animals into a number of types, illustrating various generalised patterns of age-specific mortality. His data were for time-steps of one year and seasonal effects were subsumed within each observation. Patterns of survivorship were considered to summarise age-specific traits of the species. Deevey's classification has been applied to survivorship curves of annual plants (e.g. Harper, 1977), but on such a time scale the patterns reflect a combination of age-specific and weather-specific mortality. In many cases, mortality may simply track the ambient weather either directly, or indirectly via the effects of weather on growth. As a result, survivorship curves for annuals vary considerably from year to year and with time of emergence.

It is therefore difficult to use Deevey's classification for making general qualitative statements about annual plants. In any case, the unevenness in the curves due to weather are often such that classification into to *any* of Deevey's classifications is impossible. Having said this, it may be possible to fit a curve which smoothes out deviations in the data and then to estimate rates of mortality in the same way as for seeds. Some data sets (e.g. Yoda *et al.*, 1963; Harper & Gajic, 1961) show a roughly exponential decline in numbers with time, indicating a constant per capita death rate. Other examples (e.g. Tripathi, 1985) show more complex patterns with marked disjunctions between two or more successive exponential declines with different rates. In such cases, regression can be used to estimate rates of decline over each of the separate periods identified.

One of the many examples of studies of plant survivorship is shown in Fig. 4.13. Sheppard *et al.* (1989) recorded numbers of the thistle *Carduus nutans* seasonally in southern France. In a cohort emerging in the spring, juvenile mortality was high. A small proportion of the cohort flowered and died, thus acting as summer annuals. The remainder survived throughout winter as rosettes, with little mortality. These flowered and died the following summer, thus acting as biennials. A second cohort emerged in the first autumn. Mortality in the seedling stage was lower than the spring cohort, perhaps suggesting that juvenile mortality was related to summer weather conditions. A proportion of this cohort flowered and died the following summer, acting as winter annuals; the remainder suffered mortality in summer and autumn and would presumably have behaved as biennials if the study had continued longer. Little would have been gained by classifying these curves in a simplified way using Deevey's system.

Vegetative reproductive phase

Some important weeds possess the ability to reproduce vegetatively. Vegetative structures are formed under or on the ground and act as a

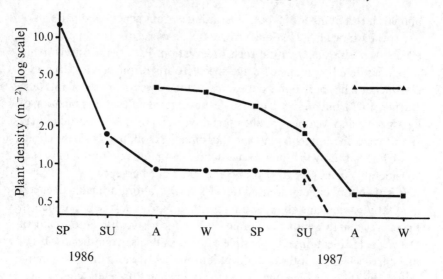

Fig. 4.13. Decline of three successive cohorts (●, ■, ▲) of *Carduus nutans* in southern France (based on data in Sheppard *et al.*, 1989). Arrows indicate times at which at least some individuals in the cohort flowered and then died; the dashed line indicates that all plants died before the next assessment date.

meristem (bud) bank akin to a seed bank. These are capable of producing shoots when stimulated to do so. Since a single shoot may produce many below-ground structures and each of these may in turn produce a number of above-ground shoots, an extensive clone of genetically similar individuals can result from a single original seed or vegetative propagule. To distinguish between a plant and a single above-ground shoot, Kays & Harper (1974) proposed a simple terminology. A single shoot, which may be only part of a clone, is termed a 'ramet', whereas the entire genetic individual, i.e. the whole clone, is termed the 'genet'. In principle, it should be possible to study separately the demography of genets and of ramets. However, unless there is considerable polymorphism amongst clones and there are distinct markers, it is often only feasible to study gains and losses of ramets.

The relative importance of sexual and vegetative reproduction to total shoot production varies, but it is often found in these species that seeds are less important to the maintenance of the shoot population than vegetative structures (e.g. Lapham *et al.*, 1985). Even if recruitment from the vegetative meristem bank is less than from the seed bank, shoots from vegetative organs usually get off to a better start, and as a result are better competitors than those from seeds, because of their greater energy reserves.

However, vegetative organs tend to be less tolerant of desiccation than seeds.

Vegetative reproductive structures are morphologically diverse and include bulbs (e.g. *Allium vineale*), tubers (e.g. *Cyperus esculentus*), pseudobulbs (e.g. *Arrhenatherum elatius*) and rhizomes (e.g. *Elymus repens*, *Sorghum halepense*). Generalisations across such diversity are likely to be difficult. Three species will therefore be discussed as examples.

Cyperus esculentus produces rhizomes from above-ground shoots. These may produce either further above-ground shoots or tubers. The tubers act as dormant meristems to ensure plant survival over the dry season (Lapham *et al.*, 1985) or over winter (Tumbleson & Kommedahl, 1961). When dormancy is broken, especially at the start of the new season, tubers will then produce new ramets. Hence, the seasonal cycle of ramet production begins with 'births' from tubers and then increasingly includes births directly from other ramets.

Tubers are generally formed in the top 15 cm of the soil; shoot production from tubers declines only below this depth (Tumbleson & Kommedahl, 1961). In Minnesota, a single tuber in the field can form up to 36 ramets in 16 weeks and up to 1900 in a year when growing from a point source in monoculture (higher rates of production over 16 weeks were recorded in pots). This same tuber had produced 6900 tubers by the end of the first season's growth. Lapham (1985) recorded mean tuber production in Zimbabwe of 17 700 in one year and 163 000 over two years from a single plant, again in monoculture.

Clearly, potential birth rates of ramets from tubers are considerable. However, as Lapham *et al.* (1985) report, ramet (as well as tuber) production is density-dependent and is also likely to be reduced when growing in mixtures with other species. Shoot and tuber production has been found to be dependent on soil type, with slower rates of increase in sand than in sandy silt loam or peat (Tumbleson & Kommedahl, 1961). Mortality rates of tubers have been estimated as 0.03 per 0.1 year if undisturbed and 0.18 per 0.1 year if ramets were removed at regular intervals (Lapham *et al.*, 1985).

Elymus repens (*Agropyron repens*) spreads by means of plagiotropic underground rhizomes. In undisturbed conditions, new ramets are formed usually by the rhizome tips turning upwards. Along the rhizomes are lateral buds which mostly remain dormant. However, when the rhizomes are cut, for example by ploughing, the most apical meristem develops and forms a new above-ground shoot. Other laterals remain dormant as a result of apical dominance.

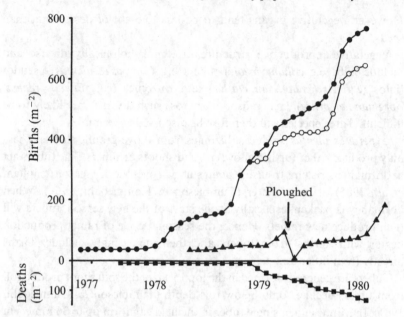

Fig. 4.14. Cumulative births (●) and deaths (■) of shoots (ramets) of *Elymus repens* (redrawn from Mortimer, 1983), developing from single rhizome fragments in monoculture. Net population size is given by open symbols (○). Net population size in a winter wheat crop is given by ▲.

McMahon & Mortimer (1980) recorded shoot emergence throughout the year in the UK. In countries with colder winters, shoots will appear only in the warmer months. McMahon & Mortimer observed peaks in shoot production in spring and autumn in monocultures, with a period of reduced production about the time of flowering and during winter. In a winter wheat crop, however, the spring peak was absent. This may indicate feedback from shoots to prevent ramet production when conditions are unfavourable, such as when there is inter- or intraspecific interference. In this way, mortality may be avoided or reduced through control of shoot production. Ramet deaths were only recorded at densities above 300 m^{-2} in monoculture and not at all within winter wheat (Fig. 4.14). In a model based on these data, the annual production of rhizome buds for a ramet present in March was set at 460; bud mortality was set at 6% per annum. Clearly, bud production will be density-dependent. It will also depend on other vegetation present: Marshall (1990), for example, recorded rhizome bud production 93% lower over 2 years on plots sown to other grasses compared with bare ground.

Sorghum halepense, like *E. repens*, spreads by rhizomes. Primary rhizomes extend horizontally and produce lateral (secondary) shoots which grow up to the surface where they each produce a ramet. Terminal buds on primary rhizomes may also produce aerial shoots. Tertiary rhizomes are produced at flowering time and are over-wintering organs from which primary rhizomes are produced the next year (Holm *et al.*, 1977). Data on population dynamics are not as extensive as for *E. repens*. However, Satorre *et al.* (1985) described the density of emerged rhizome sprouts by the equation

$$y = (11(1 - 0.86e^{-0.0034T}) - 1)^2/100 \qquad (4.16)$$

where y is the proportion of the maximum final number and T is the accumulated temperature (day degrees) above 15 °C. The minimum rhizome biomass in the soil was found to occur when 46% of the final number of shoots had emerged, after 315 day degrees. The authors proposed using this equation for timing applications of foliar herbicides so that they are most effective, when underground shoot reserves are at a minimum. Tillage, if timed so that severed rhizomes are desiccated in a dry period, has been recorded as causing up to 60% reduction in ramet numbers (Radosevich & Holt, 1984).

Conclusions

It is apparent that certain aspects of the regulation of population density have been well covered by research. For example, seed longevity and rates of seed mortality have been studied extensively for a wide range of species. As a result, it has been possible to draw some general conclusions for mortality in seed banks. Some aspects of population regulation have been studied in great detail in a small number of species, such as density dependence in *Agrostemma githago* and rhizome bud dynamics in *Elymus repens*. Thus, we have a few, detailed case studies; even most of these have only been studied in a single environment. For some components of the life-cycle, we have almost no data. For example, there are few detailed studies of plant mortality between emergence and maturity (even that resulting from herbicides). There are few studies of seed predation while still on the parent or when on the soil surface. There are few studies of density-dependence in vegetatively reproducing species: most population studies are short term studies of the progeny of single tubers or single node rhizome fragments.

If we want to model the dynamics of population density (see Chapter 6), therefore, we are faced with two options. We can restrict our modelling only to those very few species for which we have sufficient data (to date mostly monocots), or we must make (informed) guesses for many of the processes. Even in models based on many years of research, some parts (particularly below-ground events) will be based on averages of only one or two experiments. It is important when modelling (see next chapter) to recognise the limitations of the data base.

Despite the fact that we expect factors involved in the regulation of density to vary with the environment, most studies have been done in only one or a very restricted range of sites. There is little information on the effects of soil type, for example, on decline of seeds in the soil, germination rates and density-dependent plant mortality and seed production. Although we know quite well that temperatures affect breakage of seed dormancy, stimulation of rhizome development and plant growth, we have surprisingly few studies which relate population processes quantitatively to weather. It is thus dangerous to try to extrapolate from data collected at a single site, or from means of one or two years, and to predict what would happen at different sites or under different weather conditions, or even 'on average'.

With such a dearth of information, it is difficult to indicate where the most important areas are for future work. One approach which would be highly cost-effective would be more comparative studies of species, rather than so many single species case studies. Alternatively, we might target species for study which provide clear biological contrasts with previous intensive case studies.

Even within areas well covered by research, there are some major questions still to be answered. For example, how does seed decline in the soil vary with waterlogging/soil moisture levels and pH? How much do rates of decline vary with soil type? Decline of seed reserves studied in the relatively invariant year-to-year weather conditions in the UK followed remarkably smooth exponential declines; would this still be the case in more variable climates such as the wheat–sheep belt of eastern Australia? What are the specific causes of seed decline other than germination; how is it that some species are able to avoid these losses better than others? The more detailed the information we have and the more generalisations we can make, the more hypotheses we can generate for other parts of the life-cycle.

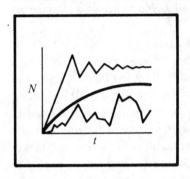

5

The intrinsic dynamics of population density

The birth and death processes discussed in the previous chapter jointly determine the dynamics of population density, assuming no (or equal) immigration and emigration. The 'path' which population density follows over time can be referred to as its *trajectory* and some examples were given in Fig. 1.5. There may be periods where population density is changing little and periods where, for whatever reasons, density is increasing or decreasing. The direction of this change is therefore of particular interest in weed management. So, too, is the rate of change, which will determine how soon a species will get out of hand, or how soon a problem will have been alleviated.

A central aim in the study of population dynamics is to understand, and therefore to predict, population trajectories. Is a species currently increasing in abundance? How high will its density go if left unchecked? How can we change our management so as to make its density steadily decrease? How fast can we force it to decrease? Is density declining because of some fortuitous series of failures in seed production, or does it reflect a successful management programme? Do we need to take further action, or will the population continue to decline? Such questions make the study of population dynamics an important aid to weed control decision-making.

By monitoring populations in an area we may be able to identify, retrospectively, particular events which coincide with changes in trajectory, such as the introduction of a new herbicide or the adoption of a different method of tillage. However, there is such a variety of farming operations which may affect weed populations and which vary from year to year that ascribing cause to effect may be difficult. For example, in a field studied for 20 years by Chancellor (1985a) there were eight different herbicide regimes used, six different crops were grown, sowing occurred in six different months, along with numerous differences in fertiliser applications, tillage,

135

and fungicide and pesticide use. There would, no doubt, have been 20 patterns of weather, several of them (if not all) being regarded as extreme or unusual in some way.

Alternatively, information on causes of change can be obtained objectively from experiments in which possible causal factors are varied systematically. Of necessity, however, experiments are of limited duration and they can also only include a small number of the vast array of possible treatment combinations. If we are to forecast the long term dynamics of populations, and to do so under conditions not included in particular experiments, we need methods by which we can take the available information and make predictions from it. Simple mathematical models have come to play a central role in the study of plant population dynamics and without models our forecasting abilities would be crude indeed. Models also give us a framework within which to collate all of our data, to see where the gaps are and to direct future experimental programmes.

In the next two chapters we will explore the dynamics of weed population density using the available empirical evidence, examining the predictions of models, searching for general conclusions, pin-pointing gaps in our knowledge, and using models to ask questions about weed management strategies.

An important first principle is that populations display their own *intrinsic* dynamic properties. If all other (*extrinsic*) factors remain constant, populations can change of their own accord. Such changes occur because of internal regulatory processes arising from interactions amongst individuals within the population. We will therefore begin by examining the population dynamics resulting from intrinsic demographic processes. What types of trajectory would we expect by intrinsic processes alone? What modelling approaches can we take to predict these trajectories? What types of information are the models capable of giving us? What data do we have to test their predictions? In Chapter 6 we will explore how this behaviour is modified by extrinsic factors.

What types of trajectory are likely?

Consider the first seed of a species dispersing into an area. If the habitat provides the resources for the species to grow and reproduce, is large enough in area to support an expanding population and is not already fully occupied by other species, then the species may increase in abundance. Its *rate of increase* (λ), or its multiplication rate, is given by the ratio of population size in successive generations, i.e. N_{t+1}/N_t where population sizes are measured at a common point in the life-cycle. This is most easily calculated for annual weeds displaying discrete generations. The trajectory of population density can be completely characterised by the way in which λ changes with time over generations of population growth.

At first, while population density is low, individual plants will have little interference with each other's growth, and the population can be expected to continue to increase at a constant rate. We will refer to the value of λ at low density, by convention referred to as the *finite rate of increase* of the species in that habitat, as R (note that this is also known as the fundamental net reproductive rate). However, as density increases the population will become crowded and both plant survival and reproduction will be reduced (see Chapter 4). We would then expect λ to decrease. Eventually, interference would become so intense that any gains to the population through reproduction would be completely cancelled out by mortality, and population density would cease to increase ($\lambda = 1$). The population would then have reached its upper density limit, the *equilibrium* level of the species (N_e), or the *carrying capacity* of the habitat. An argument such as this presumes, of course, that the resources available for population growth are limited and ultimately fixed.

The overall pattern of increase described above would be the type of trajectory shown in Fig. 5.1a. For this trajectory λ decreases consistently until it reaches 1; Fig. 5.1b shows clearly how the rate of increase is density-dependent. An alternative way of depicting the trajectory is in the form of a *generation map*, where the densities in successive generations are plotted against one another (Fig. 5.1c). A straight line can be drawn through the origin, showing the condition for no increase ($N_{t+1} = N_t$). Points above this line indicate that density is increasing; points below the line indicate a decreasing density.

Generation maps are particularly useful in that they enable graphical predictions to be made about population dynamics, without the need for any mathematics. A technique is used called 'cobwebbing' (Hoppensteadt, 1982). Starting at a given value of N_t, a line is drawn upwards until the curve is reached. A horizontal line is next drawn across to the line $N_{t+1} = N_t$. From there a line is drawn either vertically upwards or downwards to the curve, thence back to the $N_{t+1} = N_t$ line. And so on. An example, using the type of trajectory in Fig. 5.1, is shown in Fig. 5.2. The series of N_t values at each successive intersection with the curve is the predicted population dynamics. Clearly, if by chance the population starts above the equilibrium level, the generation map in Fig. 5.2 will predict a unidirectional (monotonic) decline to the equilibrium.

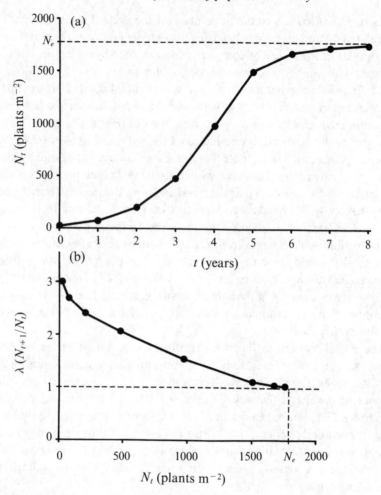

Fig. 5.1. Three ways of depicting the trajectory of an increasing population: (a) graph of population density against time; (b) rate of increase against population density; (c) 'generation map', showing the relationship between population densities in successive generations. In each case the equilibrium density (N_e) is shown, for which $N_{t+1} = N_t$, i.e. $\lambda = 1$.

So far, we have assumed that the equilibrium will be approached in an asymptotic fashion. However, individual density-dependent factors act largely during a single phase of the life-cycle (see Chapter 4), when plants interfere with one another's growth. Is it reasonable to expect that they will regulate density precisely to the asymptotic value? A good example of where this might not be the case is where most seeds do not germinate in the first year after shedding. The population may regulate seed production on the basis of the current year's plant population, only to have a large cohort

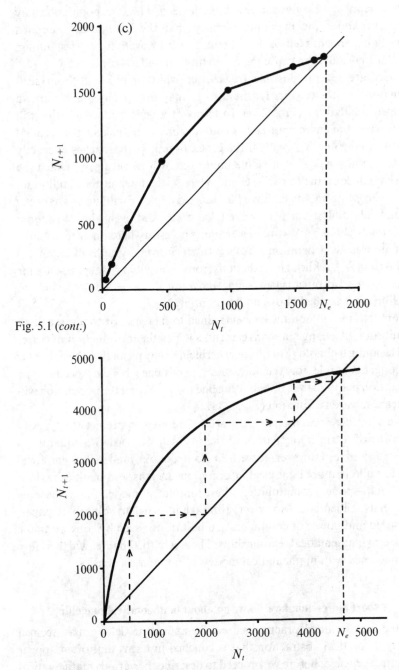

Fig. 5.1 (*cont.*)

Fig. 5.2. Illustration of how 'cobwebbing' can be used to predict a population trajectory from a generation map. The density in each generation is where the zig-zag line hits the curve.

of emergence in the following year from the seed bank. The population may then over-shoot the maximum density that the habitat can support. Density-dependent feed-back will tend to compensate for over-shooting, but this may still then cause the population to under-shoot.

There are several possible trajectories which may result from such regulatory processes. Firstly, the density may approach equilibrium via damped oscillations (Fig. 5.3a). In terms of a generation map, this will occur when the curve approaches equilibrium at an angle of between 45° and 90° to the line $N_{t+1} = N_t$. If the generation map approaches at exactly 90° to the line $N_{t+1} = N_t$, a stable limit cycle will be set up, in which the density will continue to oscillate and never settle down at the equilibrium (Fig. 5.3b). For an angle above 90°, more complex oscillations about the theoretical equilibrium will occur (Fig. 5.3c), and even almost random behaviour ('chaos'). (A word of warning is necessary here. In order to show the data clearly, generation maps are often shown as graphs of $\log(N_{t+1})$ against $\log(N_t)$. Although predictions from cobwebbing using these are the same as for untransformed data, the critical angles of approach to equilibrium described above no longer apply.)

Populations will not always be destined to increase, or to approach an equilibrium. A change in management, such as the introduction of better seed cleaning technology or of a new herbicide, may mean that the habitat is no longer suitable for the long term persistence of the species. The generation map will then be below the line $N_{t+1} = N_t$ and the population will decrease towards extinction (Fig. 5.3d).

We can, therefore, use simple graphical models to predict the types of outcome of intrinsic regulation of density. All we need is to quantify a generation map. However, if we want to make our models less empirical and to build in more biological information, such as seed bank behaviour and multi-species communities, these graphical models soon become hopelessly inadequate. Moreover, although it is easy to cobweb on paper, the speed and power of computers can best be accessed by converting the models to mathematical relationships. The rest of this chapter will therefore be concerned with mathematical models of population density.

A short digression: how do we go about mathematical modelling?

To the weed control practitioner, mathematical modelling may seem a purely theoretical abstraction that is couched in terms that often appear incomprehensible. Before we proceed to describe the various mathematical models of the dynamics of population density, we will try to explain

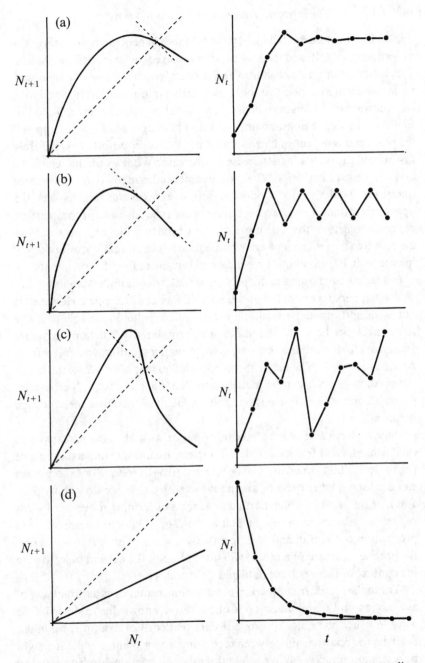

Fig. 5.3. Four further types of population trajectory, with their corresponding generation maps: (a) convergent oscillations to equilibrium; (b) non-convergent oscillations about equilibrium ('stable limit cycles'); (c) non-convergent, complex behaviour; (d) decline to extinction. Dotted lines on generation maps show the line $N_{t+1} = N_t$ and a perpendicular line through the equilibrium density, so as to emphasise the angle of the curve close to the equilibrium.

qualitatively where modelling fits into research in general, how it can assist experimental work, and the logical steps involved in the modelling process.

Whilst elucidating mechanism and process, empirical research by observation and experiment inevitably raises further questions. Say, for example, we recorded the changes in density of a weed population under a mouldboard ploughing regime and under minimum tillage: what would happen if we ploughed one year in five and minimum tilled in the other years? How fast would the weeds then increase or decrease? What would happen if we no longer used a herbicide? The most unambiguous method to answer these questions would be to conduct a set of experiments and collect the appropriate data. However, experiments are expensive and, if the question concerns long term trends, may take a long time to achieve an answer. If we are required by a manager to give immediate advice, we cannot afford to say 'please wait 10 years until I have carried out the required experiments'.

Modelling can be used to deduce, on the basis of our understanding, what *should* happen. It cannot tell us categorically what *will* happen. However, if our data and our understanding are sound, we should be able to make reasonable predictions. The difference between a prediction and pure guesswork is the reliance on data and on understanding generated from previous research. The better the quality and extent of the data, the better the predictions that can be made. Modellers have a phrase: 'garbage in, garbage out'. Experimentation and modelling therefore need to go hand-in-hand.

Once we have assembled all the information available, the first step is to establish a logical framework describing our qualitative understanding of the subject. This framework can be referred to as a *model*. For example, we can describe the life-cycle of an annual weed by a simple flow chart (Fig. 1.6a), or the ways in which particular aspects of farming move weed seeds within a field by a series of statements, or the dependence of seed production on rainfall and sowing date by a series of graphs. Necessarily, the precise structure of a model will depend upon the desired objectives in the context of the available data and time scales.

Within the model framework, we can use mathematics as a formal way of assembling and manipulating our quantitative knowledge to describe the ways in which variables or processes are related. For example, we might describe the relationship between seed number per plant and plant weight by a straight line $S = aw$. We might describe the relationship between seed number and plant density by a rectangular hyperbola (see p.96) or seed decline by an exponential function. We will also need to decide on a set of rules to connect the relationships to one another and to produce a computer program which encapsulates the entire model. Values of the parameters in the equations are typically derived directly from experimentation and statistical analysis, or extracted from the literature. The variables which describe the state of the population at any given instant, such as number of seeds in the soil or number of plants, are termed *state variables*.

Verification describes the process of testing whether the behaviour of the model is realistic. Does it predict the data which were used to derive the parameter values? Does it predict absurd outcomes? Does it, in fact, reflect the biological knowledge which was *wanted* in the model? If not, then the model will have to be changed, so that it behaves acceptably. This process of verification is often salutary and may lead to the use of a new equation to describe a particular relationship, experimentation to obtain better estimates of parameters or alteration of the entire modelling approach.

In this phase of testing the behaviour of the model, we may want to check to see how much influence particular parameters have on the predictions. This is referred to as *sensitivity analysis*. This will indicate the areas most in need of further research (if a parameter based on little data has a large influence on the model), and until that research has been carried out we should remain wary of the model predictions. In any further calculations or simulations from the model, it would be wise to give predictions based on a range of values of the sensitive parameters. For example, if few data are available on the proportion of seeds germinating, we may wish to give predictions of equilibrium densities assuming 5%, 10% and 20% emergence, so as to gain some idea of our confidence in the predictions.

In a sensitivity analysis, the aim is to compare the sensitivity of the model output to each parameter. There are two ways in which this may be done. The first method involves *varying a parameter systematically* until a change of a specified magnitude is achieved in a given state variable. For example, each parameter may be varied until a 10% reduction is reached in the size of the seed bank after 5 years. The most sensitive parameter is the one requiring the least proportional change, $\Delta P/P$ where P is the original parameter value and ΔP is the amount by which it was changed. In the second method, *a parameter is changed by a specified amount* and the impact of this on the state variable is determined. It is common to go through a process of doubling and halving each parameter. Caution is required since many parameters in population models are proportions: it is not possible to double a proportion of 0.95, and it would be nonsensical to compare doubling a proportion of 0.05 with doubling 0.50. For such parameters we might consider incrementing by a given proportion, say varying each by 1 or 2%. The sensitivity of the state variable to a parameter can be quantified by

$$[(\Delta O)/O]/[(\Delta P)/P]$$

where ΔO is the change in output caused by a ΔP change in the parameter, and O is the value of the output. Care is required whichever method is used, since different state variables are likely to be sensitive to different parameters. For example, *equilibrium density* may be most sensitive to maximum seed production per unit area, *rate of increase of plant number* in the exponential phase may depend mostly on proportional germination and seed production per plant at low density, whereas *rate of decrease of seeds* in the soil may be most sensitive to the pattern of soil inversion caused by ploughing.

Sensitivity testing can also be useful in the later stages of modelling, once there is confidence in the behaviour of the model. By knowing which parameters have the most influence, we can conclude that these may be the key aspects of the biological system. We may then use this knowledge to target future work. For example, Medd & Ridings (1989) concluded from a model of *Avena fatua* that further advances in plant control by herbicides would have less of an effect on population density than the development of agents to kill seeds in the soil.

Jeffers (1982) pointed out that the development of many models is stopped when the predictions adequately reflect the data which went into the model. Predictions outside of the experiences which were fed into the model are then made with confidence. However, work should be continued in order to test the validity of the model, by carrying out experiments to see if the predictions are upheld in real life.

Essentially, *validation* should entail comparison of model predictions with data from independent experiments (i.e. whose data were not incorporated into the model). In an efficient research programme, independent experiments could be conducted before the model has been completed; data sets may also be deliberately omitted from model formulation, to save them for later validation. However, there is always a tendency to 'peek' at the results and perhaps to modify the model slightly on that basis. Data from experiments run directly alongside those used for model formulation or from plots within the same experimental layout are highly likely to behave in a similar way to the model, since the data were obtained under virtually identical conditions. The best validation of a model will be obtained from totally independent data, preferably from different years and different locations. Unfortunately, many research programmes end before model validation. Notwithstanding the problems discussed on p.167, it is important that validation should be built into modelling programmes wherever possible, otherwise confidence in the predictions must remain weak.

The development of a model is often an irregular process, with the model gradually taking shape as the logical course of an experimental programme proceeds, and then feeding back into the experimental work as the one benefits the other. There may be no distinct end, when the model is regarded as 'finished', since any model can be constantly improved and extended.

As was pointed out earlier, modelling is not an end in itself. It is a means of asking questions about a system. The final stage of modelling is therefore the application of the model to answer those questions. This may take a variety of forms. For example, the model can be used to *simulate* the time-course of a population. Alternatively, the properties of the model may be explored *analytically*; theoretical mathematical analysis may enable deductions to be made about the intrinsic behaviour of populations. Such deductions may then be tested by further experimentation. The advantage of analytical methods is that they may show behaviour which might not be discovered from simulations, because the latter use only limited combinations of parameter values.

Mathematical models of population density

Modelling necessarily involves the making of assumptions. Since nature is inherently complex, we must begin by making a set of assertions to define what we are and what we are not trying to model. We have already stated that this chapter is concerned with the intrinsic behaviour which would occur if the habitat remained constant from one generation to the next. For simplicity, we will also assume that spatial processes have no bearing on population dynamics. This is equivalent to assuming that we are examining dynamics within the centre of an extensive, homogeneous population. Spatial dynamics and the dynamics occurring at the edge of a population will be considered in Chapter 7.

The models which we will discuss can be divided into two groups. Firstly, there are those models which consider only the density of a population (plants or seeds) at intervals of a single generation. They make no assumptions about what goes on within each generation, or what biological processes determine the density changes. They merely describe changes in densities. For a species with a single generation each year, these models will have a time-step of one year. We will refer to these as *single-stage* models. Secondly, there are those which explicitly build in the birth and death processes discussed in Chapter 4. They divide up the life-cycle of the species into a number of discrete stages and consider the gains and losses from one stage to the next. We will refer to these as *multi-stage* models or, more colloquially, *life-cycle* models.

Multi-stage models can, in turn, be separated into two categories. If we can consider that all plants emerge approximately at the same time and reach each developmental stage together, a model may need only describe the behaviour of a single *cohort* (a group of individuals whose development is synchronous). However, plants emerging much later than others may behave very differently and may be suppressed by the earlier cohorts. In a population of multiple cohorts, or even within strongly interfering single cohorts, reproduction and mortality may be highly related to plant size; the behaviour of the average individual may be far from representative of the population. For biennials, perennials and very short-lived annuals, there may be over-lapping generations, with mature and immature plants from different years or generations present at the same time. For such species, we need to recognise that the different developmental stages and sizes coexist. This can be done through the use of *matrix* models.

Single-stage models

If individual plants in a population are widely spaced, interference between them will be negligible. We may then assume that the behaviour of the population, in particular the rate of increase, will be independent of density. Earlier, we defined the finite rate of increase at such low density as R. If the density is N_t in the first generation and N_{t+1} in the next, then

$$N_{t+1} = RN_t \tag{5.1}$$

This is referred to as a *difference equation*. If we begin to simulate the dynamics of density from a starting density N_0 when no time has elapsed ($t=0$), after one generation there will be a density $N_1 = RN_0$. After two generations density will have reached $N_2 = RN_1 = R^2N_0$, and hence after t generations it will be $N_t = R^tN_0$. This is the discrete form of the *exponential* model.

This simple model has seldom been used for weed populations. One of the few examples is its use by Selman (1970) to describe changes in the density of *Avena fatua* (without herbicides) over six years with early crop sowing dates and over five years with late sowing dates. Both early and late sowings gave a reasonable fit to the exponential model (Fig. 5.4). In what was probably the first attempt to predict the dynamics of a weed population, he ran simulations using this model with sequences of values of R so as to examine the effects of different long term control programmes. High and low values of R (both greater than 1.0) were used to simulate good and bad seasons for seed production, and high and low values of R (both less than 1.0) simulated the effects of poor and good herbicide efficacy.

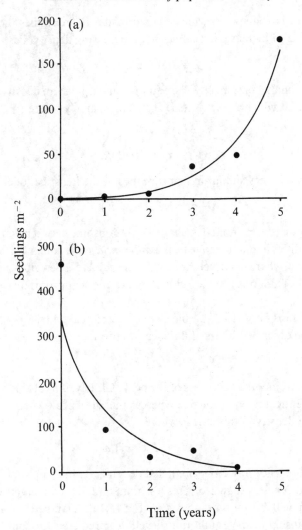

Fig. 5.4. Dynamics of *Avena fatua* growing as a weed in consecutive crops of spring barley (based on data in Selman, 1970): (a) early sown crop; (b) late sown crop. Exponential curves have been fitted by regression: estimated values of R are 2.74 and 0.40 respectively.

In order to make our models of the dynamics of population density more realistic, we need to consider what will happen as density increases to the point at which plants interfere with one another. Rate of increase will no longer be as great; as density increases, we would expect λ to decrease. For populations in which N_{t+1} asymptotically approaches an upper limit as N_t increases, the equation

$$N_{t+1} = RN_t/(1 + aN_t) \qquad (5.2)$$

is an appropriate shape, in which the parameter a determines the rate at which its rate of increase, λ, declines with increasing density, i.e.

$$\lambda = R/(1 + aN_t) \tag{5.3}$$

where a is a 'shape' parameter. The equilibrium population density is given by the point at which $N_{t+1} = N_t$ (see Fig. 5.1a), namely where

$$R = 1 + aN_t$$

or
$$N_t = N_e = (R - 1)/a \tag{5.4}$$

This model is very similar in its properties to the logistic model,

$$dN/dt = rN/(1 - N/K) \tag{5.5}$$

They both describe sigmoidal (s-shaped) relationships between density and time. They differ in that the logistic is a *differential* equation and describes a smooth curve, whereas a difference equation such as 5.2 describes a series of discrete steps from one generation to the next, more appropriate to annual plants.

Relationships between N_{t+1} and N_t which are peaked (see Fig.5.3 a–c) can often be described by the difference equation

$$N_{t+1} = RN_t/(1 + aN_t)^b \tag{5.6}$$

where b determines the acuteness of the peak. Clearly, if $b = 1$ this equation becomes identical to equation 5.2, whereas there is a peak when $b > 1$. The equilibrium density in this case is where

$$N_t = N_e = (R^{1/b} - 1)/a \tag{5.7}$$

The type of density trajectory predicted depends on the values of R and b (Fig. 5.5). If b is less than 1 or R is less than e (2.72), convergence to the equilibrium will be monotonic (as in Fig. 5.1a). Damped oscillations, finally stabilising at the equilibrium level, are predicted for some combinations of parameters where b is between 1 and 2, and where R is between e and e^2 (Mortimer *et al.*, 1989). Complex behaviour, such as chaos, can occur only if b is greater than 2 and at the same time R is greater than e^2 (i.e. 7.39).

An illustration of the utility of equation 5.6 can be seen in the following example. Mortimer (1987) used regression to estimate parameter values from the data of Manlove (1985) for *Avena fatua* in winter wheat in the absence of herbicides. The equation was

$$N_{t+1} = 96.7N_t/(1 + 0.194N_t)^{0.593} \tag{5.8}$$

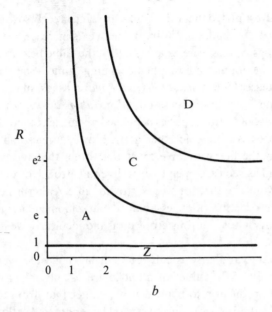

Fig. 5.5. Predicted behaviour of a population according to equation 5.6, in relation to values of the parameters R and b. Trajectories differ between the four zones of the graph: D – divergent oscillations, chaotic behaviour or complex cycles; C – convergent oscillation to equilibrium; A – asymptotic approach to equilibrium; Z – asymptotic decline to zero (after Mortimer *et al.*, 1989).

where densities were in seeds m^{-2}. Since b is less than 1.0, the predicted trajectory will be an asymptotic approach to equilibrium. From equation 5.7, the equilibrium density will be 11 484 seeds m^{-2}. When a late post-emergence herbicide was used, the fitted equation was

$$N_{t+1} = 13.2N_t/(1 + 0.228N_t)^{0.429} \qquad (5.9)$$

Again, the value of b indicates an asymptotic approach to an equilibrium, but now N_e would be 1791 seeds m^{-2}, only 16% of the level without herbicide.

What are the likely ranges of values of the parameters R and b in practice? We have already seen that changes in density in the field from one year to the next can be extremely variable. To estimate R we need to restrict our search for values to populations increasing from low density; to adequately estimate b, densities need to be extremely high. We will give just a few examples below, to illustrate the ranges which we might expect.

Wilson & Phipps (1985) recorded annual increases (λ) of up to 3 for *Avena fatua* in winter wheat, perhaps providing some indication of R.

Selman (1970) estimated that for *A. fatua* growing in early sowings of spring barley R was 2.7 overall, but for individual years increases were up to 6.0; for late sowings, values were as low as 0.14, the value below 1.0 reflecting the fact that the population was then declining. Lintell-Smith *et al.* (1991) recorded increases by *Bromus sterilis* and *Papaver rhoeas* in winter wheat of $\lambda = 30$ over the first year of their study. For *Galium aparine* λ was up to 25; rates of increase of all three species depended on nutrient conditions and the other weed species with which they grew. Many more estimates could be gleaned from the literature. We must not forget that they are all only estimates, and as such are open to experimental error. However, a general conclusion would be that for weeds growing in cereal crops and without herbicides R is probably often less than 10, but seldom greater than 50. For weeds growing in less vigorous crops or as monocultures we might expect higher values.

Because of the variability usually found in data collected over a sequence of years (see Fig. 5.7), values of a and b are usually taken from curves relating seed production to plant density, rather than from direct observations of N_{t+1} and N_t over one or several generations. Watkinson (1980) speculated that b will usually range from 1.0 to 1.8. It should be noted that few data sets have been obtained for which b is significantly greater than 2; occasional high values can be obtained, but these often have large confidence intervals and can usually be explained by statistical problems with the fitting procedure (Cousens, 1991). Most data to which equation 5.6 have been fitted have given values of b either less than 1 or not significantly different from 1 (e.g. Firbank & Watkinson, 1985; Mortimer, 1987). It would therefore appear that complex behaviour such as chaos (which requires $b > 2$) is likely to be more a mathematical property of our models than a behaviour to be expected of real populations of annual plants (Watkinson, 1980).

There are a number of ways in which the simple models described above can be extended. In order to estimate N_{t+1} from curves relating seed production to seed density sown (rather than by following complete generations) we need to incorporate a factor representing the proportion of seeds produced which germinate (g). Since some seeds may remain dormant in the soil, a persistent seed bank can also be included by incorporating seed survival (s):

$$N_{t+1} = R'gN_t/(1 + a'gN_t) + sN_t \qquad (5.10)$$

where R' and a' have been derived by fitting equation 5.2 to a seed production curve (Mortimer *et al.*, 1989). The presence of g simply acts to scale down the generation map, whereas the presence of s serves to rotate it somewhat. The incorporation of these factors is therefore unlikely to change the behaviour of the model from an asymptotic approach to equilibrium. However, if the seed production curve is peaked, and the model is adjusted accordingly to

$$N_{t+1} = R'gN_t/(1 + a'gN_t)^{b'} + sN_t \qquad (5.11)$$

the values of the parameters g and s can change the type of trajectory predicted (see Fig. 5.10).

Since the dynamics of one species is likely to be affected by population levels of another species, we can expand equation 5.6 to represent a two-species community:

$$N_{1,(t+1)} = R_1 N_{1,t}[1 + a_1(N_{1,t} + \alpha N_{2,t})]^{-b_2}$$
$$N_{2,(t+1)} = R_2 N_{2,t}[1 + a_2(N_{2,t} + \beta N_{1,t})]^{-b_1} \qquad (5.12)$$

where subscripts 1 and 2 refer to the two species (Firbank & Watkinson, 1986). If one species is a weed and the other is a crop sown at the same density each year, the dynamics of the weed can be given by

$$N_{t+1} = RN_t/[1 + a(N_t + \alpha N_c)]^b \qquad (5.13)$$

where N_c is crop density, and hence

$$N_{t+1} = KN_t/(1 + MN_t)^b \qquad (5.14)$$

where the constants $K = R/(1 + \alpha N_c a)^b$ and $M = a/(1 + \alpha N_c a)$. If we then extend this to a multi-species community of weeds in a crop (Watkinson, 1985), the density of species i will be

$$N_{t+1} = R_i N_{i,t}/[1 + a_i(\Sigma N_j \alpha_{ij} + \beta N_c)]^{b_i} \qquad (5.15)$$

where Σ refers to the summation of the effects of densities of all species on the weed of interest (including intra-specific effects). The possibilities for further elaboration are endless and it is of limited value to consider further extensions here in the absence of data.

Multi-stage single cohort models

In the previous section we examined models which subsume the demographic processes that occur within generations and that may interact to determine seed yield of the population as a whole. Thus, events within a generation are ignored. The models relate N_{t+1} to N_t by empirical functions describing the way λ changes with plant density. Because our time series data sets are both few and noisy, to parameterise these models we often have to make do with curves of seed production against density of seeds sown. As a result, in equation 5.10 we had to add factors to the models to specify for those parts of the life-cycle between seed production and germination.

To add more detail, we can deliberately set out to pull apart the life-cycle of a species into the various processes acting within a generation, and then put them back together to predict dynamics over many generations. Models constructed in this way can be as simple or as complex as the researcher decides, depending on the data available and the ways in which the processes are viewed. Studies aimed at quantifying the various steps in weed life-cycles began to appear in the early 1970s. The probabilities of an individual surviving from one stage to another and the numbers of seeds produced (together often termed 'fluxes') are measured and depicted in flow-charts (see Fig. 1.6). Perhaps the first weed life-cycles described in this way were those of three species of *Ranunculus* in a pasture (Sarukhan, 1970). A range of species life-cycles including *Avena fatua, Alopecurus myosuroides* and *Poa annua* were summarised by Sagar & Mortimer (1976). Although in neither case were the data used to simulate population dynamics over generations, it is a simple matter to incorporate their probabilities into a model and to examine iteratively changes in weed density. Recently, there has been a proliferation of multi-stage models of weeds either growing as monocultures or mixed with crops. Here, we will discuss some of the models in order to demonstrate their range of complexity and their properties; in Chapter 6 we will evaluate their utility.

Let us consider first the case of a low density population of a weed, such that transitions between one stage and the next are not dependent on density. Fig. 5.6 shows a flow-chart of a simple annual plant life-cycle. The various fluxes have been placed next to the arrows leading from one developmental stage to the next. In this example, the main causes of plant mortality identified are herbicides (k) and the combined effects of other extrinsic factors (m_p). Each plant produces s seeds. Seed mortality between production and entry into the soil seed bank is assumed to be caused by removal by harvesting machinery (r), predation and other natural agencies (q) and burning of stubble (u). Losses of seeds already in the soil will result from germination (g) and combined mortality due to fungal decay, ageing and predation (m_s). With the exception of seed production, all of the fluxes are probabilities and lie between 0 and 1.

Simulations can be made by starting with a given number of individuals at any particular developmental stage, and multiplying in turn by each of

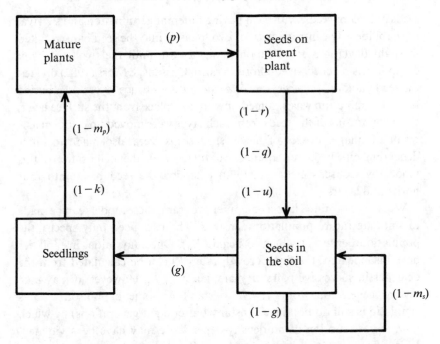

Fig. 5.6. Flow-chart illustrating the life-cycle of an annual weed. The probabilities (fluxes) of moving from one phase to the next are shown in parentheses. Letters refer to: germination (g); herbicide-induced mortality (k); natural plant mortality (m_p); seed production (p); removal by harvesting machinery (r); predation (q); burning of stubble (u); and seed mortality (m_s).

the fluxes in sequence around the flow-chart. If we begin with N_0 seeds in the seed bank, for example, after a single generation there will be

$$N_1 = gN_0(1-k)(1-m_p)s(1-r)(1-q)(1-u) + N_0(1-g-m_s) \quad (5.16)$$

seeds remaining, where $0 < (1-g-m_s) < 1$, or in general

$$N_{t+1} = gN_t(1-k)(1-m_p)s(1-r)(1-q)(1-u) + N_t(1-g-m_s) \quad (5.17)$$

Since the terms in this model are simple products of N_t, the rate of increase N_{t+1}/N_t is constant, and hence this is a discrete exponential model. In fact, we have separated R into a number of components, such that

$$R = g(1-k)(1-m_p)p(1-r)(1-q)(1-u) + (1-g-m_s) \quad (5.18)$$

Using equation 5.17 as a template, a number of elaborations have been included in models by various authors. For *Avena fatua*, Wilson *et al.* (1984) divided the seed bank into two components, the previous year's seeds and older seeds, with each having different germination and survival probabilities. The introduction of two 'phases' into the seed bank resulted in slight deviations from exponential increase until the two seed bank components reached an equilibrium ratio. Cussans & Moss (1982) divided the seed bank of *Alopecurus myosuroides* into a surface layer and a buried layer. Germination was assumed only to take place from the surface layer, and proportions of the seeds from each layer were moved up or down as a result of tillage. Losses of newly shed seeds were also made to differ depending on whether or not there was tillage. A further elaboration of this model by Cousens & Moss (1990) extended the seed bank into four horizontal layers.

When incorporated into a computer program, these models can be made to simulate future population densities. The user need only specify the population density at which to begin (N_0). Once any initial instabilities have disappeared, the finite growth rate (R) can be calculated from the densities in successive pairs of years, i.e. N_{t+1}/N_t. However, it may take several generations for N_{t+1}/N_t to stabilise (some of Wilson *et al.*'s simulations still do not have constant λ after eight generations) by which time the assumption of no density-dependence may have been violated. Wilson *et al.* (1984) calculated the overall rate of increase over the first t generations as $(N_t/N_0)^{1/t}$.

The multi-stage models discussed so far are unrealistic for high densities, since various processes will be affected by plant density as intraspecific interference increases. This can easily be taken into account by replacing the constant fluxes shown in Fig. 5.6 with equations relating the fluxes to density. For example, Firbank *et al.* (1985) made *Bromus sterilis* seedling emergence (G) a function of seed density (S):

$$\log_{10}G = a - b\log_{10}S + c(\log_{10}S)^2 \qquad (5.19)$$

Commonly, seed production per plant is replaced by a function of density. Rauber & Koch (1975) made seed production per plant decline linearly with density up to 180 *Avena fatua* plants m^{-2}, after which it was held constant. Cousens *et al.* (1986) and Doyle *et al.* (1986) assumed that seed production per plant (S) followed the hyperbolic relationship

$$S = S_{max}/(1 + aN) \qquad (5.20)$$

where N is the number of mature plants. Ballaré *et al.* (1987b) followed Watkinson (1980) by including another parameter, such that

$$S = S_{max}/(1 + aN)^b \qquad (5.21)$$

Firbank & Watkinson (1986) replaced this by two functions, one relating mature plant weight (w) to density of weed (N_w) and crop (N_c)

$$w = w_{max}/(1 + a(N_w + \alpha N_c)^b \tag{5.22}$$

and one relating plant weight to seed production

$$S = fw^g \tag{5.23}$$

where f and g are density-independent parameters.

Plant mortality between emergence and maturity is the other parameter usually made density-dependent. Rauber & Koch (1975) made mortality a quadratic function of seedling density (G), such that the surviving plant density (N) was

$$N = G - (aG^2 + bG + c) \tag{5.24}$$

Several models have followed Watkinson (1980) in assuming that

$$N = G/(1 + mG) \tag{5.25}$$

while Firbank & Watkinson (1986) introduced crop as well as weed density:

$$N_w = G_w/[1 + m(G_w + \gamma G_c)] \tag{5.26}$$

where m and γ are regression coefficients describing density-dependent mortality.

Watkinson (1980) developed a simple two-stage model from a consideration of an annual plant with no persistent seed bank and incorporating density-dependent mortality and seed production. If there are N_t seed-producing plants, from equation 5.21, the number of seeds produced will be

$$S = S_{max}N_t/(1 + aN_t)^b \tag{5.27}$$

If it is assumed that all seeds germinate, then the number of seedlings will be $G = S$. From equation 5.25, the number of mature plants produced from those will be

$$N_{t+1} = G/(1 + mG) \tag{5.28}$$

Substituting S for G in this equation, we obtain

$$N_{t+1} = S_{max}N_t/[(1 + aN_t)^b + S_{max}mN_t] \tag{5.29}$$

Parameter values, obtained from glasshouse populations, for *Agrostemma githago*, were $S_{max} = 1600$, $b = 1.06$, $m = 6 \times 10^{-5}$. This would predict convergence to an equilibrium via damped oscillations. In fact, for this data set mortality is so low that it has little effect on the predictions.

Watkinson also pointed out that it is a simple matter to extend his model to allow for seed mortality, in which case equation 5.29 becomes

$$N_{t+1} = gS_{max}N_t/[(1 + aN_t)^b + gS_{max}mN_t] \qquad (5.30)$$

where g is the proportion of seeds successfully to germinate and form viable seedlings. However, this change has little effect on the qualitative properties of the model. Watkinson ignored g in his 1981 study of *Agrostemma githago*.

Clearly, we have now come full circle, and such simple models as Watkinson's are identical in form to those we discussed at the end of the section on single-stage models. The only difference is that here we have deliberately built up a model of a generation, whereas before we brought in within-generation factors to make up for the fact that we were using seed production curves to parameterise models which attempted to span generations. The origin of our equations will not affect the predictions, provided that our particular model has been shown to adequately mimic reality.

The multi-stage models, with their modular construction, allow us to build in many aspects of weed biology and management. This is no doubt the reason for their current popularity. We can then investigate their effects by varying the parameter values of each in turn. For example, we might investigate the effects of stubble burning by setting the loss parameter (u in equation 5.17) first to zero and then to some experimentally determined value. We might rotate tillage practices or winter and spring cropping by rotating the parameter values which we use in each year. Such uses of the models will be illustrated in Chapter 6. Although it may be difficult to predict the qualitative dynamics of these more complex models analytically, it is a simple matter to do so by computer simulations.

A small number of multi-stage models of weed populations have included random, or *stochastic*, elements within the life-cycle (e.g. Rauber & Koch, 1975). Values for these are generated from frequency distributions, usually the normal distribution. Like any aspect of modelling it is important to know what is to be achieved by making such a change, and there is little point in complicating a model just for the sake of it. The reasons for making models stochastic vary. There may be a perception that the models will then more closely resemble population dynamics in a real, variable world and will not rely on the assumption of a constant environment. While this may be the case, it is unclear what additional understanding will be learned from the stochastic model that we do not learn from the deterministic version. The processes included in the model are the same apart from the random noise. The simulations have no greater chance of predicting density in a particular year of a real, variable environment, since the random noise will not necessarily be synchronous with the real weather and management variations. If a number of stochastic simulations are produced and averaged, we will usually see that the predicted densities are not the same as the densities predicted by the average parameter values. However, the qualitative behaviour of the models will often be unaffected unless population levels are extremely low.

By running a model repeatedly and producing a large number of stochastic simulations based on variable model parameters (*inputs*), it will be possible to obtain a frequency distribution (and hence confidence intervals) of the model predictions (*outputs*). This is the essence of a 'Monte Carlo' simulation. However, the results are dependent on the degree of variability placed on the parameters. We have little data on the variability of life-cycle parameters between years. We would therefore find it difficult to choose realistic variances to feed into the model or to decide which of the parameters are the more variable. Firbank & Watkinson (1986) used stochastic simulations to examine the sensitivity of their model to its parameter values. Each parameter was varied according to either its standard error obtained from the original fitting procedures (which will include experimental error) or assuming a standard error of 10% of the parameter value. The sensitivity of the model to the parameters was then assessed from the standard errors of the outputs.

Matrix models for multiple cohorts

So far in our development of models we have assumed that all plants emerge together, go through the various stages of their life-cycle together and reproduce together. In essence, we have restricted our attentions to annuals emerging in a single cohort (a single episode of emergence). This may be a realistic simplification for many weeds of cropping, where there is often a flush of emergence after crop sowing. If there are late emerging seedlings they may well be insignificant in relation to the numbers emerging earlier and, as a result of interference with the crop and other weeds, may make a negligible contribution to seed production. The assumption of a population composed of similar individuals will certainly not be valid for biennials, such as some thistles in pastures which have mature and immature plants present at the same time, for perennial shrubs, such as mesquite in rangelands, for clonally spreading herbaceous perennials, such as *Elymus repens* or *Cyperus esculentus* where new clonal shoots may be formed throughout the year, or for annuals such as *Cardamine hirsuta* or *Poa annua*, in which there may be repeated episodes of seed production and germination within a season leading to overlapping generations of plants. In these cases a simple circular flow-chart is an inappropriate way of representing the species. We therefore require a way of handling more complex population structures.

Matrix models are ideally suited for this purpose. They were developed for studies of animal populations by Leslie (1945) and Lefkovich (1965) and extended to plant populations by Usher (1973) and others. First of all, we need to divide our population into a number of 'states'; these may be seeds, various ages of plants, or definable developmental states, such as tubers, immature plants and reproductive plants. Instead of a single number N_t for total population density, the numbers in each category can be represented by a column vector (N_t) showing the numbers in each category, for example

$$N_t = \begin{Bmatrix} N_1 \\ N_2 \\ N_3 \\ N_4 \\ N_5 \end{Bmatrix} \tag{5.31}$$

where there are five states.

The various survival probabilities and fecundities for each state can be summarised in a 'projection' or 'transition' matrix, where the columns represent the current states, and the rows give the probabilities of surviving in the same state, becoming a plant of another category, or producing offspring within the time interval. For example, McMahon & Mortimer (1980) divided the life-cycle of *Elymus repens* into five states: seeds, rhizome buds, immature adults and mature adults. Their transition matrix was:

$$M = \begin{Bmatrix} P_s & 0 & 0 & s & 0 \\ a & P_{sl} & 0 & 0 & 0 \\ 0 & c & P_i & 0 & b \\ 0 & 0 & d & P_m & 0 \\ 0 & 0 & 0 & f & P_b \end{Bmatrix} \tag{5.32}$$

where P_s is the probability of a seed surviving as a seed

s is the number of seeds produced by a mature plant

a is the proportion of seeds becoming seedlings

P_{sl} is the proportion of seedlings surviving as seedlings

c is the proportion of seedlings surviving to produce immature plants

P_i is the proportion of immature plants remaining as immature plants

b is the proportion of rhizome buds producing immature plants

d is the proportion of immature plants surviving to become mature plants

P_m is the proportion of mature plants surviving as mature plants

f is the number of rhizome buds produced by a mature plant

P_b is the proportion of rhizome buds surviving as dormant buds

The corresponding column vector is

$$N_t = \left\{ \begin{array}{c} N_s \\ N_{sl} \\ N_i \\ N_m \\ N_b \end{array} \right\} \tag{5.33}$$

Matrix algebra allows us to multiply the matrix of probabilities and fecundities by the column vector at the start of the time interval, so as to obtain a column vector of the numbers in each state at the end of the time interval:

$$N_{t+1} = MN_t \tag{5.34}$$

This calculation is achieved by multiplying each number in the column vector by its corresponding probability or fecundity in a row of the transition matrix. For example, the number of seeds at time $t+1$ will be

$$(N_s \times P_s) + (N_{sl} \times 0) + (N_i \times 0) + (N_m \times s) + (N_b \times 0)$$

and the number of seedlings at time $t+1$ will be

$$(N_s \times a) + (N_{sl} \times P_{sl}) + (N_i \times 0) + (N_m \times 0) + (N_b \times 0)$$

so that for all states

$$N_{t+1} = \left\{ \begin{array}{c} N_s P_s + s N_m \\ N_s a + N_{sl} P_{sl} \\ c N_{sl} + N_i P_i + b N_b \\ d N_i + N_m P_m \\ f N_m + N_b P_b \end{array} \right\} \tag{5.35}$$

If we repeat this calculation over several time intervals, starting from a population N_0, at time t we will have a population size of size $N_t = M^t N_0$. This is similar in structure to the exponential equation and, since the values in the matrix do not depend on density, the predicted population dynamics will be exponential (once initial instabilities have abated). In fact, after several multiplications, the proportions of the numbers in the different states will stabilise, so that $N_{t+1} = R(N_t)$. The value of the rate of increase, R, can be calculated by matrix algebra techniques (for further details see Pielou, 1977).

As with other multi-stage models, we can introduce the effects of herbicides. If a herbicide kills a proportion k of seedlings, we can replace P_{sl} in the matrix by $(1-k)P_{sl}$ and c by $(1-k)c$. For a translocated herbicide, we might multiply bud survival similarly by a herbicide kill factor. We can also have different matrices for different crop management scenarios. We could then simulate the effects of crop rotations by using different matrices on different occasions.

So far, we have not considered the length of the time-step between successive multiplications. For populations of perennials, this might be an interval of one year, as in the discrete exponential model (equation 5.1). However, for overlapping generations of annuals or for herbaceous perennials, we would expect the probabilities and fecundities to change significantly during the year. We can then produce transition matrices for different parts of the year and apply each transition matrix in turn. Thus, Sarukhan & Gadgil (1974) divided the year into four seasons, each with its own transition matrix; McMahon & Mortimer (1980) used a new transition matrix for every two months. If we begin with a column vector N_t, after the first 'season' the column vector will have become $M_1 N_t$, after the second season it will be $M_2 M_1 N_t$, and after the s seasons within a year and measuring time t in intervals of one year, we obtain

$$N_{t+1} = M_s M_{s-1} \ldots M_1 N_t \qquad (5.36)$$

By matrix algebra we can replace the product of the series of matrices by a single matrix.

A very simple model of this type was developed by Andujar et al. (1986) for *Avena sterilis*. They divided a population into three states: seeds, seedlings and mature plants (although since the plant phase consisted primarily of a single cohort, seedlings and mature plants did not occur at the same time). Their transition matrix for any part (i) of the year was

$$M_i = \begin{Bmatrix} P_s & 0 & b \\ a & 0 & 0 \\ 0 & d & 0 \end{Bmatrix} \qquad (5.37)$$

The values assigned to the transition probabilities for each season were

$$\text{autumn–spring:} \quad M_1 = \begin{Bmatrix} 0.4 & 0 & 0 \\ 0.3 & 0 & 0 \\ 0 & 0 & 0 \end{Bmatrix} \qquad (5.38)$$

$$\text{spring–summer:} \quad M_2 = \begin{Bmatrix} 0.9 & 0 & 0 \\ 0 & 0 & 0 \\ 0 & 0.75 & 0 \end{Bmatrix} \quad (5.39)$$

$$\text{summer–autumn:} \quad M_3 = \begin{Bmatrix} 0.5 & 0 & 30 \\ 0 & 0 & 0 \\ 0 & 0 & 0 \end{Bmatrix} \quad (5.40)$$

This model predicts an annual rate of increase of the seed bank of $R = 6.93$.

As before with the simpler multi-stage models, the single-value elements in the matrix models can be replaced by equations, so that the models include density-dependent processes. They could also be replaced by functions of environmental factors, such as soil water availability (Maxwell *et al.*, 1988). In a matrix model produced by Law (1975; described in detail by Begon & Mortimer, 1986) *Poa annua* seed production was given by the equation

$$s = v \exp(-hN) \quad (5.41)$$

and seedling survival by the equation

$$d = p - q \exp(rN) \quad (5.42)$$

unless N exceeded a threshold value, after which $d = 0$. All other matrix elements remained independent of density; v, p and q are arbitrary regression parameters.

In recent years matrix modelling has been the subject of considerable theoretical interest. Useful methods have been developed for exploring their sensitivity, particularly for exponential models. However, rather than give a far too brief description here, we refer the reader to Caswell (1989) and Silvertown *et al.* (1993) for detailed discussion.

What types of trajectory occur in practice?

Earlier we described a suite of intuitively *possible* intrinsic population trajectories. We have seen that the same trajectories can be predicted from mathematical models as well as graphical models. But do the trajectories occur in practice? Which of them have been observed, and which are only theoretical possibilities? No model, whether graphical or mathematical, is of any use to us unless it can be shown to mimic reality.

Immediately, we face three major problems. Firstly, there are very few long-term data sets; traditional research funding tends to inhibit studies long enough to record population changes over several generations. To establish that a population is indeed undergoing stable cycles, for example, would require data from at least four generations and preferably six or more. Studies of six years duration do exist (e.g. Selman, 1970; Wilson & Phipps, 1985), and even some of eight years (Dessaint *et al.*, 1990), but they are very rare.

Secondly, we are considering intrinsic population changes and assuming that all else except density remains constant. But all else does not remain constant. Weather changes constantly; farm management practices change from year to year. These changes would repeatedly perturb the population away from equilibrium, as well as altering the equilibrium towards which the population is heading. We would therefore expect that any intrinsic density-dependent behaviour will be masked in the field by a large amount of extrinsic 'noise'. How could we distinguish intrinsic density-driven chaos, for example, from environmentally driven random fluctuations of what would otherwise be an asymptotic approach to equilibrium? Or from any of the other possible trajectories?

Thirdly, populations of weeds are seldom followed from low enough densities to describe the full range of density-dependent population behaviour. Studies are usually restricted to established weed populations, or to populations sown at a reasonably high density. We would therefore only expect the populations to follow a restricted trajectory. The more restricted the range, the less likely it is that intrinsic population behaviour will be detected.

To illustrate the problem with background noise, Fig. 5.7 shows an attempt to construct a generation map from data obtained in a six year study by Wilson & Phipps (1985). They sowed plots with *Avena fatua* at a density of 373 seeds m^{-2}, and applied (as near as possible) identical husbandry in every year. Seedling densities were monitored each year. The result is a disappointing scatter of points. It would be a brave person indeed who would suggest a curve to smooth out these data! No confident predictions could be made about the intrinsic population dynamics of *Avena fatua*.

A clearer, though still very limited, picture of a trajectory emerges from a three year study by Sutton (1988) of *Bromus sterilis*, *Galium aparine* and *Avena fatua*. His *B. sterilis* data are presented in Fig. 5.8. Even after as little as one year from a sowing of only 5 seeds m^{-2} (thinned to 3 seedlings m^{-2}), *B. sterilis* appeared as if it might have reached an asymptote. This conclusion is supported by the fact that population levels were then similar to those which were achieved over the same period from a sowing of 1000 m^{-2}. The slight fall from the second to the third year appears small on a

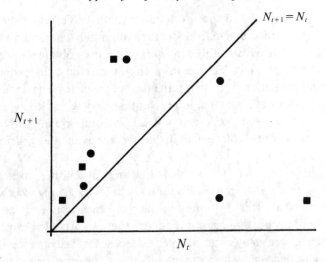

Fig. 5.7. An attempt to construct a generation map from data collected over a number of years. *Avena fatua* seedling densities were recorded in plots tilled with a tine cultivator and sprayed with barban each year (based on data in Wilson & Phipps, 1985): in wheat (■) or in barley (●).

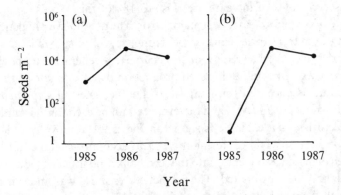

Fig. 5.8. Changes in the seed bank of *Bromus sterilis* sown into a winter wheat crop in 1985 (Mortimer *et al.*, 1993): (a) high initial density, (b) low initial density. No herbicide was applied. Note that after only one year population densities were similar.

log-transformed axis, but is in reality quite large (though not statistically significant).

Rather than follow population density through several generations, we could set up treatments of a range of densities and observe each of them *over a single generation*. In this way, we can establish a generation map by plotting the values of N_1 against N_0 for all treatments. This can then be used to predict population dynamics under the assumption that those conditions occurred every year. Hopefully, the generation map obtained in this way would have less variance than one from a long term study. One problem will be that the populations are sown at the start of the generation under study; if seed dormancy and germination percentage vary with seed age, the age structure of the seed bank will not be at equilibrium. This lack of equilibrium will not be allowed for in the predictions and this will introduce some error.

There have been very few studies of a range of densities over a single *complete generation*. Density studies usually begin by sowing seeds and end at plant maturity. It is tempting to construct generation maps merely by plotting seed production against number of seeds sown. However, significant losses of seeds may occur between dehiscence and emergence; many of the seeds produced may be dormant and may not germinate the following year. If these factors are ignored, the quantitative predictions of population trajectories will be incorrect. For *Avena fatua*, the number of seeds produced by a single plant at low weed density (s_{max})in a wheat crop may commonly be of the order of 100 to 250 (Mortimer, 1987, and Table 4.2), yet observed rates of increase are close to 1% of this. To use s_{max} as an estimate of R would clearly be a major error. The qualitative predictions of the long term trajectories can also be wrong, as will be shown later.

In order to convert a seed production curve to a generation map, we can scale the y-axis by multiplying seed production by an estimate of survival between dehiscence and germination. It then becomes an *estimated N_{t+1}* axis (note not *observed N_{t+1}* and therefore cannot be used for the validation of trajectories). An example is shown in Fig. 5.9. Manlove *et al.* (1982) assumed 30% survival of dormant seed; no account was taken of losses between dehiscence and entry into the seed bank, and thus N_{t+1} may still be considerably over-estimated. The predicted dynamics was an asymptotic approach from low densities to equilibrium. There was some suggestion that high density sowings would not fall, but would instead maintain a higher equilibrium than that achieved from low densities.

An equivalent approach is to plot seed production (y) against seeds sown (x), but to cobweb about the line $y = x/g$, where g is the proportion of seeds produced which germinate in the next year, rather than about $y = x$ (Symonides *et al.*, 1986). If the seed production curve has no peak to the left of the line $y = x$, asymptotic behaviour will be predicted for all values of g. Fig. 5.10 shows a hypothetical seed production curve with a peak to the left of $y = x$. Lines have been drawn for a range of values of g. In this case, for

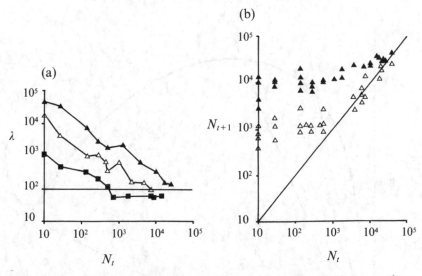

Fig. 5.9. Rate of population increase, λ (a) and generation map (b) for *Avena fatua* seed banks, calculated from data obtained within a single generation (redrawn from Manlove, 1985): in monoculture ($\blacktriangle$); in mixture with winter wheat ($\triangle$); in winter wheat sprayed with 1–flamprop-isopropyl ($\blacksquare$).

$g = 1.0$ (i.e. no post-dehiscence mortality) cobwebbing about the line $y = x$ will lead to a prediction of a cyclic trajectory. As g becomes smaller, the predictions tend towards asymptotic trajectories. Knowledge of the value of g is therefore critical; the wrong assumption can lead to a false prediction. Unfortunately, g *is usually poorly estimated* and our confidence in our predictions must therefore be poor. In many cases g appears to be well below 1.0, perhaps often less than 0.1. The use of the seed production curve as a surrogate for a generation map, without allowing for mortality or seed dormancy, can therefore be very misleading.

For cyclic or other complex trajectories even to be possible, the seed production curve must have a peak to the left of the line $y = x$. Is this common? Again, there are problems with the extent and quality of the data. Firstly, few researchers on weeds include extremely high densities. For pragmatic reasons they restrict their experimental densities to those which would be found in a typical farmer's field. As a result, the line $y = x$ is seldom reached. Maximum seed production per unit area is not achieved in any of the treatments and it is impossible to deduce whether or not a peak exists. Secondly, the limited data on seed production at high densities of weeds tend to suggest that peaked curves are unusual (see Chapter 4). This would therefore indicate that trajectories will usually be asymptotic. Extreme

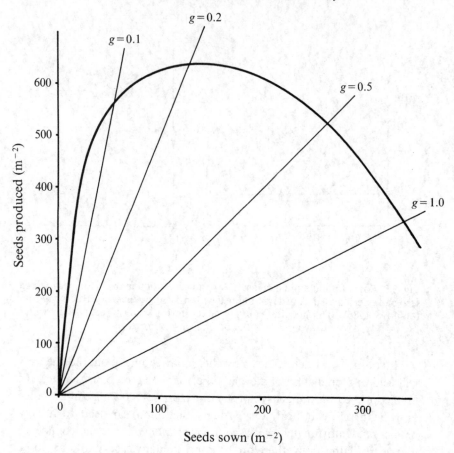

Fig. 5.10. Illustration of how a seed production curve can be used as a generation map, by assuming an appropriate level of germination (g). For $g = 1.0$ stable cycles are predicted; at values of g of 0.1 and 0.2 an asymptotic approach to equilibrium would be predicted (note axes not to same scale).

peaks of the type depicted in Fig. 5.3c have rarely been recorded for seed production curves. Thrall *et al.* (1989) present a possible example for *Abutilon theophrasti*, but they do not have data at low enough densities to be able to describe the whole curve. Finally, much of our data are for weeds growing as monocultures, not growing within crops: it may well be that the types of density trajectory will differ in the two cases. Watkinson (1981) suggested that the addition of a crop to *Agrostemma githago* would change the trajectory from damped oscillations to an asymptotic approach to equilibrium. We clearly need more and better data on weed seed production.

How, then, can we validate the long term trajectories predicted by our theoretical models? There are several possible courses of action, but few of them are practical:

1. We could try to validate our predictions over several generations under strictly controlled environments, such as in 'phytotrons'. This has not been done to date, and may prove too expensive to do on the scale required for higher plants.

2. We could build more complex models, which explicitly incorporate environmental factors. Given data on all of these factors, we would then be able to filter out the environmental noise from our clouds of points. However, the data required to do this would be extensive and beyond the limit of most budgets. Also, the more assumptions that are built into models, the greater the scope for excuses that it is the data which are poor and not the model which is wrong.

3. We could obtain data from as wide a range of habitats and years as possible and average them in the hope that variability will cancel out to reveal the fundamental population behaviour. However, it is also possible for real, interesting behaviour to cancel out, such as two oscillations not in synchrony. If a number of peaked generation curves are combined, the result may appear almost as a monotonic curve.

4. It should be possible to distinguish between behaviour driven *primarily* by the environment and behaviour driven *mainly* by density. If we set up plots at a range of densities, their changes over generations should show a marked degree of synchrony if weather is primarily driving dynamics, whereas if dynamics is driven by density their trajectories will vary independently in a manner predicted by the model for that density (see Symonides *et al.*, 1986).

5. The most pragmatic attitude would be to argue that we can *never* adequately validate our predictions of intrinsic behaviour, so that they must remain a summary of our current (theoretical) understanding. Any predictions from our models will remain a best guess of what might/ could possibly happen. Their usefulness would thus be very restricted. In any case, the assumption of a constant environment in which only density drives population changes is totally artificial. Perhaps a long term validation would be pointless.

Conclusions

We have a wide range of models at our disposal for predicting the intrinsic dynamics of populations. Their properties have been thoroughly explored and include both simple and complex trajectories. The limited range of experimentally derived parameter values for the models indicates that complex trajectories would be uncommon, and that most populations would reach their equilibrium density either asymptotically or by damped oscillations. However, most of these predictions are, and will remain, poorly validated because of the considerable background variability experienced in the field. We must appreciate that neither seed production curves nor true generation maps constitute observed trajectories. They do not demonstrate that a particular dynamics *does* occur, only that it *could* occur. Our predictions from unvalidated models, whether graphical models or mathematical models, must be treated as no more than best guesses. However, the models described in this chapter have been invaluable in focussing the minds of researchers on the likely dynamics which may arise from density-dependence. They have also provided the necessary building blocks on which to base our attempts to understand the ways in which management affects weed populations. This will be the subject of the next chapter.

6

Extrinsic factors affecting population density

In the previous chapter we considered what would happen to populations of weeds in an invariant habitat, where population dynamics would be driven solely by intrinsic processes such as intraspecific interference. However, the environment of a weed population is usually far from being constant, either within or between generations. The management of a field, for example, may vary widely from one generation to the next, causing changes both in a weed species' population growth rate and in its potential equilibrium population density. Weather patterns will vary both between and within generations. An understanding of the dynamics of weed populations in the 'real world' thus depends on a knowledge of the effects of the various extrinsic factors and their interaction with intrinsically regulatory ones.

In weed population ecology the key extrinsic factors can be divided conveniently into three groups (Fig. 6.1): management factors, weather factors and interactions with other organisms (including insects, pathogens, large herbivores and other plant species). The relative importance of the different factors will vary with year, geographic location and habitat type. Erratic rainfall patterns, for example, will cause both crop and weed growth to vary considerably from year to year, leading to variation in weed seed production. In more equable climates, crop management may be more intensive, and changes in management may have a greater influence than weather in determining weed dynamics. In perennial crop communities and those involving grazing by animals, weed population dynamics may be particularly affected by the dynamics of other organisms present.

It will be extremely unlikely that each extrinsic factor will act in isolation. For example, the timing of rains may affect the number and timing of cultivations, the crop species chosen and the management regime of the crop. Together, these weather and management factors are likely to affect

169

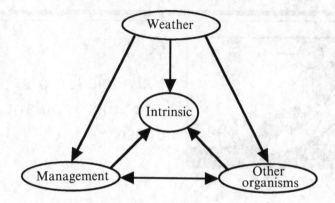

Fig. 6.1. Diagrammatic representation of the interaction between intrinsic population processes and extrinsic factors.

the relative seedling densities of different weed species. In turn, temperature, rainfall, the crop species, sowing density and level of fertiliser application may affect the vigour of the weeds and the abundance of herbivores. The net result will be an effect on weed seed production and hence population growth rate.

In this chapter we will discuss the three major groups of extrinsic factors in turn. We will examine experimental data and the predictions of the population models described in Chapter 5. Of necessity, most attention will be given to management factors, since these are the ones which we can manipulate most easily in experiments and which we can use as tools to control weeds in practice.

The effects of agricultural management factors

To an agronomist, management is the suite of deliberate actions taken to affect crop or animal production and/or quality. This will include the preparations for crop and pasture seeding, crop husbandry, harvesting method and grazing management. Husbandry will include actions which directly benefit the crop, such as fertiliser application, and those which protect the crop from present or future interference by other species, namely weeds, pests and diseases. To a weed, crop management constitutes a variety of forms of habitat modification, which will affect the life history of a weed species. Those species that are pre-adapted to a particular crop management system will be those that have an inevitable tendency to exhibit a positive rate of increase when introduced to that system.

Farm management is constantly changing. Technological improvements, fashions and economic conditions result in changes in the way crops are grown and pastures managed. Changes may be abrupt, as with the introduction of a new herbicide which rapidly becomes widely adopted, or more gradual, as in the introduction of a new crop species. But seldom will one aspect of management change at a time. Fertiliser use, crop cultivars, sowing dates, rotations and herbicide use have all changed markedly since World War II, but not independently of one another. As a result, it will be difficult to separate out the individual effects of the different factors, even with the most complex analytical tools (Cousens *et al.*, 1988b). For example, in one of the most exhaustive studies of changes in weed communities, Haas & Streibig (1982) analysed data from 4600 plots. Although they identified some management factors loosely correlated with the species present (see below), considerable subjectivity was required to interpret the data and their conclusions remain largely speculative. Surveys of weed densities in fields or regions over several years are therefore of very limited use in determining causes of changes in populations and are really only useful to generate hypotheses rather than to determine cause and effect. Most of our information on factors determining population dynamics therefore comes from experiments in which each aspect of management is varied as a separate treatment or as part of a 'factorial' design.

We will discuss the various management factors roughly in the order of the seasonal growth cycle of a crop in an intensive agricultural system. It will become apparent, however, that our knowledge of the effect of extrinsic management factors on the population dynamics of weeds is very incomplete and in many instances the effects have only been measured over a short time scale. Our approach has therefore been to choose examples to illustrate some of the likely consequences that may result from change in management and to discuss these in the context of long term dynamics.

Cultivation

The way in which land is prepared for crop sowing will determine the distributions of weed seeds in the soil profile. Different methods of cultivation will disturb the soil to different depths and will cause different types of soil movement. For example, a mouldboard plough will tend to turn the soil over in blocks, thus burying surface material (including recently shed weed seeds). The efficiency of inversion (i.e. the proportion of surface material buried) will depend on such factors as the texture of the soil, its wetness and the speed of operation of the cultivator. Seeds from as much as 30 cm may be moved to the soil surface, while seeds from the surface may be buried to that depth. Tine cultivators result in less vertical movement. Because of their raking action, they tend to stir the soil, mixing material from different depths. The amount of mixing will decrease with depth; also, the deeper that the tines are set to penetrate the soil, the deeper the mixing. Harrows disturb only the soil surface and are used mainly in the later stages of seed bed preparation. Seed bed preparation by cultivation is usually not restricted to a single implement, but consists of a programme of operations often using several types of machinery. Some of the implements will loosen the soil, some will break up soil clods, and others will be used to kill weeds or to encourage weed germination for later control.

In some countries in recent years, direct-drilling and sod-seeding have increased in popularity, where crop or pasture seeds are sown without prior cultivation. With the introduction of non-selective herbicides, such as paraquat and glyphosate, it was realised that cultivation to kill weeds was not essential. As a result, time and money could be saved, soil erosion could be reduced and soil moisture retained. A programme of cultivation can be replaced by a pre-sowing application of a herbicide, followed by sowing directly into undisturbed soil. This innovation, and the floristic changes resulting from it, has generated a considerable amount of research on weed population dynamics. Results are usually summarised according to whether the weeds are annual grasses, annual broad-leaved weeds or perennials. However, these groupings are taxonomically diverse, and it is of little surprise to find very different responses within the groups, reflecting varying species biology. The response by a species to tillage will depend on seed dormancy mechanisms and germination behaviour (Mohler, 1993, and see Chapter 4).

The most noticeable responses to tillage have been recorded in annual grass weeds. In one example, Moss (1979) compared the densities of *Alopecurus myosuroides* at three times of the year under different cultivation methods (Table 6.1). Approximately two weeks after cultivation there were far fewer seedlings in ploughed plots than tine cultivated plots. All plots were sprayed with paraquat and then the crop was sown. At maturity, there were once again lower densities of *A. myosuroides* on ploughed than on tine cultivated or direct drilled plots. In experiments reported by Froud-Williams (1983), complete eradication of *Bromus sterilis* was achieved after only one year of mouldboard ploughing, whereas high densities remained on direct drilled plots.

For annual grasses with low levels of seed dormancy, these short term effects will inevitably be reflected in long term population dynamics. Some

Table 6.1. *Effect of tillage on the density (plants* m^{-2} *) of* Alopecurus myosuroides *in the first year after treatment. All plots were sprayed with paraquat prior to sowing of the wheat crop. No other herbicides were applied. Data given are averages of two straw disposal methods*

	Plough	Tine	Direct drill
Density pre-spraying:			
Before tillage	2657	2607	—
After tillage	20	557	3615
Density post-spraying:			
At plant maturity	51	322	351

Source: After Moss (1979).

grass weeds will only germinate close to the soil surface and will not survive for long when buried in the soil; hence their abundance will be favoured by direct drilling and impaired by ploughing. *Bromus sterilis* seeds will germinate and fail to reach the soil surface if buried too deeply (see p.105), and the seed bank will decline rapidly. Greater germination from seeds left at the soil surface will lead to greater seed production, shedding even more seeds to the soil surface, and so on. Increased annual grass weed problems under reduced tillage systems have been reported in the USA, Europe and Australia (Medd, 1987a) and are one of the major problems encountered when trying to minimise farmer work-loads and erosion by this technique. The greater rates of population increase will raise the levels of control required from herbicides (see p.198) to keep the weed populations in check.

Most experimental studies of the effects of tillage on weed populations have been only short term, seldom exceeding three or four years, illustrating *potential* population dynamics rather than long term observed dynamics. In a rare study, however, Wicks *et al.* (1971) grew winter wheat every year for 12 years under three cultivation treatments and examined changes in the abundance of *Bromus tectorum*. Sweep plough, one-way disc and mouldboard plough cultivations were applied each year between harvest and crop sowing, representing an increasing scale of soil inversion. At the end of the study, the densities of *B. tectorum* were 93, 24 and 0 plants m^{-2} respectively. Forcella & Lindstrom (1988) reported seed and seedling densities (mostly *Setaria* spp.) after 8 years of cropping, with either conventional or ridge cultivation. There were greater densities of both seeds and seedlings under ridge tillage (Table 6.2).

Although most annual grass weeds stand out as responding positively to

Table 6.2. *Seed and seedling densities (m^{-2}) of the weed flora in maize after 8 years of cropping, as affected by tillage method and cropping sequence*

	Conventional tillage	Ridge tillage
Seed bank:		
Continuous maize	960	2980
Maize/soybean rotation	580	680
Seedlings:		
Continuous maize	102	181
Maize/soybean rotation	64	81

Source: From Forcella & Lindstrom, 1988.

reduced tillage, some species show either no obvious response, or they respond unpredictably. Derksen *et al.* (1991) found in a 5 year study that the weed species responding positively to tillage varied between sites and years. *Setaria viridis*, for example, displayed a strong association with conventional tillage in only one year and at only one site. Perennial weeds in this study were associated with zero tillage at one site, but not at others.

Unlike annual grasses, the responses of annual broad-leaved weeds to tillage method are far more varied, and it is difficult to generalise about the types of species which respond in a given situation. In one experiment (Pollard & Cussans, 1976), annual broad-leaved weed seedlings were counted on plots receiving four depths of cultivation, from mouldboard ploughing (to a depth of 20–25 cm) to direct drilling (no cultivation). Over 5 years, there was generally a higher density of broad-leaved weeds in plots receiving the deepest cultivations (Fig. 6.2). The dominant species in ploughed plots were *Polygonum aviculare*, *Raphanus raphanistrum* and *Sinapis arvensis*. There was no clear trend in total population density over time. This was interpreted as being due to the buffering effect of a large, persistent seed bank and good weed control by herbicides. Similar results were recorded by Bachthaler (1974) in Germany. Few broad-leaved species, if any, in Pollard & Cussans' study were most abundant under reduced tillage. Most species showed no response to tillage method.

Some perennial weeds have been observed to increase under reduced tillage systems. Bachthaler (1974) found an increase in *Elymus repens* to such a density that direct drilling had to be discontinued. Perennial broad-leaved weeds also increased. Other species observed to increase under minimum tillage include *Cirsium arvense* in North America (Staniforth & Wiese, 1985) and *Rumex crispus*, *Taraxacum officinale*, *Trifolium repens*,

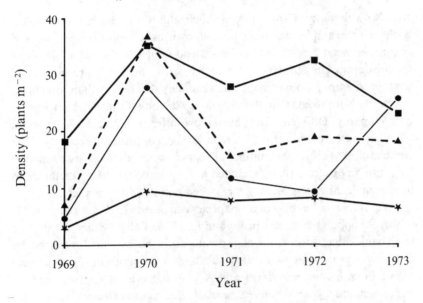

Fig. 6.2. Dynamics of broad-leaved weed seedlings (all species combined) under four cultivation regimes in England (based on data in Pollard & Cussans, 1976): direct drill (★); shallow tine (●); deep tine (▲); mouldboard plough (■).

Sonchus arvensis and *Convolvulus arvensis* in the UK (Bachthaler, 1974; Pollard & Cussans, 1976). One reason often given for the decline of perennials under ploughing (and of the increase under direct drilling) is that rhizomes and other organs are brought to the soil surface by cultivation, where they desiccate and die. In species with tap roots, cutting by the plough may cause a delay in regrowth from sprouts, reducing their ability to interfere with a crop. However, cultivation of the rhizomatous *Solanum elaeagnifolium* in Australia caused an increase in shoot density, perhaps because buds on rhizomes were released from apical dominance (A. R. Leys, pers. commun.). This species seems to be extremely tolerant of desiccation.

It is clear that, apart from perhaps the annual grasses, we have little ability to predict the direction of response which species may show to cultivation treatments. This is clearly an area deserving closer attention.

Crop selection and crop rotation

Crop species vary in the times of year at which they are sown. The degree to which the sowing date of the crop matches the germination period of the weed species will determine which species of weeds successfully emerge and grow with the crop. Crops differ in their ability to interfere with weeds. Barley, for example, has long been known as a 'smothering crop' and suppresses weed growth better than wheat (when compared at the same sowing dates). The genotype of the crop is also critical in determining weed seed production. Some crop cultivars may be better than others at interfering with weeds (Challaiah *et al.*, 1986), either through competition or allelopathy. Different crops also allow different herbicides to be used, allowing different spectra of weeds to be controlled effectively. The combination of crops, the number of years in which each is grown, and the length of any pasture phase included in the rotation, is likely therefore to have significant effects on long term weed population dynamics.

An example of the effect of crop species can be seen from a study of weeds in three cropping sequences in Wyoming, USA (Ball & Miller, 1990). The seed bank after three years of growing pinto beans was dominated by *Solanum sarrachoides*, with also substantial amounts of *Chenopodium album, Amaranthus retroflexus* and *Kochia scoparia*. After two years of sugar beet and a year of maize *S. sarrachoides* was at extremely low levels; the main species in the seed bank were *C. album, A. retroflexus* and *K. scoparia*, with some *Eragrostis cilianensis*. After three years of maize, all weed populations were at much lower densities, but with *C. album, Setaria viridis* and *A. retroflexus* being the most abundant species. Despite the clear differences in weed populations between cropping sequences, it remains unclear whether the primary cause was crop sowing date, the different herbicide programmes used in each crop, or some other aspect of crop husbandry. However, the initial management decision which led to the divergence of the weed floras was the choice of which crop to grow. Another example of the effect of crop rotation can be seen in Table 6.2, where the total weed density was greater in a continuous maize sequence than in a maize/soybean rotation.

The effects of growing a particular crop can be carried through for several generations. Lotz *et al.* (1991) grew five different crops in the same year and then monitored the density of *Cyperus esculentus* through three further maize crops. Clear differences were still present at the end of the study (Fig. 6.3). The least increase in population density was shown following a hemp crop. The greater suppression of the weed by hemp than by maize was attributed to the earlier closure of the crop canopy of the former species.

Some of the most profound effects of arable crop selection come with the choice of winter versus spring crop species or cultivars. In a winter crop, sown in autumn and harvested in summer, summer germinating weeds (if

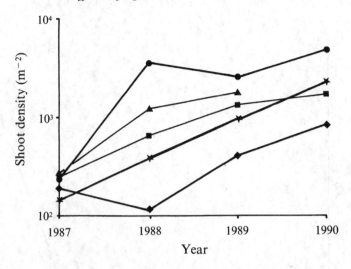

Fig. 6.3. Dynamics of *Cyperus esculentus* in the years following growth of four different crops in The Netherlands (redrawn from Lotz *et al.*, 1991): hemp (◆); winter barley (★); maize (■); winter rye (▲); no crop (●). In the succeeding years maize was grown on all plots.

they germinate at all) must emerge into an established crop canopy and will grow poorly if at all. In a summer crop, sown in spring or early summer and harvested in summer or autumn, weeds which germinated in the previous winter will usually have been killed by cultivation. Hence, winter and summer crops tend to develop distinctive floras. For example, Zemanek (1976) found in a 4 year study that *Sinapis arvensis* increased in spring barley and decreased in winter wheat. The change in Britain over the past few decades towards predominantly winter sown grain crops has probably been partly responsible for the rise in dominance of *Galium aparine* and the decrease in *Polygonum aviculare*.

In eastern Australia, most wheat is grown in rotation with pastures, fallows or summer crops. The frequency of each in the rotation varies with climate and market prices for the various commodities, not (in general) for reasons of weed control. An annually regenerating legume pasture is the most common break from cereals in the southern part of the region, providing feed for sheep as well as increasing soil nitrogen content. At present, the cropping phase is the most valuable part of the rotation. Palatable broad-leaved species such as *Arctotheca calendula* and grasses such as *Lolium rigidum* may be tolerated in the pasture (as either beneficial species or as only minor weeds), but deliberately killed by the use of

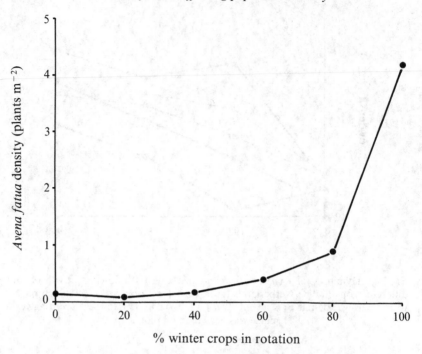

Fig. 6.4. The density of *Avena fatua* plants in winter wheat crops in relation to the frequency of winter cropping in the crop rotation. Data were obtained from northern New South Wales, Australia, by sampling the fields of respondents to a random questionnaire (redrawn from Martin & McMillan, 1984).

herbicides in the crop (where they may reduce yields considerably). The relative lengths of wheat and pasture phase must therefore be expected to affect their population dynamics. Summer crops are grown in more northerly areas, where there is summer rainfall, or under irrigation. In a survey of a summer rainfall area which included winter wheat, long fallow, summer crops and pasture, Martin & McMillan (1984) found a marked increase in the density of *Avena* spp. with the frequency of winter cropping in the rotation (Fig. 6.4).

Survival of weed seed banks under the pasture phase of a rotation tends to be greater than under cultivated crops (see Chapter 4). Anecdotes abound of fields in the UK which, when ploughed in World War II for the first time in living memory, became infested with *Sinapis arvensis*, probably from a dormant seed bank. On a shorter time-scale, Wilson & Phipps (1985) found that 1% of seeds of *Avena fatua* survived in the soil after three years under a grass ley, whereas none appeared to survive three years of cropping for arable (spring barley) silage. At the end of the experiment the numbers of buried seeds of *A. fatua* under wheat and barley grown for grain production (with herbicides) was similar to that under pasture, but this may have been a result of plants that escaped spraying and produced seed.

Warnes & Andersen (1984) found slower decline of *Sinapis arvensis* seed banks under a grass ley than under cropping; decline of *Abutilon theophrasti* was slower under a (perennial) alfalfa crop than under oats, but similar to the decline under a two-year maize/soybean rotation (Lueschen & Andersen, 1980). The rate of decline will clearly depend on whether species are able to set seed in the pasture: seeds may be set if grazing pressure is low or concentrated outside the flowering period. Cropping for silage or hay may, or may not, prevent weed seed production; the timing of cutting relative to weed phenological development will be critical.

Sowing date

Each species of weed will respond in a characteristic way to temperature and rainfall, so that they will tend to display seasonal patterns of emergence. The date around which a farmer intends to sow a crop and the recent rainfall pattern together dictate when final seed bed preparations take place. These preparations will kill most of those weeds that have already emerged. If the main flush of emergence of a weed occurs just after the crop has been sown, then its density in the crop will be high; if emergence occurs just before cultivation or sowing, then many seedlings will be killed and its density will be lower. Moreover, those weed seedlings emerging well after the crop will interfere poorly and will produce fewer seeds than those emerging before or with the crop.

A traditional way to minimise weed problems was to delay cultivations until after emergence had occurred, then cultivate and sow. The increased use of herbicides has allowed sowing of winter crops to take place earlier, thereby achieving higher yields, but such systems rely on chemicals rather than tillage to kill weeds. Farmers returning to 'organic' methods regard timeliness of cultivations and sowing in relation to weed emergence as one of their primary methods of weed control (Wookey, 1987).

In Chapter 5 we described the study by Selman (1970), who found that *Avena fatua* increased in spring barley when sowings were early, but decreased steadily when sowing was delayed. For early sowing $\lambda = 2.74$, whereas with late sowing $\lambda = 0.40$. Whybrew (1964) also presented results of normal and delayed sowing of spring barley from an experiment at the same location, showing that repeated late sowing could almost eliminate a high

Table 6.3. *Effect of sowing date on the density of* Avena fatua *seedlings in long-term plots cropped with spring barley each year. Data are averages of six straw disposal and tillage treatments. Seedling densities* (m^{-2}) *were assessed in late autumn. 'Normal' sowings were between 7 March and 13 April; 'late' sowings were between the end of April and early May, 3–4 weeks after the main spring germination period of* A. fatua

	1958	1959	1960	1961	1962	1963
Normal sowing	3.22	0.74	2.88	2.17	6.44	29.17
Late sowing	1.37	0.03	0.01	0.07	0.00	0.08

Source: From Whybrew, 1964.

density population of *Avena fatua* (Table 6.3). A shorter study by Jarvis (1981) showed that considerable reductions in *Alopecurus myosuroides* density could be achieved by late sowing of winter wheat. In one year, a delay of sowing from 12 November to mid-December reduced weed density by 25%; in the next year a delay from 8 October to mid-December caused a weed reduction of 47%. However, the economic benefits of delaying sowing may be slight or even negative, since the advantages in terms of weed control may be inversely correlated with potential crop yield. If herbicides are available, many farmers will prefer to rely on them, rather than delay sowing and face uncertain weather conditions.

Sowing density

Seed production by weeds, and hence population rate of increase, is likely to be affected by any factor which changes the competitive ability of the crop stand. Crop density is known to be one such factor (see Chapter 4). For example, Cussans & Wilson (1975) found that seed production by *Avena fatua* was increased greatly when crop seed rate was halved. From response surface equations, such as equation 4.3, we might expect a greater effect on seed production from halving crop sowing rate than from doubling it (see also Wright, 1993). We can examine the likely effect of crop density on rate of population increase through the simple models described in Chapter 5. For example, equation 5.13 expressed the rate of increase of a population in terms of crop and weed density, the rate of increase of the weed at extremely low densities in monoculture (R), and three other parameters. Let us assume parameter values of $\alpha = 1$, $a = 0.5$, $b = 1$ and $R = 500$ (where densities are expressed per m^2). If there is one weed m^{-2} and 200 crop plants m^{-2}, the rate of population increase will be $\lambda = 4.93$. If crop density is halved, $\lambda = 9.71$; if crop density is doubled, $\lambda = 2.48$. However, there have been no direct field studies of the long term effects of crop seed rate.

Crop sowing patterns and inter-row distances can affect yield losses resulting from weeds (e.g. Medd *et al.*, 1985), but their effects on weed seed production (or long term population dynamics) have seldom been studied.

Fertilisers (and soil ameliorants)

Species vary in their demands for nutrients and in their tolerances to soil conditions. Some species are argued to be nitrophilous, although not necessarily responding only to nitrogen. Nitrogen application rates increased steadily in most arable crops from the 1950s to the 1980s. Haas & Streibig (1982) considered that this has resulted in the increase of some arable weeds, particularly grasses, in Denmark since 1950. Liming was speculated to be the cause of a reduction in the abundance of calcifuge species, such as *Rumex acetosella*, *Erodium cicutarium* and *Spergula arvensis*, and an increase in calcicole species, such as *Veronica persica*, *Anagallis arvensis* and *Lamium purpureum*. However, so many practices have changed with the intensification of agriculture that it is difficult to ascribe specific causes to changes in weed abundance without careful experimentation.

Of the many studies of interference between *Avena fatua* and cereals, some have shown that nitrogenous fertilisers increase yield loss, some show a decrease and others show no effect. Whether the effect is positive or negative can also vary from one year to the next within a single site. This is likely to indicate that effects on weed population dynamics will also be variable. The response to fertiliser is likely to vary with the rate applied. For example, Gonzalez Ponce *et al.* (1988) found that the size of plants of *Avena sterilis* growing in a wheat crop was greatest at an intermediate fertiliser rate. Ampong-Nyarko & De Datta (1993) found that, of four weed species growing in rice, *Amaranthus spinosus* was the least vigorous at low fertiliser rates, but the most vigorous at high rates. Clearly, the response will vary with the relative physiologies of the species involved and the background fertility of the soil. However, such short term studies can be useful in assessing likely long term dynamics (although this has never, to our knowledge, been tested rigorously).

Perhaps the longest-running fertiliser trial in the world was started on Broadbalk field at Rothamsted, UK, in 1843. Since then, some plots have received annual applications of farmyard manure only, while other plots have received a range of different levels of inorganic nitrogen fertiliser in combination with magnesium, phosphorus and potassium. In some areas, winter wheat has been the only crop grown since 1843 (though occasionally interspersed with fallows), while in other areas winter wheat has been rotated with other crops, such as potatoes. In 1964, Thurston reported that *Alopecurus myosuroides* density had become greatest on continually fertilised plots.

From 1991 to 1993, S. R. Moss (pers. commun.) surveyed those areas of Broadbalk which have not received herbicides. Some species, such as *Medicago lupulina*, *Equisetum arvense* and *Aphanes arvensis*, showed clear preferences for low nitrogen plots. *Tripleurospermum inodorum* and *Ranunculus arvensis* were most frequent at intermediate levels of nitrogen, whereas *Stellaria media* favoured plots with high nitrogen or farmyard manure. *A. myosuroides* and *Papaver* spp. occurred abundantly in all treatments, but their density increased with nitrogen level. Although the data do not reveal the detail of the dynamics over the 150 years, it is clear that in the absence of herbicides the use of different fertiliser levels has caused particular species to increase and others to decrease. The final weed flora in the high nitrogen plots was typical of modern cereal fields, whereas the low fertiliser plots included greater proportions of species which have now become rare in the UK. This suggests strongly that fertiliser use has had a large influence on community composition, perhaps even more so than herbicides. For 'high nutrient' species, the reason for their current abundance may well be some physiological response to a high supply of nutrients; 'low nutrient' species may simply be intolerant of high nutrient levels or they may be suppressed by other species which respond better.

In Australia, pasture production has been increased by the introduction of *Trifolium subterraneum*, which fixes nitrogen, and by applications of superphosphate. However, as well as benefiting desirable species the resulting increase in soil fertility has favoured weeds, in particular thistles (Sindel, 1991). On a farm owned by the University of Sydney, some 'improved' fields were so badly infested with *Cirsium vulgare*, *Carduus nutans* and *Onopordum acanthium* that grazing was severely limited. In an attempt to reduce the fertility and hopefully therefore to decrease the thistle problem, the farm manager sowed a crop of oats (sod-seeded after glyphosate application). The result the following year was a monoculture of *Carthamus lanatus*, a thistle previously at low abundance and which has been shown to prefer lower levels of fertility than the other species (Austin *et al.*, 1985)! However, it was not determined whether the reduced nutrient levels were directly responsible for the increase in *C. lanatus*, or whether optimal germination conditions for that species simply happened to coincide with the senescence of the oat crop.

Irrigation

In many tropical and sub-tropical areas, water is used in rice paddies deliberately to alter the habitat in a major way, from a dryland to an aquatic environment. One of the major reasons for this is so that the dryland weeds will be killed by waterlogging. Most rice does not require standing water for growth. The paddy conditions allow certain aquatic and semi-aquatic species to become abundant, and hence paddy rice has a characteristic weed flora. *Echinochloa* spp. are especially well adapted to waterlogging and are some of the major international rice weeds. In Australia, *Echinochloa crus-galli* remains the major problem in crops where rice is sod-sown and young seedlings are flushed periodically with water, before being flooded permanently. A recent development is the sowing of pre-germinated rice seeds by air into already flooded fields. Under these conditions, the semi-aquatic *E. crus-galli* has become less of a problem, and the true aquatics have become dominant, such as *Cyperus difformis* and members of the Alismataceae (McIntyre *et al.*, 1991). Although quantitative data are not available, there have been marked changes in density as a result of irrigation management.

The effect of irrigation will depend on how much the previous habitat is changed, i.e. by how much extra water is supplied (the same will be true in the reverse process, when water is removed by drainage). The addition of water into a habitat may be expected to have a minor effect if it is being used to supplement rainfall in occasional dry spells, such as on temperate vegetable farms. Irrigation of areas previously desert is likely to have a major effect on endemic plant populations; it may be that few of them will be able to take advantage of the abundance of water, so that invaders better adapted will increase and interfere with them.

Herbicides

The use of herbicides since the Second World War has been one of the greatest changes to the cropping environment in agriculture and horticulture. The effectiveness of modern herbicides in controlling weeds can be seen most vividly in years when weather conditions prevent spraying. The brightly flowered *Papaver rhoeas* in Europe is easily killed by broad-leaved herbicides in cereals, but it has a very persistent seed bank. When wet weather occurs at the critical time of year for spraying and herbicide use is therefore abandoned, fields can be turned scarlet by the weed, as they used to be before herbicides.

The trend in recent decades in intensive agricultural systems has been for increased frequency of herbicide use, to the point at which weeds in many crops are being sprayed prophylactically with pre-emergence herbicides (i.e. before the weed plant population can even be seen). Many intensive cereal producers spray broad-leaved weeds every year, without any confirmation of whether expenditure is justified, since the herbicides are cheap and effective. It is hardly surprising that there has been a general decrease in the abundance of weeds within intensive cropping (Haas & Streibig, 1982), probably as a result of the slow but steady decline in the abundance of seeds in the soil. The selectivity of herbicides varies, such that each chemical has its unique spectrum of weeds which it kills or retards and those which it does not. As herbicides come and go in the market place, so different weeds are likely to increase and decrease.

Despite repeated changes in the herbicides being used by farmers, some species have shown clear long term trends in abundance. Haas & Streibig (1982) attributed the decline of *Sinapis arvensis* and *Sonchus arvensis* since 1945 to their susceptibility to the still commonly used phenoxy acid herbicides (such as 2,4-D and MCPA). Weaver (1985) stated that increased use of pre-plant and pre-emergence herbicides in maize and soybeans has resulted in an increased abundance of late-germinating, large-seeded weeds. Ervio & Salonen (1986), however, could find little influence of herbicides on changes in the Finnish weed flora from the 1960s to the 1980s. In their study, drainage and liming proved to be more significant. In all of the above cases the causes of the dynamics have been inferred rather than demonstrated experimentally.

Experimental evidence suggests that while herbicides may affect the relative abundances of species, they will seldom lead to the disappearance of a species altogether. Thurston (1964) found that seven years of annual applications of phenoxy acid herbicides caused a decrease in abundance of sensitive species, but none were eliminated. *Polygonum aviculare* and *Stellaria media* decreased, but others, such as *Tripleurospermum maritimum*, *Veronica arvensis*, *Aphanes arvensis* and *Euphorbia exigua* increased. Rademacher *et al.* (1970) found that the relative abundances of different broad-leaved weeds after 12 years depended on the particular herbicide used (Table 6.4). *Sinapis arvensis* had declined markedly in abundance under all herbicide treatments even though the species possesses dormant seeds which may survive for long periods in the seed bank. Most species had

Table 6.4. *Effect of the repeated use of four herbicides over 12 years on the weed flora of an arable field. Seedling densities of each species prior to spraying are shown as a percentage of their density in a no-herbicide control*

	MCPA	2,4-D	DNOC	Calcium cyanamide
Thlaspi arvense	32	27	107	148
Sinapis arvensis	16	9	11	19
Veronica hederifolia	72	126	54	66
Lamium purpureum	55	41	48	33
Polygonum convolvulus	83	32	25	35
Vicia hirsuta	19	7	66	96
Stellaria media	127	134	107	92
Matricaria chamomilla	83	110	56	83
Alopecurus myosuroides	88	104	121	83
Galium aparine	1067	500	133	100
Rumex crispus	18	12	45	83
Cirsium arvense	0	0	61	94

Source: After Rademacher *et al.* 1970.

declined under those herbicides to which they were susceptible. With the use of MCPA and 2,4-D, one species, *Cirsium arvense*, had become too infrequent to be recorded. *Galium aparine* was the only species to show a marked increase, primarily in the two phenoxy acid herbicide treatments.

Even after 36 years of annual applications of 2,4-D amine in a Canadian experiment begun in 1947, Hume (1987) found that no species sensitive to the herbicide had been eradicated. Although most sensitive species had declined and tolerant species had increased, some sensitive species (e.g. *Chenopodium album* and *Thlaspi arvense*) had not decreased in abundance relative to more tolerant species. The failure to eradicate species is probably a result of persistent seed banks, dispersal from neighbouring areas and the fact that few herbicides achieve 100% weed control in every year. Whether the reduced populations have declined to a lower stable equilibrium (see p.202) or whether they will continue to decline to extinction as the experiment continues remains to be seen.

Seed cleaning

Combine harvesters take in weed material containing seeds, along with the crop seeds being harvested. Through a combination of sieving and air blowing, crop seeds are separated from larger, smaller, heavier and lighter material. Even though most weed seeds are removed, some get through to contaminate the grain. In many countries there are allowable limits to the level of impurities in grain offered for sale, depending on its intended use. Farmers producing wheat as 'certified' seed for sale to other growers, for example, have to observe strict in-crop hygiene restrictions to minimise any wild oat and other weed seed contamination. Wheat produced for feed, on the other hand, may be accepted with considerable contaminants (up to a limit dependent on the buyer). Farmers saving their own seed for planting of future crops, however, would aim to have as clean a crop as possible. Too many weed seeds may mean that the grain has to be cleaned.

Monospecific crop stands have not always been as achievable as they (almost) are today, and harvesting machinery has not always been as good in removing unwanted weed seeds. In the past, weeds such as *Agrostemma githago* and *Lolium temulentum* contaminated harvested cereal grain due to their similarity in size to the crop seeds (Salisbury, 1961). Not only did they have toxic effects on those consuming cereal products, they were also unavoidably resown with the crop each year. The introduction of better seed-cleaning methods resulted in a dramatic decrease in these species to extinction in many regions. Indeed, the absence of seed dormancy in *A. githago* means that it apparently cannot survive in fields in northern Europe without being re-sown each year. Firbank & Watkinson (1986) used experimental data on interference between winter wheat and *A. githago* to simulate the population dynamics of the weed (Fig. 6.5). They found that in order for the weed to start to decline in density, the efficiency of removal of *A. githago* seeds from winter wheat in the cleaning process would have to be at least 91%.

Stubble management

During the growth of the crop and as a direct result of harvesting, weed seeds will be left lying on the soil surface. So too will crop material, ejected from the harvester, which will contain weed seeds (Wilson, 1981; Fogelfors, 1982). Stubble, the plant remains not cut by the harvester, will be left standing. In order to prepare for sowing a future crop, the farmer is faced with a number of decisions and options. The straw can be baled and removed, or it may be burnt. The straw and/or stubble may be chopped up and then incorporated into the soil by cultivation. Each of these operations has implications for the fate of weed seeds and may be expected to affect population dynamics. Burning will kill many weed seeds and may stimulate

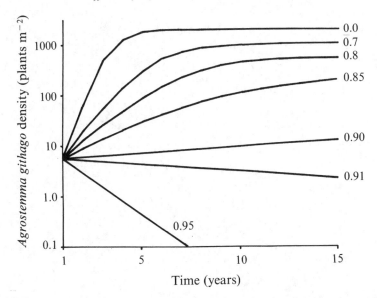

Fig. 6.5. Simulated dynamics of *Agrostemma githago* in a continuous winter wheat production system (after Firbank & Watkinson, 1986). Trajectories are shown for different proportions of seed removal by a combination of harvesting and cleaning.

some to germinate (Moss, 1987); straw and ash from burnt straw on the soil surface may intercept soil-acting herbicides and result in poorer weed control (Moss, 1979).

Several British studies have examined the effects of straw burning on annual grass populations; the longest time-series is given by Whybrew (1964). He found that burnt plots had the higher densities of *Avena fatua* seedlings in the autumn following harvest, but by the end of the next spring barley crop they had on average only about half the density of plots from which the straw was physically removed. Over a period of 6 years, density on plots where straw had been burnt remained lower (Fig. 6.6).

Burning has long been recognised as a tool in the management of rangelands. Woody weeds (shrubs) and unpalatable grasses can be reduced in biomass (if not in numerical abundance), allowing more beneficial species to increase. However, this topic will not be discussed further.

Fallow management

A fallow is any period in a cropping programme when land is not being used for production. The length of this period will depend on the crops being grown, crop sequence and the environment. In some dryland areas a long

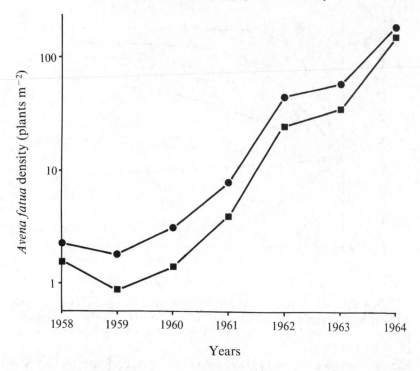

Fig. 6.6. Effect of straw burning on the population density of *Avena fatua* in repeated years of spring barley in England (based on data in Whybrew, 1964). Data are means of three tillage methods: ● straw removed; ■ straw burnt.

fallow (12–18 months) is used to build up soil moisture for a following crop. A shorter fallow will, out of necessity, come between the harvesting of one crop and the sowing of another. The duration of this will depend on the successive crops; six months may elapse between consecutive summer crops, whereas there may be only one or two months between winter crops. A farmer must either manage these fallow periods or allow weed seeds to be produced and added to the soil. Stubbles may be left standing throughout; cut straw may be left as a mulch; weeds may be controlled by cultivation, of varying frequency and by a range of implements, or by herbicides.

Various studies have reported greater rates of loss of seeds under a cultivated fallow than under a chemical fallow (no cultivation, weeds controlled by herbicides) or ley (Warnes & Andersen, 1984; Lueschen & Andersen, 1980). Good weed control in a fallow can be expected to have its greatest long term effect on the density of species with short-lived seeds. For example, Brenchley & Warrington (1936) found that a 2.5 year fallow reduced *Alopecurus myosuroides* seed populations to only 4–5% of the original density. The half-life of *Sinapis arvensis* seeds in a chemical fallow was approximately 7 years, compared with 1–2 years in a ploughed fallow or under cropping (Warnes & Andersen, 1984). The half-life of *Abutilon theophrasti* seeds (see Table 4.3) was about 2 years in a chemical fallow and less than 1 year in a 'two-plough continuous fallow' (Lueschen & Andersen, 1980).

Young *et al.* (1969) estimated the 'reproductive capacity' (plant density plus seed bank density) of *Bromus tectorum* under several fallow management regimes. Two years of an atrazine/paraquat fallow reduced reproductive capacity in the third year to an average of 11% of unsprayed plots. A one year chemical fallow followed by one year without chemicals reduced the reproductive capacity to 68% of unsprayed plots.

In some countries, such as those in the European Union, farmers are now being paid to take fields temporarily out of production. Rules have been laid down for the management of these 'set-aside' fields, such as when and how often they can be mown and whether herbicides can be applied. As with other management regimes, it is likely that set-aside will affect weed populations, at least in the short term. It is likely that the effects on seed decline will be similar to 'no-till' fallows or to pasture leys, but the effects on seed production will depend on the timing of management operations relative to the phenology of the species.

The effects of weather

Of the various meteorological factors which together constitute 'the weather', temperature and rainfall are likely to have the greatest effect on the dynamics of weed populations. Although the typical seasonal pattern of weather conditions at a location is a habitat characteristic and will therefore partly determine the rate of population increase and the potential equilibrium density, we will here consider only those weather events which can be regarded as departures from the expected (average) pattern. Unfortunately, weather has been regarded as background noise by most agronomists and ecologists – unpredictable and beyond our control. It has therefore received little attention in its own right. However, for populations of weeds on the edges of their geographic range or in highly variable climates, weather may be critical in driving population dynamics.

There are numerous anecdotes of population explosions during extreme weather patterns. For example, Salisbury (1961) recorded that *Antirrhinum orontium*, a summer annual, decreased in abundance in the UK following wet, cold summers due to poor seed production. It is likely that summer rains, droughts and temperatures will have implications for growth and seed production, particularly of summer annuals, affecting the sizes of future seed banks and hence future densities. However, there are very few hard data to support conjectured weather-driven dynamics. Some agronomists in eastern Australia believe that certain pasture weeds, such as *Vulpia* spp., have increased since, and in some way as a result of, the 1982/3 drought. However, this may also be related to changes in management in response to technical innovations and reduced profit margins during the 1980s, as much as to any consequence of the drought.

In temperate regions, winter frosts will be important regulators of population density: their frequency and intensity will affect the number and sizes of plants. For example, *Avena fatua* is susceptible to frosts. Although it germinates mainly in spring, autumn germination can be considerable. A hard winter will kill or retard these autumn emergers (Davies, 1985); a mild winter will allow them to survive, become more vigorous than those emerging in the spring and result in greater overall population seed production. *Sinapis arvensis* may similarly germinate in autumn or spring; although normally killed by frosts (Mulligan & Bailey, 1975), a mild winter could cause an increase in population density.

Soil conditions, as modified by temperature and rainfall, will affect seed survival, breakage of dormancy, and hence the proportion of a seed bank germinating. They will also determine the rates of emergence of both weed and crop seeds (Weaver *et al.*, 1988), setting up a hierarchy of plant sizes which will influence the intensity of later interference. Inevitably, this will affect seed production, additions to seed banks, and hence future population sizes. Several researchers have related weed growth to temperature and daylength using heat unit systems (e.g. Nussbaum *et al.*, 1985); these methods may be useful in the future for studying the implications of particular weather patterns on population dynamics, as may growth simulation modelling (Kropff, 1988).

There is currently a proliferation of research on the implications of future global changes to both natural vegetation and agriculture. Although not strictly concerned with weather changes, there has been at least one study of the effects of elevated CO_2 levels on weed population dynamics. Bazzaz *et al.* (1992) studied seed production by *Abutilon theophrasti* grown in shallow trays in controlled environment cabinets at two levels of CO_2. They used a simple population model, assuming 100% germination every year, to explore the long term consequences for the weed. They predicted that complex cycles of abundance or chaos would be more likely under high CO_2

levels; however, when they incorporated more realistic assumptions about germination and seed persistence into their model, they found that complex dynamics was unlikely in both high and low CO_2.

Interactions with other organisms

The other organisms with which a weed comes into contact consist primarily of other plants (weeds, crops, pasture or species of conservation value), plant pathogens, soil micro-organisms and animals (notably herbivores). The animals can in turn be divided into vertebrates, both domestic and wild, feeding on the vegetation (plants and seeds) within which the weeds are growing, endemic invertebrates and deliberately introduced invertebrates for biological control. The plant pathogens may also be either endemic or deliberately introduced. In this section we will give examples of just some of the many possible interactions between populations of weeds and other organisms.

Other plants

For convenience, we usually study the population dynamics of only a single weed species at a time. However, there may be as many as 20 species of weeds, sometimes even more, within a single arable field. If the various weed species are at low densities such that they seldom interact, then their dynamics may be virtually independent of one another. If they are intermixed at densities within which interactions are very common, however, increased vigour of one species or poor control of another may affect seed production by the remainder, and hence lead to long term changes in relative abundance.

Most experimental studies of the interactions between weeds and other plants have established simple two-species mixtures. Marshall (1990) planted rhizome segments of *Elymus repens* either in bare ground or surrounded by a hexagonal arrangement of stands of six other grass species. He recorded the positions of plantlets (ramets) of *E. repens* as the weed spread outwards. Not surprisingly, he found that the rate of increase was greater when the species was growing alone. After 2.5 years, rhizome bud density was 14.5 times higher on average in bare plots than in grass plots. He was able to rank the six species in their ability to withstand invasion by *E. repens. Arrhenatherum elatius* and *Agrostis stolonifera* were the most able and *Holcus lanatus* and *Poa trivialis* the least able to prevent invasion.

Interactions between *Galium aparine*, *Bromus sterilis* and *Papaver rhoeas* were examined by Lintell-Smith *et al.* (1991) in plots of winter wheat containing pairs of the weed species. The rate of increase of *B. sterilis* (measured over a single generation) was affected little by the other weeds. *G aparine* increased faster ($\lambda \approx 25$) on its own than in mixture with *B. sterilis* ($\lambda \approx 10$). The rate of increase of *P. rhoeas* was decreased considerably by both *B. sterilis* and *G. aparine*.

Such studies are able to show that interactions between species occur. However, it would be an unrealistic task to try to quantify experimentally all of the first-, second- and higher-order interactions between the many species within a typical field. An alternative approach, multiple regression of complex (multi-species) natural communities of weeds, inevitably only detects the effects of a few species as statistically significant and most interactions remain unquantified. We are therefore left with difficult methodological problems in how to proceed with the study of multi-species interactions. While accepting the reality of complex communities, most researchers have concentrated on their single target species and have regarded the other species as simply part of the background habitat.

Biological control agents

Biological control aims to reduce the abundance of a weed to a level at which it can be either tolerated or managed by other measures (Fig. 6.7). It is unrealistic to expect that biocontrol will eradicate a weed, although the final equilibrium populations of pest and weed may be so low that the weed seems to most untrained observers to have disappeared. Unfortunately, the successful biological control of a small number of major weeds has led to undue public expectations from the technique. For example, releases of the moth *Cactoblastis cactorum* drastically reduced the populations of *Opuntia* spp. in Australia which had been making vast areas un-grazable (see Chapter 2). This early success has led to the belief that similar successes are possible with *all* other weeds. In Australia, over $10 million is being invested annually on biological control programmes. Introductions of new control organisms for an array of weeds are continually taking place, but with few further major successes to date. It remains to be seen how many (or how few) releases result in significant reductions in weed populations.

Although *Opuntia* control is the classic biocontrol success story, there have been others worthy of note. Large areas of water in various countries have been cleared of *Eichhornia crassipes* through the introduction of the weevil *Neochetina eichhorniae* (Cullen & Delfosse, 1990). *Hypericum perforatum* has been reduced in abundance in California by *Chrysolina* spp.,

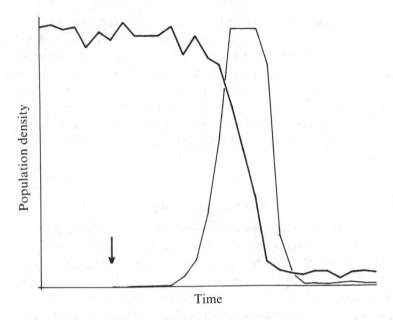

Fig. 6.7. Likely effect of the introduction of a biological control agent on the dynamics of a weed. The expected trajectories of both weed and biocontrol agent are shown. The introduction of the biocontrol agent is shown by an arrow.

but only in open parts of fields, since the insect avoids the shade under trees (Harper, 1969).

In Australia the rust *Puccinia chondrillina* was introduced in 1971 to control *Chondrilla juncea*, a weed from the Mediterranean. A narrow-leafed form of *C. juncea* was controlled to great effect (Burdon *et al.*, 1981); its density decreased, although its geographic distribution did not. However, intermediate and broad-leaved forms of the weed, unaffected by that strain of rust, have since increased. Their abundance within their distribution increased by over 100% between 1977 and 1980. Other races of *P. chondrillina* are being sought which will affect these other forms of *C. juncea*. A gall midge, capable of attacking all three forms has also released. For some target weeds, successful control may therefore involve the release of several control agents, perhaps each with different climatic preferences and genotypic specificity.

An outcome of the biocontrol story is the realisation of the part played by predators and diseases in native plant communities. Although we may study a species in its native habitat and assume that it is not regulated by predation or disease, when it is released into a new location without its predators and pathogens may become extraordinarily abundant. We may interpret this as the invasive weed being more fit in some way than endemic species, or preferring the new edaphic and climatic environment which it finds there. Only when we introduce organisms from the plant's native habitat, which then reduce its abundance, do we see that predation or disease may have been a potent force after all. We must accept from this evidence that interactions between organisms can be important regulators of abundance in natural systems even though it is difficult to observe them taking place.

Grazing animals

Farm animals can have considerable effects on weed populations, particularly if they graze selectively. Although cattle and horses may graze paddocks heavily, they will leave most of the poisonous and unpalatable species, causing them to increase in abundance. The actions of animal's feet, especially 'poaching' in wet conditions, will create gaps in perennial pastures which can be colonised by weeds. Prevention of weed population increase in pastures is facilitated by a dense, close sward with few gaps (Panetta *et al.*, 1993).

Goats are often recommended for the control of woody (shrub) weeds and thistles in pastures and rangelands (e.g. Wood, 1987); however, although this may provide a 'quick-fix' the goats may cause damage by also eating the more palatable species and by trampling. There is only a limited market for goat products and it is unrealistic to expect farmers to use the large numbers of goats that would be required for weed control in some extensively weed-infested regions. As part of an integrated management programme, however, selective grazing of pastures may play a key part in reduction of weed problems.

The effect of grazing animals on a weed population is likely to be a function of both the timing of the grazing relative to the life-cycle of the weed and the intensity of the grazing pressure. Table 6.5 shows an example of a sheep grazing trial in New Zealand. An increase in the duration of grazing from 2 to 3 days per fortnight in spring combined with an increase from 2 to 3 days per month in summer was sufficient to send a population of *Cirsium arvense* from an increase into a decline. In a continuously stocked sheep system, the density of *Lolium rigidum* was shown to decline with higher stocking rates, whereas the total number of weed plants increased (Sharkey *et al.*, 1964). Unfortunately, a great many grazing studies record ground cover rather than plant numbers; quantitative population data are rare.

Table 6.5. *Effect of various rotational sheep grazing management systems on the density of* Cirsium arvense *in New Zealand. Stocking densities for the grazing period are given as sheep* ha^{-1}; *the number of days grazed per cycle and the cycle length in days are also shown (e.g. 2/14)*

Treatment	Spring management	Summer management	Thistle shoot density as percentage of density 11 months previously
1	231 ha^{-1} (2/14)	182 ha^{-1} (2/28)	169
2	231 ha^{-1} (3/14)	182 ha^{-1} (3/28)	54
3	231 ha^{-1} (4/14)	182 ha^{-1} (4/28)	5
4	462 ha^{-1} (4/28)	364 ha^{-1} (4/56)	29

Source: From Hartley *et al.* 1984.

Using models to explore weed control options

We clearly have a good empirical understanding of how some extrinsic factors (particularly those concerned with crop management) affect weed populations. How can we use this knowledge to develop future, and hopefully better, weed control strategies? Can we predict what would happen if we change a management strategy in a particular way? What strategy should we adopt to obtain the most effective and most economic control of a particular weed?

The most powerful technique at our disposal for answering such questions is mathematical modelling in conjuction with experimental verification. Various models for simulating the intrinsic dynamics of populations were described in Chapter 5. If we know how the parameters of a model change under different management practices (i.e. if we collect the relevant experimental data), we can run simulations to examine the potential long term behaviour of populations under a range of management scenarios. We may even calculate the economic impact of the weed population in each year of a simulation, and hence predict the net financial benefits of different management options. In this section we will describe some examples of the use of models to answer questions about the management of weed populations.

Control at different stages of the weed life-cycle

In the previous chapter we described multi-stage population models, where the life history of a weed was described by a series of survival probabilities and fecundity parameters. These models can be used to examine the effects of changing specific causes of mortality. In this section we will examine the long term consequences of varying tillage methods, stubble burning and seed collection by combine harvesters. We will also show how models can determine the efficacy required of a herbicide and the cost-effectiveness of developing new control technologies.

Cussans & Moss (1982) produced an exponential multi-stage model of the dynamics of the annual grass *Alopecurus myosuroides* in winter cereals. From an extensive experimental programme they were able to estimate the parameters of their model for both mouldboard ploughing, tine cultivations and direct drilling (see p.154). Cousens & Moss (1990) extended the model to include density-dependent plant mortality and seed production. They also divided the soil into four depth horizons, with tillage determining movement of seeds between each depth. Seeds were assumed only to emerge from the top-most soil layer.

Predictions from this model, assuming no use of herbicide, are shown in Fig. 6.8. Both the rate of increase and the equilibrium density were greatest for direct drilling and least for mouldboard ploughing. This is in accordance with farmers' experience in eastern England, where *A. myosuroides* became very much more of a problem when direct drilling was adopted. *A. myosuroides* seeds have a short life span in the soil and decline quickly after burial by ploughing; by leaving seeds at the soil surface, direct drilling allows more of them to establish seedlings. In all simulations plant density (as well as total seed density) approached the equilibrium density asymptotically. The frequencies of seeds in the different soil depth classes followed damped oscillations for ploughing and a monotonic approach to an equilibrium for rigid tine cultivations. The predicted rate of increase at low weed density using the original Cussans & Moss model (Table 6.6) was over four times higher under direct drilling than under ploughing.

A mortality factor can be introduced into models to represent weed control by a herbicide. Cussans & Moss (1982) used iterative simulations to find the level of herbicide performance required to just prevent populations from increasing ($\lambda = 1$). Very much greater herbicidal control was required under direct drilling (Table 6.6). In the absence of stubble burning, for example, 92% control was predicted to prevent population increase under direct drilling, whereas only 65% control was required with ploughing. For a herbicide giving a weed control of about 90%, populations would therefore increase slowly under direct drilling, but decrease rapidly under ploughing. Although stubble burning, killing an average of 33% of seeds, helped to reduce the rate of population increase, the requirements for

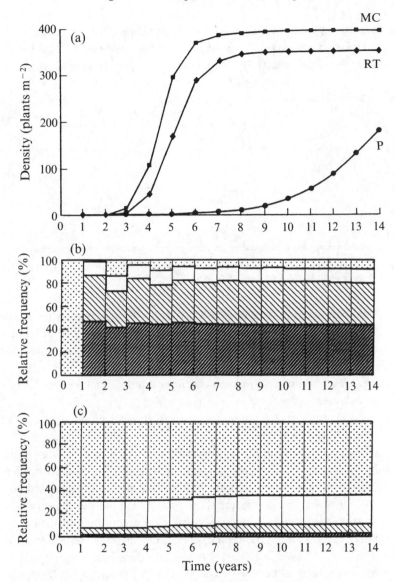

Fig. 6.8. Predicted changes in *Alopecurus myosuroides* populations under different tillage regimes (after Cousens & Moss, 1990): (a) mature plant density under minimum tillage (■), rigid tine tillage (♦) and mouldboard ploughing (●); (b) seed depth distribution under ploughing; (c) seed depth distribution under rigid tine tillage. Depth horizons are shown as: (⬚) 0–5 cm; (□) 5–10 cm; (◩) 10–15 cm; (▨) 15–20 cm.

Table 6.6. *Predictions of the annual rate of population increase at low weed density in the absence of herbicides (λ) and the minimum percentage control by herbicide required for no increase (k'), for two annual grasses in winter cereals grown under two tillage methods*

	Straw burnt		Straw removed	
	λ	k'	λ	k'
Alopecurus myosuroides				
Plough	1.5	50	1.9	65
Direct drill	6.3	88	9.3	92
Avena fatua				
Plough	1.6	72	1.9	81
Shallow tine	2.1	73	2.9	82

Source: From Cousens *et al.* (1987), who present this table containing the results of Wilson *et al.* (1984) and Cussans & Moss (1982).

herbicide performance were only slightly reduced under direct drilling. In contrast to these results, Wilson *et al.* (1984) found very little difference between ploughing and shallow tine cultivations using a similar model for *Avena fatua* (Table 6.6), although again stubble burning was able to relieve the pressure on herbicides somewhat. *A. fatua* is able to emerge successfully from much greater depths than *A. myosuroides*.

Given the problem of *A. myosuroides* under direct drilling but not under ploughing, Cussans & Moss (1982) used their model to investigate the frequency of ploughing in an otherwise continuous direct drilling regime required just to keep the population from increasing in the long term. They also built into their model a decline in herbicide efficacy with consecutive years of direct drilling; research had shown that herbicides could be adsorbed on to ash and straw residues building up on the soil surface. By iteration, they found that ploughing was required at least one year in every five (Fig. 6.9). Coincidentally, this is roughly the frequency adopted by farmers on the basis of intuition/experience. In a similar way, Wilson *et al.* (1984) investigated the introduction of a spring cereal break crop into an otherwise continuous winter cropping pattern. A spring barley break crop every four years reduced the required herbicide performance to only 50%, well within the capabilities of even mediocre herbicides.

The above examples primarily considered the effects of changing seed burial characteristics, survival parameters in the soil and herbicide perfor-mance. Now that more weeds are becoming resistant to herbicides (see

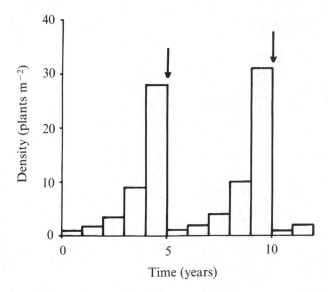

Fig. 6.9. Predicted dynamics of *Alopecurus myosuroides* in winter wheat when ploughing (indicated by arrows) is used once every five years in an otherwise continuous direct drilling system (after Cussans & Moss, 1982). Herbicide was assumed to control 95% of plants initially after each ploughing, but decreasing by 10% for each year of direct drilling.

Chapter 8), weed scientists are asking whether increased mortality can be achieved in other parts of the life-cycle. In particular, it has been pointed out that for some species a large proportion of seeds may be taken up into the combine harvester. Rather than return them to the soil surface, why not find ways of retaining and destroying them? What level of seed retention is required in order to cause population decline in the absence of herbicides? In the case of *Agrostemma githago*, modelling has suggested that at least 91% had to be removed by seed cleaning to result in its decline (see p.186).

To stabilise a weed population without a herbicide will require an equivalent level of control from alternative means (assuming exponential growth). For example, if 90% control from a herbicide is required for $\lambda = 1$, a level of 90% removal and retention of seeds by a combine harvester would be needed to maintain $\lambda = 1$ without a herbicide. Alternatively, if only 50% of seeds can be removed by the combine, 80% control will still need to be achieved through some additional non-chemical means (calculated by multiplying survival probabilities: $(1 - 0.5)(1 - x) = (1 - 0.9)$). Fig. 6.10 shows predictions made using the Cussans & Moss (1982) model for *A. myosuroides*. Although this species sheds almost all of its seeds prior to the

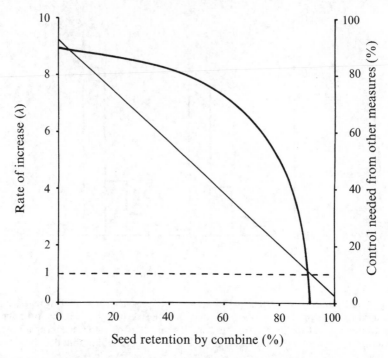

Fig. 6.10. Effect of percentage of seed removed and retained by a combine harvester on the dynamics and control of a hypothetical species (based on a late maturing population of *Alopecurus myosuroides* – see text for details); rate of population increase (thin line), showing the level of seed removal required to achieve $\lambda = 1$ (dashed line); control required from other sources of mortality to achieve $\lambda = 1$ (thick line).

harvest of a cereal crop, the model can be used to examine the implications of seed removal for a similar species whose seeds do not completely dehisce before combining commences. The rate of population increase declines linearly as the percentage uptake and retention of seeds increases.

Medd & Ridings (1989) produced a density-dependent three-cohort model of the life-cycle of *Avena fatua* and made similar calculations to those above. They used their model to simulate the effects of different levels of plant control with herbicides and seed control by new technologies, assuming that the latter could be achieved. They argued that the marginal costs of further improvements in plant kill from herbicides would be low, given the already high levels of control obtained. It would be more cost-effective to fund future research into ways of controlling either seeds in the soil or seed production. Relatively small increases in seed kill, in conjunction with present herbicide use, could result in significant improvements in rate of decline of weed populations. Attention was drawn to the possibilities of 'crop-topping', using herbicides late on in the crop to prevent weed seed formation/maturity, and of using seed pathogens as biological herbicides.

Pandey & Medd (1990) went on to combine the technique of 'dynamic programming' with the population model, to examine the economics of the various control strategies. They found that, in order for seed control methods to be cost-effective, costs of use (in Australian dollars) could not exceed \$16 ha^{-1} for 90% seed kill, \$12 ha^{-1} for 60% kill and \$5 ha^{-1} for only 30% kill. The optimum strategy was to use seed control measures in all years.

Biological control agents are another means of controlling weed populations. Smith *et al.* (1993) simulated the dynamics of *Striga hermonthica*, a parasitic weed of millet in Africa. They used their model to determine the level of seed predation required from the potential biocontrol agent *Smicronyx umbrinus* so as to reduce the equilibrium weed density by 50%. The result, 95% seed predation, was regarded by the researchers as unobtainable from this insect alone. If the biocontrol agent was used in conjunction with a crop rotation in which millet was grown only one year in four, a seed reduction of only 22% would be required from the insect for the same effect on the predicted weed equilibrium density.

In the above examples, multi-stage models were used to explore the implications of *specified* control options at particular points in the life-cycle. However, an expectation held by many researchers has been that the use of such models, through sensitivity analyses, can tell us the weak points in a species' life-cycle, and hence show us where to direct our efforts. We should be able to find the 'Achilles' Heel' of the weed. If a small change in a parameter results in a large change in rate of increase of the weed, small increases in mortality at that point in the life-cycle should result in much better weed control. However, there have been very few, if any, examples where such insight has been achieved. Our models are mostly simple exponential models which have very predictable behaviour, with (in general) no one parameter being more influential than another. Arguably, by the time we have enough data to run a model, we have already discovered by other means enough about the species to tell how best to control it.

Effects of herbicides on population dynamics

The effects of herbicide use on weed population dynamics have been explored extensively with the aid of models. As we saw above, a factor can be introduced into models to represent herbicide mortality and/or sub-lethal effects of a herbicide on seed production. By varying this factor, questions can be asked about the long-term implications of herbicide use. In this section we explore the effects of herbicides on population dynamics in more detail.

Let the proportion of plants killed by a herbicide be k; a proportion $(1-k)$ will survive herbicide application. Then, in the simple exponential model (equation 5.1)

$$N_{t+1} = R(1-k)N_t \qquad (6.1)$$

The level of kill (k') required so as to just prevent further population increase (i.e. $N_{t+1} = N_t$) is thus

$$k' = 1 - 1/R \qquad (6.2)$$

Hence, given a low density of a weed (such that the assumptions of the exponential model are reasonable) and knowledge of its potential rate of increase, we can calculate the minimum performance required of a herbicide. For example, if the potential rate of increase of the weed is three-fold per year ($R=3$), a plant kill of 67% is required; if density increases by a factor of $R=10$, then 90% of plants need to be controlled. This, of course, assumes that reproduction by plants surviving or escaping the herbicide will be unaffected by the chemical.

In a similar way, we can introduce the effect of a herbicide into the density-dependent model of equation 5.6. Let the number of individuals at the end of a generation be reduced by a proportion k (hence k implicitly includes both plant mortality and sub-lethal effects on seed production). Equation 5.6 then becomes

$$N_{t+1} = (1-k)RN_t/(1+aN_t)^b \qquad (6.3)$$

which has an equilibrium at

$$N_e = \{[(1-k)R]^{1/b} - 1\}/a \qquad (6.4)$$

Clearly, in this case the herbicide control required to stabilise a population $(N_{t+1} = N_t)$ will vary with the density of the population at the time of application, i.e.

$$k' = 1 - (1 + aN_t)^b/R \qquad (6.5)$$

For $b=1$, this model will produce a series of generation maps as shown in Fig. 6.11, with the equilibrium density decreasing as k increases.

Any factor which increases mortality will act to reduce N_{t+1}, and hence will lower the trajectory on the generation map. Increasing the kill achieved by herbicides will thus reduce the equilibrium density (Fig. 6.11). If

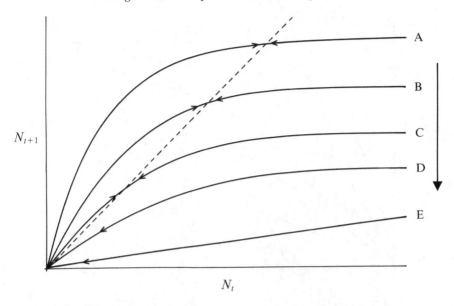

Fig. 6.11. Trajectories predicted by equation 6.3 for different levels of control by a herbicide (after Mortimer, 1985). Weed control is assumed to reduce the population by a fixed proportion of the seeds produced. Increasing level of weed control is indicated by a vertical arrow. Dashed line shows $N_{t+1} = N_t$.

eradication of a weed is to occur, mortality (by whatever means) must be increased to a level such that there is no cross-over of the trajectory and the line $N_{t+1} = N_t$, and it must be maintained at that level (line D, Fig. 6.11). If there is complete kill of plants and therefore no seed production, the population will decline at a rate set solely by the rate of mortality of seeds in the soil. There will be no intersection with the line $N_{t+1} = N_t$ and the trajectory will be a straight line on the plot of N_{t+1} against N_t (see Fig. 5.3d).

Control of grass weeds is often expressed in terms of reduction of inflorescence number or seed production at maturity (Doyle *et al.*, 1986; Cousens *et al.*, 1986), and the above way of expressing herbicide control is thus appropriate to some types of data. However, if we record the proportional mortality of seedlings before interference takes place, and assuming that surviving plants are potentially as fecund as unsprayed plants, herbicide effects could be incorporated thus:

$$N_{t+1} = RN_t(1-k)/[1 + aN_t(1-k)]^b \qquad (6.6)$$

with an equilibrium at

$$N_e = \{[R(1-k)]^{1/b} - 1\}/[a(1-k)] \qquad (6.7)$$

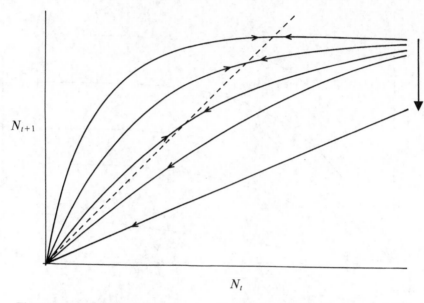

Fig. 6.12. Trajectories predicted by equation 6.6 for different levels of control by a herbicide, assuming that the herbicide reduces seedling density by a fixed proportion. Increasing level of weed control is indicated by a vertical arrow. Dashed line shows $N_{t+1} = N_t$.

Although this also predicts a decreasing equilibrium with increasing k, the shapes of the generation maps for a range of ks differ from those for the previous model (for $b = 1$, see Fig. 6.12).

An alternative approach is to assume that herbicides act to reduce R by a fixed proportion (Mortimer, 1987), such that

$$N_{t+1} = RN_t/(1 + aN_t)^b - kRN_t \qquad (6.8)$$

This has an equilibrium at

$$N_e = \{[R/(1 + kR)]^{1/b} - 1\}/a \qquad (6.9)$$

The predicted generation maps for this model are similar in form to those of equation 6.3. Like the original model without herbicides, this model can predict either asymptotic, convergent oscillatory or more complex approaches to equilibrium (Mortimer, 1987; Mortimer *et al.*, 1989), depending on the parameter values (Fig. 6.13). However, it is restricted in that if populations are initiated at high densities, negative values of the rate of increase (λ) can be predicted (which are impossible). The value of kR must therefore be constrained to within fixed limits.

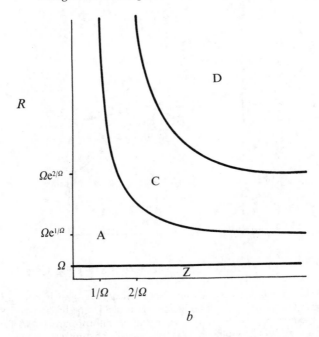

Fig. 6.13. Types of trajectory predicted from different values of the parameters R and b in equation 6.8 (redrawn from Mortimer, 1987). The model assumes that herbicides reduce the intrinsic rate of increase (R) by a fixed proportion. Trajectories differ between the four zones of the graph: D – divergent oscillations, chaotic behaviour or complex cycles; C – convergent oscillation to equilibrium; A – asymptotic approach to equilibrium; Z – asymptotic decline to zero. Limiting values are shown on the axes, where $\Omega = (1 + kR)$.

Using single-stage models, there are therefore a range of ways in which herbicide effects can be incorporated. It is possible to explore the conditions for equilibrium and the types of dynamics predicted by each. The particular method selected should reflect the way in which herbicide efficacy is being assessed in experiments.

Although the predictions thus far have been that the application of a herbicide will cause a simple decrease in the equilibrium density or a decline to extinction, more complex behaviour is possible if the mortality from a chemical is density-dependent. Ulf-Hansen (1989) obtained estimates of rates of population increase from a number of experimental densities of a biotype of *Alopecurus myosuroides* susceptible to chlorotoluron. Fig. 6.14 shows that for the recommended application rate, all densities would decline ($\lambda < 1$). For unsprayed plots an equilibrium ($\lambda = 1$) would be approached as density becomes very high ($> 10^4$ seeds m^{-2}). However,

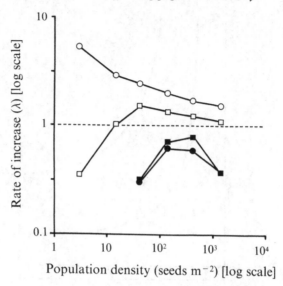

Fig. 6.14. The rate of increase of *Alopecurus myosuroides* growing in winter wheat when sprayed with phenyl-urea herbicides (redrawn from Ulf-Hansen, 1989): (○) unsprayed control; (□) half application rate of chlorotoluron; (■) full rate of chlorotoluron; (●) full rate of isoproturon.

using half the application rate there would be an additional equilibrium at a lower density (close to 15 seeds m^{-2}). This equilibrium would be unstable: at very slightly lower densities the population would decline, while at slightly higher densities the population would increase. The upper equilibrium would be stable, with slight deviations from that density resulting in trajectories back towards it. None of the current population models allow for such a possibility.

In Chapter 5, we introduced the subject of analytical methods to aid understanding of weed population dynamics and to predict changes in trajectories. The basis of the approach is to select a relevant equation for population dynamics and to analyse mathematically its properties. For specific difference equation models it has proved possible to answer certain questions about the long term dynamics of two species mixtures of plants. The methodology was originally proposed by Hassell & Comins (1976) and uses equation 6.8 as the starting point. The growth rate of a single species can be expressed as

$$\lambda = R/[1 + aN_t]^b - \Lambda \qquad (6.10)$$

where Λ represents the *rate of extrinsic control* kR (Mortimer *et al.*,1989). If the rate of control Λ exceeds $(R-1)$ in successive generations, then

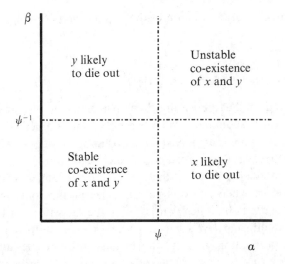

Fig. 6.15. Phase plane analysis of a model of two interacting species (see text for details), showing the combinations of parameters α and β leading to different dynamical behaviour. ψ is defined by equation 6.12.

population extinction will result. For the population to show long term persistence and to achieve an equilibrium size under extrinsic control, Λ will need to lie in the range $0 \leq \Lambda \leq R - 1$.

Using the two-species equations described in Chapter 5 (p.151), we can model the growth rates of the seed banks of two competing weeds x and y as

$$\lambda(x) = R_1/[1 + a_1(X_t + \alpha Y_t)]^{b_1} - \Lambda_1 + d_1$$
$$\lambda(y) = R_2/[1 + a_2(\beta X_t + Y_t)]^{b_2} - \Lambda_2 + d_2 \qquad (6.11)$$

where $\lambda(x)$ is the population growth rate of the seed bank of species X in the presence of species Y and $\lambda(y)$ the corresponding growth rate of Y. The parameters R_i, a_i, b_i, α and β have been defined earlier. The parameters d_1 and d_2 are the contributions to population growth rate from the seed bank, the proportion of the seed population that remains dormant and viable over a generation.

Predicting whether either of the species in a mixture will decline in the face of the combined effects of interference between species and external control may be achieved by 'phase plane analysis'. Pairs of parameters are examined and the limiting values for particular population outcomes are determined. Fig. 6.15 shows four regions of parameter space (in this case α and β) for the two species. The precise domains of each outcome are determined by ψ, the calculation of which includes measures of the rate of extrinsic control and the fraction of the population persisting in the seed bank:

$$\psi = \left\{ a_2 \left[\left(\frac{R_1}{(1 + \Lambda_1 - d_1)} \right)^{1/b_1} - 1 \right] \right\} \Big/ \left\{ a_1 \left[\left(\frac{R_2}{(1 + \Lambda_2 - d_2)} \right)^{1/b_2} - 1 \right] \right\} (6.12)$$

The model predicts that there is one domain in which stable co-existence of both species will exist but that in all others either one or other will tend to be driven to extinction or that an unstable association may occur. However, it is important to realise the true meaning of such predictions. In the lower left domain, the model predicts that there is an intrinsic tendency for the two species to coexist when one or both are experiencing some measure of extrinsic regulation. In other words, both species will persist at lower theoretical equilibrium levels which arise from the combined effects of reduction in population growth rate (through interspecific interference) and through extrinsic control. In two other domains one species will tend to drive the other to extinction.

What is the relevance of such predictions? The model answers the two questions 'What are the long term effects of reducing population growth rate(s) on the population dynamics of two competing plant species?' and 'to what degree must the population growth rate of one species suffering interspecific interference be reduced by extrinsic control measures theoretically to drive it to extinction?'. As such these are of strategic rather than practical significance in weed science. The answers depend on several major assumptions. The first is that equation 6.11 is an appropriate description of population growth. The second is that the predictions are based on fixed parameter values. As we have already argued, parameter values will change, for example in relation to weather. However, if we have knowledge of the range in values that parameters might span we may still use the approach to determine whether or not all possible combinations of parameter values lead to the same predictions. The third assumption is, of course, that we are only concerned with two-species interactions.

The use of this approach (Mortimer *et al.*,1990, 1993), remains in its infancy due in part to a lack of data but also of understanding of the interactions between weed species experiencing weed control measures.

Weed control decision thresholds

A particularly successful application of plant population models has been their use by weed scientists to study the economics of various weed management strategies. Before describing the details of the modelling, we need first to consider the types of control strategy being considered.

One approach to weed and pest management is prophylactic spraying, where chemicals are used on a regular basis so as to prevent losses from occurring. The perceived threat by the pest is considered large enough, or the cost of the chemical is small enough, to make it worthwhile taking measures to avoid the risk of damage (a 'risk aversion' strategy). At no time is the actual threat (expected yield loss) assessed. An alternative approach is to calculate the likely damage and the benefits from control each time spraying is considered. If the difference between the costs and the benefits indicates that spraying is economically justified, then (and only then) a spray will be applied. Since potential losses from a weed population will be correlated with the density of the infestation, we can conceive of an 'economic threshold' density; above this density the benefits from spraying will exceed the combined costs of the operation and of any toxic effects on the crop.

The calculation of the threshold density requires a knowledge of the relationship between crop yield loss and weed density. Although sometimes depicted in older text books as a sigmoidal function, field data consistently follow a monotonic curve (Fig. 6.16). An equation used commonly for describing this damage function was proposed by Cousens (1985a)

$$Y_L = iN_w/(1 + sN_w) \qquad (6.13)$$

where Y_L is the proportion of yield lost, N_w is weed density and i and s are parameters estimated by regression from appropriate data. If weed densities remain low, the relationship

$$Y_L = iN_w \qquad (6.14)$$

may be appropriate; the parameter i is the proportion of yield lost per unit weed density. A range of other published yield–density regression curves have been summarised by Cousens (1985a,b), but equations 6.13 and 6.14 seem to fit most data sets adequately.

If the yield loss function is assumed to be linear (at least at densities up to the threshold), the economic threshold density (N_w*) can be calculated as

$$N_w* = (C_h + C_a)/iY_{wf}Pk \qquad (6.15)$$

where C_h and C_a are the costs of the chemical and its application respectively, Y_{wf} is the weed-free crop yield, P is the sale price obtained per unit of produce and k is the proportion of weeds killed by the herbicide. If the units of N_w, Y_{wf} and P are plants m^{-2}, t ha^{-1} and \$ t^{-1} respectively, C_h and C_a are in units of \$ ha^{-1} and i is in (plants m^{-2})$^{-1}$. More complicated expressions for the threshold can be derived if a curved damage function is assumed (Cousens, 1987). Using equation 6.15, thresholds have been

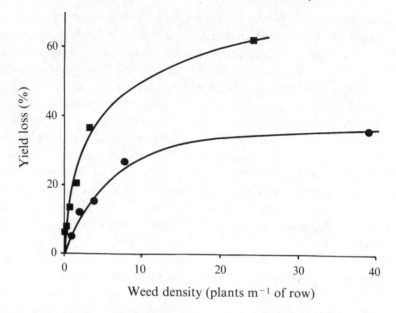

Fig. 6.16. Examples of the relationship between crop yield loss and weed density: (■) *Polygonum pensylvanicum* in soybean, (●) *Amaranthus hybridus* in maize (based on data in Coble & Ritter, 1978, and Moolani *et al.*, 1964, respectively). Curves are equation 6.13 fitted by non-linear regression.

calculated for many weeds. It is common to assume that the herbicide gives complete control ($k = 1$) and that only a single target species is controlled.

In the calculation of the economic threshold, costs and benefits are calculated only for the current crop. However, if weeds are not killed they will set seed, thereby affecting population levels (and hence cost/benefit analyses) in future years. The financial implications of a control decision are therefore not confined to the current year. The appropriate economic analysis should consider a sequence of spray decisions over a number of years. A population model can be used to simulate the long term dynamics which will result from particular strategies. In each annual cycle of the simulation the population can be 'controlled' or not, according to some pre-determined decision rule, and the next year's population level calculated. For each strategy we can therefore generate a time-series of simulated weed densities. The crop yield loss which would result in each year can be calculated from an appropriate equation.

In the same way that weed control affects future population levels, financial decisions made in the present year also have implications on future use of money. It is necessary to discount the net benefits of control according to the year in which they occur. For example, both costs and benefits can be assumed to depreciate by $r\%$ per year. If an anticipated benefit B_t is predicted to occur in t years time, its 'present value' will be only

$$V = B_t/(1 + r/100)^t \qquad (6.16)$$

The present values of costs and benefits can be summed over all years of the simulation. The strategy giving the highest net profits over that period can be referred to as the long term economic optimum strategy; the weed density at which control decisions should be made has been referred to as the 'economic optimum threshold' (Cousens, 1987).

Doyle *et al.* (1986) and Cousens *et al.* (1986) used this approach to examine the long term benefits of spraying herbicides according to different threshold values. They ran simulations using control thresholds from zero up to the 'economic threshold' in 0.1 plant m^{-2} increments. They found that the economically optimum threshold in winter wheat from the population model was considerably lower than that derived from calculations based solely on costs and benefits in a single year: 7.5 plants m^{-2} as opposed to 30 plants m^{-2} for *Alopecurus myosuroides* and 2–3 plants m^{-2} as opposed to 10 plants m^{-2} for *Avena fatua* (see Fig. 6.17). Hence, the more traditional thresholds for these species would be about four times too high for maximum profit in the long term. Although the use of thresholds reduced the number of occasions on which herbicides would be used, the financial gains over prophylactic herbicide use were predicted to be small.

Murdoch (1988) used a similar approach to examine the economics of controlling *A. fatua* in continuous spring barley production. He found that the long term economic optimum threshold was 2.1 plants m^{-2} for the recommended rate of an unspecified herbicide and only 1.2 plants m^{-2} for half that rate. The single year economic threshold densities were seven to eight times higher than these values. He also found that over a 15 year period the use of a herbicide and hand-roguing in an attempt to eradicate the weed would be more cost-effective than the use of even the best weed control threshold strategy. For *Cyperus esculentus*, Lapham (1987) found that the single year economic threshold was two to three times higher than the long term optimum threshold.

The calculation of long term economic optimum thresholds, taking into account the consequences of allowing weeds to set seeds, may appear to be a much more sound basis on which to develop a weed management plan than thresholds based only on current year economics. However, they have at least one major drawback: the models used to run the simulations include several more parameters. More data are therefore required, which may only

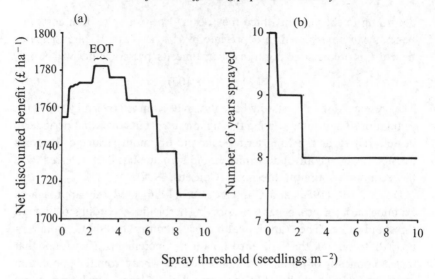

Fig. 6.17. Simulated net discounted benefit (a) and frequency of herbicide application (b), over a period of 10 years, in relation to the threshold density used for control of *Avena fatua* (redrawn from Cousens *et al.*, 1986). The economic optimum threshold, EOT (in this case a range of values), is shown by a bracket. See text for details.

be obtained from an extensive and perhaps lengthy experimental programme. Hence, the necessary data are available for very few weeds. Even for those, there may be very high variability in the data (e.g. Moss, 1990a), leading to extremely high uncertainty in the model predictions. It would be unwise to give precise advice of a threshold to a farmer on the basis such imprecise predictions. All that can be certain from our modelling is that the single year economic threshold is too high, perhaps by as much as ten-fold, but perhaps only two-fold. The practical benefits of all weed thresholds has been questioned by Cousens (1987).

King *et al.* (1986) modelled the economics of controlling mixed broadleaved and grass weed populations in continuous maize production. Their yield–weed density function, however, was an unrealistic sigmoidal curve (see Cousens, 1985a, 1988). They found that control only when the population exceeded single year economic thresholds was more cost-effective than annual prophylactic applications of pre-emergence herbicides. However, they also concluded that it 'would cost producers more to acquire the information required to implement the flexible strategy than would be gained by using it'.

Optimal timing of herbicide applications

In the previous section we explored optimal strategies for controlling weeds over several generations, where herbicides would be applied at some unspecified time each year. Another question requiring the use of optimisation techniques is when to apply a herbicide within a year. Many species exhibit a 'staggered' germination over a considerable period. If we spray a single application of a foliar acting herbicide too soon, a large number of later emerging seedlings will escape damage. If the herbicide is sprayed too late, the early emerging seedlings may not be killed because they have become tolerant; these are likely to produce more seeds per plant than later emergers. When should we spray in order to minimise population increase (note that this problem is usually optimised according to minimisation of yield loss in the year of spraying)?

Mortimer (1983) described an analysis of the timing of a herbicide application to control *Avena fatua*. Emerging seedlings were divided into a number of convenient groups (cohorts) determined by the time interval in which they first appeared. The number of seeds produced by cohort *i* will be the product of the proportion of the seed bank establishing seedlings into that cohort (g_i), the probability of a seedling surviving to maturity (P_i), the fecundity of the mature plants (F_i) and the density of the seed bank at the start of the generation (N_t). Hence, the population density in the next generation will be given by

$$N_{t+1} = bN_t + \sum N_t g_i P_i F_i \qquad (6.17)$$

where b is the proportion of the seed bank which survives. This calculation implicitly assumes that the census of the population density (N_t) is made directly after seed production. If a herbicide kills a proportion k_i of a cohort,

$$N_{t+1} = bN_t + \sum N_t g_i P_i F_i (1 - k_i) \qquad (6.18)$$

Hence

$$\lambda = b + \sum g_i P_i F_i (1 - k_i) \qquad (6.19)$$

The performance of the herbicide was made a function of the age of the cohort at the time of spraying. The rate of population increase was calculated for a range of herbicide application times, in order to see which gave the lowest value of λ. The optimal time of application was found to be in the period 31–40 days after crop drilling (Fig. 6.18).

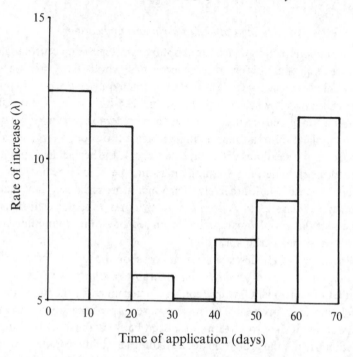

Fig. 6.18. Estimated rate of population increase for a range of times of herbicide application, using equation 6.19 (based on data in Mortimer, 1983). Data were for a population of *Avena fatua* divided into cohorts emerging within 10 day intervals. The optimum timing in order to minimise λ is 31–40 days.

Postscript on validation

In the past two chapters we have described a range of models, we have explored the intrinsic dynamics of populations and we have shown how various extrinsic factors can affect that behaviour. In most cases, the model predictions sound highly plausible. However, since models are often formulated at the end of a research programme, their predictions have rarely been tested (validated) in the field. For example, in Chapter 5 we could present very little data to show the types of intrinsic population trajectories found in practice. In the present chapter, the model predictions for different management factors merely reflect the data which went into them: if we increase a seed production parameter to reflect greater growth under nutrient application, the rate of increase will necessarily be greater. However, the prediction that a rotation of one year ploughing for every four years of direct drilling would keep *Alopecurus myosuroides* in check is supported anecdotally by farmers.

It was argued in Chapter 5 that even an unvalidated model is better than none at all. It allows us, on the basis of our current (perhaps very limited) knowledge, to make informed guesses as to the effects that a particular change in management might have, or how best to tackle a particular problem. However, it would be far better if our models were thoroughly validated! What attempts have been made to do this? How good have our models been found to be? It is not possible to generalise, because there have been so few examples. Our discussion here is, or necessity, based on just two examples.

Firbank *et al.* (1985) used equations derived in one year to predict population densities of *Bromus sterilis* in the next. When growing in wheat, the weed density was consistently underestimated by 15%; it was claimed that this was an acceptable margin of error. For *B. sterilis* monocultures, densities were overestimated by a factor of three. Of course, from one year's data we would not expect concordance in every year: one or other year may be atypical (is there ever a typical year?). Even if the parameters were based on mean data over several years, we would expect the actual parameters in a particular year to be greater than their means in roughly 50% of years and below their means in the other 50%. The departure of predictions in a given year will therefore depend on the variances of the parameters.

Short term forecasting (e.g. one annual time step of a model) is all very well, but we must remember that our models are being recommended for examining long term weed management. As we know from weather forecasting, long term predictions are likely to be much less accurate and small early errors can be compounded to result in very large errors over longer periods. Debaeke & Barralis (1988) examined their predictions over time series of 8 to 17 years. Their model parameters were estimated from data collected at one site for two years, and then annual predictions were compared with observed densities at another site over all years. They concluded that good predictions were made in 66% of cases (though they did not define what constituted a good prediction). There were, not surprisingly, greater fluctuations in the real weed populations over the course of the rotations than were predicted by the fixed value crop-dependent model parameters. In particular, densities were often greatly underestimated (by a factor of more than two) by the model; these events corresponded to single years in which seed production was underestimated in less vigorous crops.

Really, whether a model's degree of predictability is good, mediocre or poor depends entirely on personal perception. If any of our current models are used to persuade farmers not to control weeds and it turns out to be one of the poorly predicted years, they will lose a considerable income as a result of our advice. Anyone aiming to use a population model to advise farmers should therefore take out ample personal liability insurance, or ensure that copious written disclaimers accompany every recommendation! One bad prediction, however, and it is unlikely that your advice will ever be sought again.

Conclusions

Much of our information on the dynamics of population density comes from observing which species change in density under particular treatments. We perturb a population, then observe the effect. A posteriori, we interpret the reasons for this, perhaps leading to further confirmatory experiments. At the present time, we cannot *predict* which species in a community will increase and which will decrease, except for one or two particular species. Also, our observations are restricted to such a small set of conditions that we cannot extrapolate with any confidence to a new type of management.

There is a wealth of information on population responses to particular aspects of habitat modification, especially tillage. However, even for tillage, there are few long term studies, most lasting a maximum of three or four years. The correspondence between real populations and model predictions is therefore unclear. Although our models mostly predict that different management regimes will cause population densities to steadily diverge from each other, few experiments show this: most show an immediate effect in the first year, but little divergence thereafter. In the study of the effects of burning by Whybrew (1964), the ratio of densities on burnt and unburnt plots remained fairly constant as density increased. This apparent departure from model predictions needs further examination in longer term experiments.

There is almost no information on the effect of weather on weed populations. Year-to-year variations in weather are usually treated as background noise and are ignored. Although weather patterns cannot be modified by management, a knowledge of the scale of their effects can help in making decisions and in determining the viability of management systems: risk avoidance in the real (variable) world is of as much relevance to farmers as profit maximisation.

7

The spatial dynamics
of weed populations

In the previous two chapters we dealt with populations as if they were spread evenly over a uniform area, such that the behaviour of a population could be described by its dynamics within any sub-area, such as a randomly placed quadrat. We made the assumption that rates of immigration and emigration were equal and their effects cancelled out one another: the dynamics of population density were effectively dependent only on factors affecting birth and death rates. This led to an understanding of the processes involved in determining weed density.

But a population is neither spread evenly, nor is its environment spatially homogeneous. In addition, populations have edges (though the exact location of the boundary may be fuzzy) across which dispersal will be predominantly in one direction (Fig. 7.1). Although it may be *convenient* to study only the centres of homogeneous patches of species, populations do have a spatial dimension. To understand fully the dynamics of weed populations, we must examine how dispersal, demography and habitat characteristics interact. The net dispersal outwards depicted in Fig. 7.1 has two important implications for the way in which we view (and study) the dynamics of weed patches. The first is that since populations can expand at their periphery we must be concerned with the rate of diffusion into the neighbourhood immediately around a patch. However, such diffusion may be prohibited by a barrier of unfavourable habitat (not suitable for sustained survival and reproduction), resulting in discrete sub-populations. We must therefore also be concerned also with inter-patch dispersal and hence with 'meta-population' dynamics (Levins, 1970), a subject currently at the forefront of ecological research.

Although models of the diffusion of organisms outwards from a point source were published at least as long ago as 1937, the spatial dynamics of populations (and meta-populations) has only emerged comparatively

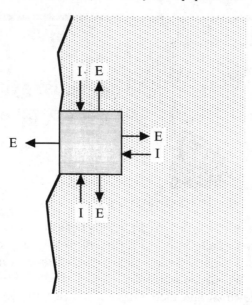

Fig. 7.1. Illustration of the importance of dispersal to dynamics at the edge of a population (shaded). Immigration (I) into and emigration (E) in each of four directions from a hypothetical quadrat is shown. If seed production in the edge quadrat was zero, population density could be maintained there by immigration from adjacent quadrats.

recently as a major issue. Indeed, there are few relevant studies concerning weeds. The discussion in this chapter will therefore be discursive and somewhat speculative. Our intention is to establish a framework for future studies rather than to provide a factual description of what we already know: more questions will be raised than will be answered!

We will begin by discussing the qualitative nature of the area being dispersed into (e.g. a field) and the way that this might affect the developing spatial distribution of a species. We will then review both the empirical data on the spatial dynamics of weeds and the suggestions which models can give us about the processes involved.

The implications of habitat mosaics within fields

Consider a species as it first invades suitable habitats at a new location. Its spread within the site will depend, as with geographic spread, on the suitability and contiguity of habitat types and on the species' ability to disperse. A site may be spatially heterogeneous, consisting of an aggregate of distinct habitat types each differing in the density of population which it will support. There may be discrete regions containing a single habitat type, for example caused by poor drainage in hollows. Crop management may create linear strips, such as where there is cultivation between rows in vineyards, where herbicides are applied in strips in orchards, or where there are hedgerows between fields. There may also be gradients of variation in soil type and drainage associated with slopes. The resulting mosaic of habitats will include routes along which a species can spread easily ('corridors') and barriers, across which spread is more difficult.

At a much finer scale, even in the most apparently homogeneous of habitats there will be a mosaic of micro-sites which will affect establishment and growth of individual plants. Consider a cereal crop, often regarded as the epitome of homogeneity. If the crop is sown by the tractor moving repeatedly up and down the field, the seed drill must turn at the ends of the field, depositing seed at a higher density on the inside of its turning curve than on the outside. It is common then to sow around the 'headlands' afterwards, resulting in overlap with previous manoeuvres (as well as sometimes small areas being missed completely). Towards the edge of the field, then, the density of the competing crop may range over short distances from zero to double the intended density (Fig. 7.2). If fields are sown in a rectangular spiral, farmers often finish by sowing across the diagonals, resulting in regions of higher density within the field. Similar patterns in fertility will also arise from fertiliser applications, and patchiness in the patterns of pesticide deposition from spraying operations. In turn, all of these farming operations will cause spatial variation in compaction (and especially so along the compacted 'tramlines' used in some countries).

Although we might expect the positions of some of the coarser-scale habitat units to be relatively invariant, such as patches of a particular soil type or locations of hedgerows, at a finer scale we would expect micro-sites to be dynamic. For example, although the suite of micro-habitats within a cereal field may be present each year that crop is grown, their exact positions will be extremely unpredictable. Different crops grown in success-ive years, such as maize and wheat, will create patterns of habitat types on different scales. Also, the clonal spread of some perennial plants in pastures may make it impossible for an annual weed to re-establish from seed in micro-sites which it formerly occupied. A field therefore constitutes a mosaic of habitat types, both in space and time.

Such habitat mosaics will confront invading weed species as they disperse into fields and will affect their subsequent dynamics. The species will first form one or more founder populations in suitable habitats. Each successful

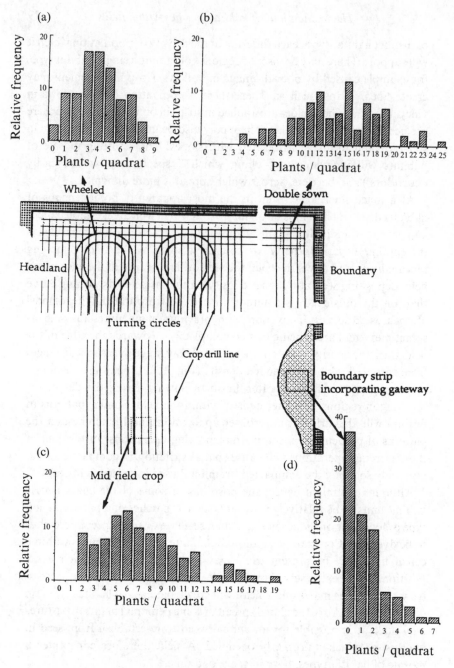

Fig. 7.2. Survey of winter wheat seedling densities in different parts of a commercial field, illustrating how an apparently uniform crop may constitute a mosaic of habitat types at a scale relevant to a weed plant (from Howard, 1991). Seedling numbers were counted in 92 quadrats, each 0.0225 m², in four regions of the field; sowing rate was 200 kg ha⁻¹.

founder will form a patch which will spread outwards at a rate and in a manner determined by its dispersal characteristics, its fecundity and the spatial pattern of micro-sites of different degrees of suitability surrounding it. These initial patches will probably be followed, as a result of chance longer distance dispersal, by the founding of discrete satellite foci, which will in turn (in those parts of the field in which the mosaic consists entirely of suitable micro-sites) coalesce. This is the same process that we described on a geographic scale in Chapter 2 and has been referred to as 'infiltration invasion' by Bastow Wilson & Lee (1989). Once the species has accessed all potential micro-sites, its distribution will remain patchy, reflecting spatial variability in the habitat.

Patchiness in the spatial distribution of a weed will therefore be prevalent both during and after the invasion. This has been revealed in surveys of both plants and seed banks; it is commonly found that frequency distributions of numbers of weeds in quadrats or of weed seeds in soil cores follows the negative binomial distribution, indicating aggregation (Chauvel *et al.*, 1989; Marshall, 1988). Such spatial patterns within weed populations are of more than academic interest: they will affect estimates of crop losses and thereby weed control decisions made in anticipation of a given level of loss (e.g. Brain & Cousens, 1990), the dynamics of beneficial and pest organisms and the abundance of rare weeds of conservation value which may be surviving only in particular micro-sites. Herbicide resistant weeds may first occur in patches: their spread throughout a population may be considered analogous to a new species entering and spreading within a field. However, with the exception of vegetatively spreading clones, the dynamics of weed patches has been given only passing attention by field ecologists.

Field observations of the spread of weed populations

Information on the invasion of fields by weeds has been largely anecdotal, such as in the example of the spread of *Thlaspi arvense* at Butser Hill (see p.61). There are only a small number of exceptions to this. Of particular note is the work of Chancellor, who for 20 years mapped the occurrences of all weeds along grids of points in the fields of the Weed Research Organization, UK (e.g. Chancellor, 1985b). During this time he was able to observe the decline of some weeds (notably those from former pastures) and the spread of others, particularly in one 5.6 ha field. Some species suddenly became widespread throughout the field, this being interpreted as the result of sowing contaminated seed. Others, such as *Chamomilla recutita*, spread steadily over the course of 6–8 years from a restricted distribution to

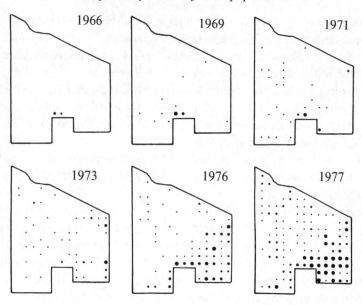

Fig. 7.3. Spread of *Chamomilla recutita* in a field at the former Weed Research Organization, UK (from Chancellor, 1985b). Size of dots indicates weed density: 1–2 per 30 m², 3–6 per 30 m², 7–14 per 30 m². Recording was carried out at 18.3 m intervals.

eventually occupy most of the field (Fig. 7.3). *Aethusa cynapium* (see Fig. 7.4) and *Fumaria officinalis* spread little and remained largely confined to their original distributions (probably coinciding with former field boundaries). The differences between species may reflect preferences for particular edaphic factors in parts of the field and spread of some species may have been limited by this. However, it is more likely that the differences relate to adaptations and opportunities for dispersal. *A. cynapium* and *F. officinalis* are known to disperse poorly; they possess persistent seed banks and perhaps rely on very rare events for their spread. It has been suggested that seeds of *C. recutita* are spread on the outside of animals (although farm animals were not present at any time during this study); their seeds survive well, although they have a shorter half-life in the soil than seeds of the other two species (Chancellor, 1986).

None of the species in Chancellor's study spread in a systematic way along a single front of occupation. It is therefore not possible to estimate rates of linear spread on the scale at which they were recorded. In principle, some idea of relative rates of spread could be obtained by plotting the cumulative number of locations in which the species was recorded against

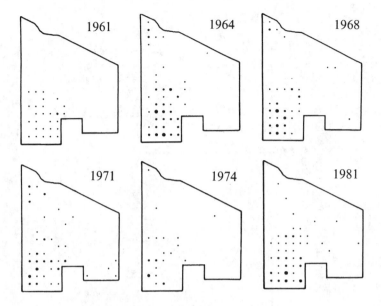

Fig. 7.4. Spread of *Aethusa cynapium* in a field at the former Weed Research Organization, UK (from Chancellor, 1985b). Details are given in the legend to Fig. 7.3.

time, in the same way that the number of counties occupied through time has been used in studies of geographic spread. Those species with the greatest rates of spread might be expected to have the greatest rates of increase in site occupancy.

Unfortunately, however, the interpretation of such graphs is equivocal. Consider a species occurring at very low density; it will not be recorded at every location within its distribution in every year simply because of a low sampling frequency. Cumulative occupancy of locations will therefore appear to increase even in the absence of spread, simply due to the low probability of being recorded. A change in the rate of increase in the number of recorded locations will result either from an increase in density throughout its range (such as by a change in management or by the introduction of new seeds in contaminated grain) or from an increase in rate of spread. Fig. 7.5 shows the cumulative site recordings for the eight species which were restricted in their distributions at the start of the study. Three species showed a steady increase throughout the 20 year study: recordings of *Fumaria officinalis* and *Aethusa cynapium* increased more rapidly than for *Papaver rhoeas*. From Chancellor's original maps *P. rhoeas* certainly appeared to have spread the least of the three species over the 20 year

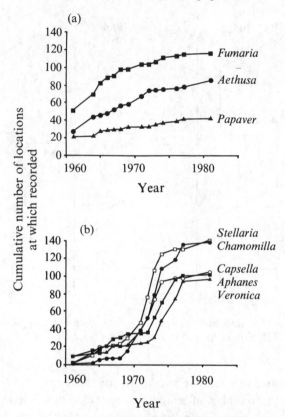

Fig. 7.5. Cumulative number of locations within a field at which each of eight species were recorded (based on data from Chancellor, 1985b). Species have been grouped according to whether they showed (a) a steady increase in cumulative number of locations, or (b) a sharp change in the frequency of records.

period. The other five species all increased slowly at first, but then began rapidly to increase. The increase for *Capsella bursa-pastoris*, *Stellaria media* and *Chamomilla* spp. appeared to begin around 1968; for *Aphanes arvensis* and *Veronica* spp. increase began around 1972. The phase of rapid increase may, as Chancellor suggested, have been initiated by the sowing of a contaminated seed lot, but without samples of the seed it is difficult to be conclusive. There were no obvious changes in crop management around that time. No data on the types of machinery used (possibly influencing dispersal) have been published. Hence, despite the existence of this unique data set, it is still very difficult to interpret the species' rates of spread within the field. However, persistence of a high frequency of records for several years after sowing of contaminated grain would indicate that dispersal had previously been limiting to the populations rather than habitat suitability.

It is clear that analysis of the spatial dynamics of weed populations may not be a simple matter, even for species newly invading a site. For species already distributed throughout a field, dynamics may be even harder to discern unless the correct sampling scale for the study is chosen. Rather than rely on weight of empirical evidence from a large number of studies (which in any case do not exist at this time), the development of a theoretical framework is an essential pre-requisite to understanding weed spatial dynamics. In the following section we will describe briefly some of the models which have been developed to describe the spatial expansion of populations. This will then enable us to review the available field observations in light of the model predictions.

Models of patch expansion

Consider a weed population expanding in a homogeneous area of habitat in which the centre of the weed patch is separated from the periphery by more than the maximum distance of dispersal (d_{max} in Chapter 3). Individual plants complete their life-cycle in one year and disperse propagules randomly in all directions of the compass. This expansion may be visualised as a set of annual concentric rings, each marking the 'edge' of the population after a generation of growth. A sub-population close to the edge of the patch will be at a lower density than at the patch centre, owing to dispersal into unoccupied habitat. The sub-population of plants growing in the centre of the patch may experience more intense density-dependent regulation than that at the periphery. The process of spread thus has two components: (i) dispersal of propagules and (ii) reproduction by plants arising from those propagules which generate further dispersal. We considered dispersal in Chapter 3 and reviewed the different forms of seed dispersal curve that may result from an individual plant and subsequent dispersal agencies. The dispersal curve relevant to the present chapter and which must be used as input into the models to be described here is the frequency distribution of distances moved by propagules over a period of time (usually a single generation), modified by their probability of surviving to reproduce (i.e. 'effective dispersal').

The conventional modelling approaches for population spread use partial differential equations to describe changes in density over space and through time. In this chapter we introduce these models sufficiently to explain the rationale and predictions arising from them; formal definitions are given elsewhere (Skellam, 1951; Okubo, 1980). Most of the models start with two basic assumptions: firstly, that organisms disperse by 'Brownian motion' at a constant rate over generations and in space, and secondly, that the population shows exponential growth at any point in space.

For a population initially comprising a single reproductive individual at an origin ($x=0$), undergoing exponential growth and expanding in unbounded space by random motion, the population density, N, x distance units away from the origin at time t, is

$$N = \frac{1}{2(\pi Dt)^{0.5}} \exp\left(\alpha t - \frac{x^2}{4Dt}\right) \tag{7.1}$$

where α is the intrinsic population growth rate (i.e. $\log_e R$), and D is a coefficient which measures the dispersal rate in units of distance^2 per unit of time (Okubo, 1980). D is related to the mean squared displacement (MSD) of dispersing individuals, the exact relationship being determined by the specific dispersal pattern that is followed. In the specific case of a Normal dispersal curve, the MSD is simply the variance (σ^2) of the dispersal distribution, which over an annual cycle is equal to $4D$. In equation 7.1 it is assumed that individuals are diffusing in a random manner and it is easy to consider time in 'snap-shots' of generations of dispersal and population growth of an annual species. Spread of the population along a transect appears as successive Normal distributions increasing in mode and variance (Fig. 7.6).

The Normal distribution describes just one shape of dispersal curve and is a specific case of a more general model which can describe curves of various shapes. The general model is given by

$$G_{(x,t)} = \exp\left[-(x/a)^b\right] \tag{7.2}$$

where $G_{(x,t)}$ is the fraction of individuals having dispersed a distance x, a is the root mean square displacement of individuals, b determines the rate of decrease with distance and t is the duration over which dispersal is measured. With random (Normally distributed) dispersal, $G_{(x,t)} = \exp[-(x/\sigma)^2]$. Mean squared displacement may be described by a variety of models which may take into account specific directionality of dispersal (Holmes, 1993).

Equation 7.1 predicts (i) that a plot of the boundary of a population expanding radially against time is a linear one (or in other words, $\sqrt{\text{area}}$

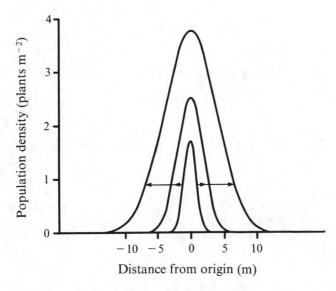

Fig. 7.6. Spread of a population along a transect outward from a point source as predicted by equation 7.1, which assumes dispersal by a process analogous to Brownian motion and exponential increase in density. The distribution of densities follows a Normal distribution which increases in mode and variance through time.

occupied by the population is linearly related to time) and (ii) that the ultimate velocity of spread is $(4\alpha D)^{0.5}$ (Skellam, 1951). The former prediction has been shown to be true for the spread of animal species on a geographic scale (e.g. muskrats in Europe), but it has not been examined extensively in plants.

But how do the predictions change if we alter our assumptions about population growth rate and dispersal pattern? As the density at a point in space becomes greater, interference between plants will result in a pattern of population growth which is no longer exponential. We can envisage that populations showing a logistic pattern of growth will disperse outward as shown in Fig. 7.7, spreading as a wave in which density-dependent regulation limits wave 'height'. Analysis shows that whilst the incorporation of such regulation slows the initial expansion rate of the population, iso-density population waves asymptotically achieve the same velocity, namely $(4\alpha D)^{0.5}$ (Kendall, 1948). This is not true, however, if we depart from the assumption of a Normal dispersal. The velocity of spread and shape of the travelling wave front depends on the underlying dispersal function (Mollison, 1977, and Holmes *et al.*, 1994, give further examples).

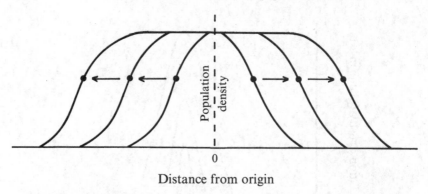

Distance from origin

Fig. 7.7. Spread of a population in which density increases in a logistic-like manner. A single density near the periphery of the population is highlighted (●), illustrating that 'waves' of any density are predicted to move at the same (constant) rate.

Implications for patch formation

The modelling approach just described provides us with a theoretical basis for addressing questions of rate of patch expansion and persistence of weed species, the two essential parameters being the rate of population growth and the effective dispersal coefficient, D. The expected rate of patch expansion can be calculated as outlined above, given the implicit assumptions of homogeneity of habitat over time and space.

The data for *Bromus sterilis* in Chapter 3 enable us to give an example. Natural dissemination of seeds in a wheat crop resulted in a Normal dispersal curve with a standard deviation of 0.312 m and virtually no drift of the population centre. Let us consider for simplicity that there are no further causes of dispersal: all seeds dehisce before harvest and there is no subsequent cultivation. From field experiments, Howard (1991) estimated that the finite rate of increase (R) of *B. sterilis* under minimum tillage and with no chemical control may be as high as 50-fold per annum in uncrowded populations. Thus, the intrinsic rate of increase (α) would be $\log_e 50$ and $D = 0.0243$ per generation. From this, the rate of patch expansion would be $(4\alpha D)^{0.5} = 0.62$ m per year.

Whilst it may be argued that the prediction for *B. sterilis* is at best notional and is difficult to evaluate because of the absence of comparative data for other species, the model suggests that patches of this weed would expand by an advancing wave relatively slowly in a winter wheat cropping system under minimum tillage without the aid of other dispersal agencies. The velocity of the expanding patch is an upper estimate, since exponential growth is assumed: in density regulated populations the same velocity will be approached only after several generations of growth. The estimate is also highly dependent on the many other assumptions which were made. For example, in strongly competitive wheat crops, population growth rates of *B. sterilis* may be only five-fold per annum and the estimated rate of radial spread of a patch consequently will then be reduced to only 0.40 m per year.

A critical assumption is the Normal dispersal curve of individuals by the end of a cycle of population growth. While this may be appropriate for some species, we know that a range of more complex dispersal distributions can occur, such as when the effects of several dispersal vectors are superimposed upon one another (see p.84). At present, we know too little about dispersal functions of weeds in agriculture to make generalisations. Mechanical dispersal agents, such as combine harvesters and cultivators, may produce distributions following the Bessel function (an exponential distribution superimposed over a Normal) or patterns so complex that they can be represented by no one simple function. For simplicity, we assumed above that no mechanical agents were involved, whereas it is known that *B. sterilis* can be carried along on the outside of a combine or its seeds taken up and distributed with the crop residues. The choice of the appropriate dispersal function will therefore affect the rate of spread, as will the values of the function's parameters. The value of the dispersal coefficient, D, of 0.0243 m^2 per generation was calculated for a population of erect plants disseminating seeds into the crop. Lodging of culms may play a significant role in increasing the variance of the dispersal curve, since culm height may be 20–30 cm.

All of our discussion to date has been based on the assumption that the weed patch is in a homogeneous habitat. Field observations suggest that this is often not the case. Some weeds may be confined to areas which are poorly managed, poorly drained or less frequently disturbed. These weed patches may be envisaged as populations occurring in a favourable habitat but dispersing propagules into surrounding unfavourable habitat. Newly arriving propagules may encounter such an area of favourable habitat and establish within it. Whether or not the new patch will persist will depend both on the area and geometry of the favourable habitat in succeeding generations, the dynamics of the population within it, the rate of loss of individuals by dispersal and the degree of 'unfavourability' of the surrounding area. There is thus a critical habitat size which will enable the population to persist.

Models of the type described by equation 7.1 may be further developed to determine this critical habitat size. Again, a starting assumption is that there is exponential population growth, but now we constrain the population to be within a circular area of habitat. Dispersal again occurs randomly within and outside the patch, but now into lethal surrounding habitat. In this case, the critical habitat area is given by $1.84 \pi^2 D/\alpha$ (Okubo, 1980). The parameter values assumed earlier would predict a critical patch area of 0.11 m^2 for *B. sterilis* in winter wheat if $R = 50$ and 0.27 m^2 if $R = 5$: areas of habitat smaller than this would not allow a population to persist.

Ecological research into critical habitat patch sizes has been focused in entomology and marine biology, from which three pertinent generalisations have emerged. The first is that density-dependent growth rates, which regulate population size within the patch, do not affect the critical patch size (unless there is a threshold density needed for positive population growth). Thus, whilst dense monospecific weed patches, arising perhaps from failure of chemical control, may exhibit intraspecific interference, this will not determine the minimum area required for patch persistence, but only the density of the population in the patch. Secondly, and perhaps the most obvious, critical patch size decreases as the boundary habitat becomes less hostile. A strongly competitive crop surrounding a patch may ensure suitable habitat for weed population growth, but yearly variation in crop competitiveness will result in fluctuation in critical patch sizes. Thirdly, factors that increase movement out of a patch will lead to larger critical patch sizes. Thus, the effects of tillage practices in moving propagules away from a patch into hostile habitat will have the effect of reducing the likelihood of patch persistence, since they will lead to a requirement for larger areas of suitable habitat.

Field observations of the dynamics of weed patches

The information available for testing the predictions of models of patch spread for weeds is extremely limited. There have been few attempts to record the margins of patches over time, and still fewer which have monitored density throughout patches as they expand. Indeed, the patchiness of weeds has often been regarded as an impediment to doing weed control trials, for which a homogeneous area is preferred, rather than something to be studied in its own right.

Most of the examples of the spread of weed patches are for clonal perennials, in which it is relatively easy to identify a distinct patch boundary. In the example of *Cirsium arvense* discussed on p. 77, there appeared to be a distinct increase in the rate of spread over the first few years of patch establishment. This is probably because a clone establishing from a seedling has to build up an energy reserve before it can achieve its maximum rate of spread. The data given by Horowitz (1973) show that over an annual cycle *Sorghum halepense* clones grown from single rhizome fragments expanded slightly faster over the second year of growth than over the first. However, at the height of the growing season the rate of expansion was greater in the first year. It is not possible to determine whether these differences were due to age *per se* or to the weather in the particular year. Overall, however, it is to be expected that, given uniform weather conditions, a constant rate of expansion will not occur for at least the first year or two in the life of a patch of a clonal perennial.

Studies of patches of annual weeds are more difficult, since the boundary is often less distinct. Although there have been few detailed studies, anecdotes abound amongst researchers concerning the spread of weeds from plots which they have sown at high density (i.e. artificial patches) when studying weed–crop interference. A recurring theme is the absence of significant spread in many cases, despite an apparent array of potential dispersal agents. For example, G. W. Cussans (pers. commun.) noted that the clear shape of plots of *Alopecurus myosuroides* in the UK could still be seen several years after experiments. In research involving *Avena fatua* at the University of Saskatchewan, B. Frick (pers. commun.) noted that this species only spread by about 0.5 m from one year to the next, even in the direction of cultivation. Such observations suggest that many species may be limited by dispersal and will spread only slowly unless they are sown as contaminants in crop seed. The timing of seed dehiscence in relation to harvesting may be an important factor in this.

Few farmers or managers would be eager to allow the introduction of any weed so far absent from a field. An experiment by Auld (1988), however, was mentioned in Chapter 3 in which he introduced plants of *Avena fatua*, *Carduus tenuiflorus* and *Onopordum acanthium* to two pastures (on a research station) in which they had not been recorded previously. One year after introduction, plants of the three species were found up to 5, 7 and 5 m respectively from their parent plants, but their abundance declined rapidly with distance. After two years, they had spread further, but not as far as might have been expected from the first year's data.

Harradine (1985) recorded the spread of the wind-dispersed *Carduus pycnocephalus* along three contrasting strips of habitat: bare ground, annual pasture and perennial cocksfoot (*Dactylis glomerata*) pasture. He found that throughout the three years of study population densities were highest on bare ground and least in the cocksfoot pasture. Within a year seedlings were found as far as 10 m from their parents in bare and annual

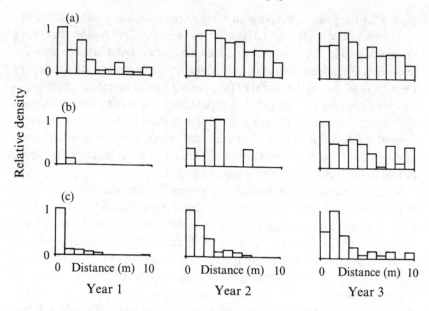

Fig. 7.8. Spread of *Carduus pycnocephalus* in strips of three contrasting habitats: (a) bare ground; (b) *Dactylis glomerata* dominated pasture; (c) annual grass dominated pasture. Densities in each 1 m band along each strip have been standardised relative to the highest density observed in that habitat type in that year. Based on data from Harradine (1985).

grass strips, but only 1.5 m from their source in cocksfoot strips. This rate of spread for the leading edge of the population is greater than would be predicted from simple models of dispersal through the air (without turbulence), and was greater than any of the species in Auld's study. However, the median distance spread from the source in each habitat type was only approximately 2 m, 1 m and 1 m respectively. By the end of three years, the weed had reached the end of the 10 m plots even in the cocksfoot pasture. Whereas thistle density at this time was still greatest close to the source in cocksfoot and annual grass plots, plants were spread almost evenly throughout the bare plots (Fig. 7.8). Although the picture is far from clear, it would appear in this study that the leading edge of the expanding populations moved at a greater rate than the region of highest density. If this was indeed the case, such a pattern of spread would not be predicted by the models described earlier in the chapter.

All of the examples given above for annual weeds are from observations of individual patches, mostly established artificially. Studies of the dynamics of naturally occurring weed patches are uncommon at any scale. Here, we describe two examples of the spatial dynamics of natural populations at different spatial and temporal scales, one based on observations within a part of a field over a period of months, the other based on monitoring of a whole farm over several years.

The first example concerns the distribution of seeds of the cosmopolitan summer weed *Chenopodium album* in a maize crop (Benoit *et al.*, 1992). Soil cores were collected at 8 m intervals in an area 40 m by 40 m at three times during the growing season. The soil was spread out in containers and incubated so as to ensure maximum germination. Seed bank density was estimated from the number of seedlings emerging. Fig. 7.9 shows that there was a considerable change in the distribution of *C. album* over the season. There were three distinct patches prior to sowing in May. After the crop had been sown, and presumably considerable emergence had taken place, there appeared to have been a slight shift in the positions of the patches. However, after dissemination of the seeds produced during the season, two new patches dominated. The final distribution was very different from the initial seed bank. (Some care should be taken when interpreting Fig. 7.9, since there are four pixels between each sampling location.) For this species, then, it would appear that patches are extremely transient, at least on this scale.

The second example concerns the grass weed *Alopecurus myosuroides*, a problem in cropping in northern Europe, especially on heavy soils. Wilson & Brain (1990) analysed data from a survey of the weed on a farm in southern England, collected over a 10 year period. Each summer the density of inflorescences was assessed on a rectangular grid of points, with distances between adjacent points of 36 m and 40 m. It should be noted that the densities represented the weed populations surviving herbicides. A total of 938 points were assessed over the 173 ha farm. Fig. 7.10 shows the results of the survey. It can be seen that there are distinct regions in which the weed was recorded repeatedly, suggesting that patches, at least on this scale, can be consistent in their positions (see especially field F). It may be that these patches correspond to areas of the farm with a particular soil type or drainage condition. Alternatively, they may represent populations expanding from a point introduction (this species usually sheds its seeds before harvest and is likely to disperse poorly). There are clear implications for herbicide use. If the areas in which the weed occurs are known (or can be determined by survey or remote sensing) only those areas need be sprayed, avoiding wastage of herbicide on parts of the field where it is not a major problem.

The previous two examples illustrate contrasting outcomes. The long

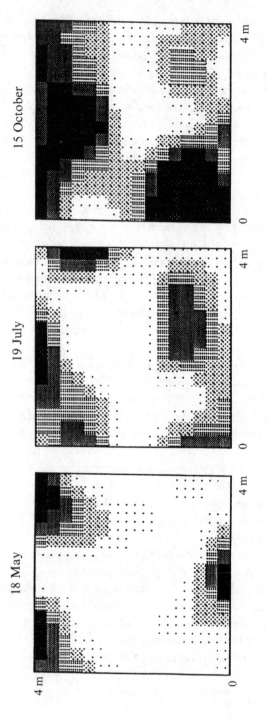

Fig. 7.9. Seasonal distributions of *Chenopodium album* seed bank densities in a maize field in southern Canada (from Benoit *et al.*, 1992). The area sampled was 40.3 m by 40.3 m; soil cores were sampled at 8.06 m intervals. Levels of shading represent different densities.

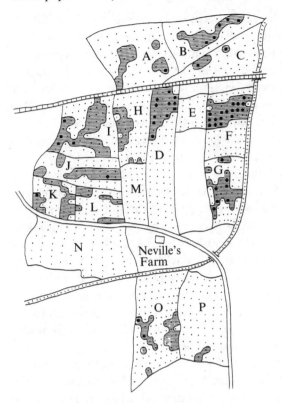

Fig. 7.10. Map of a farm in central England on which the distribution of *Alopecurus myosuroides* was mapped over the period 1977 to 1986 (from Wilson & Brain, 1990). Shaded areas include locations in which the species was recorded for more than half of the years when the field was in cereals; large dots show occurrences in more than 80% of cereal crop years; small dots show all surveyed points.

term field-scale survey indicates that patches can be relatively stable, whilst the other, admittedly on a smaller scale and within a year, suggests that patches can be very dynamic. At present it is impossible to judge which is the more common situation, or whether patch stability is related to mechanism of dispersal.

Meta-population dynamics and weed populations: a conceptual view

From an ecological point of view, a group of weed patches set in a fragmented habitat constitutes a *meta-population,* so long as the group is interconnected by dispersing propagules (Levins, 1970). Dispersal is the process by which connectivity amongst sub-populations is achieved and the meta-population is defined at this higher scale of organisation. Certainly, the modern agricultural landscape is spatially fragmented for non-crop species. It may be conceived variously as an archipelago of favourable habitats for weeds within a field, or as habitat patches fringing a hostile ocean of competitive crop. Whatever metaphor is invoked, the important point in assessing whether a weed species persists as a meta-population in this habitat mosaic is in demonstrating the importance and role of dispersal to species persistence in the sub-populations (patches).

To explore the relevance of the meta-population concept to weed population dynamics, it is convenient to consider two scenarios for the persistence of a species in a field. In the first, persistence of a weed in sub-populations is simply determined by the occurrence of favourable habitat conditions, patchily distributed across the field. Sub-populations behave independently of one another at locations in the field and the species shows no spatial dynamics at the sub-population level. Periods of extreme habitat unfavourability at a location may be bridged over generations if the species exhibits seed dormancy or can perennate from buds; however, if it is annual with a transient seed bank then extended periods of poor habitat may lead to the extinction of the sub-population.

In the alternative scenario, we can envisage the patches of the species as a meta-population in which persistence is functionally dependent on propagule dispersal. In a hypothetical field, where fragments of habitat for the weed are of equal quality, size and equi-distant from one another, then chance processes may cause some sub-populations to become extinct but others to occur by recolonisation of unoccupied habitat as a consequence of local propagule dispersal. In this model, the meta-population will move randomly over time within the field. However, in actual agricultural landscapes differences in size, location and quality of habitat will generate noticeable variation in the extinction and recolonisation rates. Thus, local dispersal of propagules from sub-populations may be insufficient to ensure habitat colonisation. If the average rate of extinction of sub-populations exceeds the rate of recolonisation over a sufficiently long time span then the meta-population will become extinct unless there is propagule immigration from sources external to it.

To what extent, then, does a weed species behave as a meta-population in agriculture and how important is this in considering the management of weeds? The heart of the answer lies in demonstrating the relative import-ance of the various dispersal processes to both rate of colonisation of habitat and the spread (and hence size) of sub-populations. Theoretically,

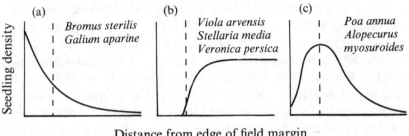

Fig. 7.11. Three typical patterns of weed seedling distribution along a transect through the margin of an arable field (after Marshall & Hopkins, 1990). The dashed lines indicate the position of the crop edge.

the chance of a sub-population becoming extinct decreases exponentially with numerical size (Goodman, 1987) and isolated habitats are less likely to be recolonised quickly than those surrounded by similar type or those interconnected by dispersal corridors (Opdam, 1990). At a regional scale, a weed species may occur as meta-population as a result of seed dispersal through long distance vehicle (combine and tillage implement) movement amongst farms. Conversely, at the field scale the same implements may act as short distance dispersal agents ensuring spread of weed sub-populations as well as re-colonisation of habitat.

The distances which propagules may travel were reviewed in Chapter 3; however, such data do not in themselves demonstrate the *importance of dispersal to the maintenance* of sub-populations. Analysis of the role of dispersal in the persistence of weed species has been limited to date and the approaches used have mostly been either inferential on the basis of long term observations or from simulation modelling. There have also been attempts to infer the role of dispersal in spatial dynamics from a 'snap-shot' of populations at a single point in time. Such a procedure is fraught with dangers. One example is given here to illustrate the interpretative problems.

At first glance, the distribution of weed seedlings at the margins of arable land might be thought to yield some evidence of weed dispersal. Marshall (1985) mapped the distributions of species along transects from hedgerows into cereal fields. He described three main patterns of abundance (Fig. 7.11). One of these (Fig. 7.11a) was where a species was most abundant in the hedgerow and decreased (but was still present) out into the field. Is the species steadily spreading into the field from the hedgerow, with the field margin fuelling the spread by supplying copious numbers of seeds and/or rhizomes? Is the weed population in a dynamic equilibrium, each year dispersing into the field, but being killed in that otherwise favourable habitat by the herbicides used in the field? Or is it a stable distribution reflecting the species' preference for the hedgerow habitat as opposed to the field habitat, and offering no threat? In each case the management strategy would be different. In the first case, a different herbicide should be tried in the field, and the field margin should perhaps also be sprayed. In the second case no immediate action is warranted, but care should be taken to prevent invasion if the herbicide used in the field is changed. In the final case, no change in strategy is necessary and spraying of field margins would be needless. Hence, the development of management strategies from single surveys is entirely a subjective judgement. An unequivocal statement about weed dispersal and its role in dynamics cannot be made.

Suggestions from simulation models

Analytical models, such as those for individual patches, do not allow the investigation of the implications of meta-population dynamics. They cannot, for example, easily be extended to the consideration of multiple weed patches, irregular patch shapes and complex habitat mosaics. The mathematical solutions can only be determined when the assumptions are simple. Where population behaviour does not conform to one of the standard mathematical functions, such as when there are multiple dispersal agents in a cropping system, either the mathematics becomes intractable or suitable functions cannot be found to describe each process.

The consideration of realistic, rather than just mathematically tractable, systems is central to developing an understanding of the dynamics of weed spread at a whole-field scale. The solution to the problem is to use cellular simulation models. Instead of assuming a continuous habitat, we can divide a field up into a grid of a large number of small cells. Each one can be considered as capable of supporting its own small weed population, which is able to disperse seeds into surrounding habitat cells. For each cell we can apply the difference equation models described in Chapters 5 and 6 to calculate seed production. Each cell can be assigned its own population parameters to reflect the local habitat conditions. Seed dispersal can be described by a grid of probability values, obtained from experimental work, centred on each cell. An initial patch of weeds, or a complex of patches, can be set up within the grid of cells. Using the iterative power of a computer, the gains and losses to and from each cell can be calculated in each generation, thus simulating the spatial dynamics of population density.

A model such as this was used by Ballaré *et al.* (1987b) to predict the

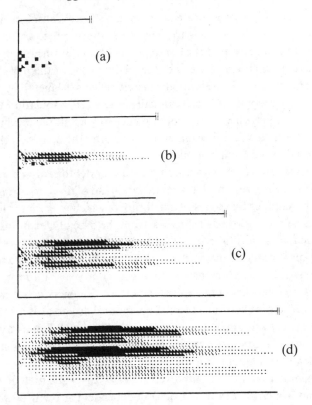

Fig. 7.12. Spread of *Datura ferox* over 3 years, as predicted by a discrete cellular model (from Ballaré *et al.*, 1987b). The initial seed bank is shown in (a); the degree of shading represents seed bank density. Dispersal was primarily by combine harvester and population density was determined by a multi-stage density-dependent model.

spread of *Datura ferox* in successive soybean crops. They divided their (edgeless) field into 0.7 m by 0.7 m cells. Seed production and mortality within a cell were described by a multi-stage population model (see Chapter 5). It was assumed that all seeds produced in a given year were taken up into the combine harvester. These seeds were distributed by the harvester according to probabilities calculated from dispersal studies (Fig. 3.16a); a proportion were also retained in the combine as grain contaminants. The harvester was assumed always to travel in the same direction and along exactly the same path. Once on the ground, seeds were further dispersed by cultivation; it was assumed that 10% of the seeds in a cell at that time were moved to a neighbouring cell (presumably in the direction of cultivation). The simulated spread of *D. ferox* is shown in Fig. 7.12. The initial population was 500 seeds located in a small cluster of cells. Considering the dispersal pattern, it is not surprising that the population formed long patches in the direction of the combine. It is also not surprising that, after the first year, the rate of spread of the leading edge of the population was approximately constant. The model merely reflects the fact that, if seeds are dispersed by a combine, they can cause the weed population to quickly spread over a large area. As a consequence of spread, the total number of plants in the field will still be increasing long after the density in the original patch has reached equilibrium.

Schippers *et al.* (1993) used a simulation model to examine the importance of three dispersal mechanisms in the spread of *Cyperus esculentus*. The species perennates by means of small tubers which overwinter to give rise during the growing season to shoots, from the base of which rhizomes and subsequently small tubers are propagated. Six partially overlapping generations of shoots may be produced in a season and on average one tuber may give rise to a further 15 tubers in a year in a crop of maize, densities of 1000 tubers m^{-2} being reported (Groenendael & Habekotté, 1988). Spread of tubers may result from natural dispersal as a consequence of rhizome growth and tubers may be placed up to 0.7 m away from the parent shoot. Dispersal over longer distances may occur as a result of mechanical farming operations.

The simulation model employed a series of sequential deterministic equations which incorporated density-dependent growth of shoots and tuber production, and described the spread of tubers by natural growth, the horizontal and vertical redistribution of tubers in the soil due to ploughing and harrowing and the horizontal redistribution of tubers by adherence and transportation in soil on farm machinery. Distribution of tubers was modelled in three dimensions. Soil tillage was assumed to redistribute tubers horizontally according to a Normal distribution (Sibbesen *et al.*, 1985) whilst distribution in the vertical plane was simulated in 5 cm layers of the soil profile down to a depth of 45 cm using the transition matrix described by Cousens & Moss (1990) (see p. 196). Soil adhering to farm machinery was assumed to have an exchange rate with that in the surface 5 cm in the range of 0–2 litres for every 1 m^2 of surface area.

To examine the relative importance of redistribution of tubers by soil mixing and by soil adherence to farm implements, a simulation was conducted from a starting population of 1000 tubers at the centre of a 50 m by 50 m grid. Population size and spread were assessed after a simulated period of ten years growth. In the first simulation, the exchange rate of soil between implements and ground was assumed to be 0.875 litres m^{-2} and the

Table 7.1. *The effects of soil adhering to machinery and soil mixing on a simulated population of* Cyperus esculentus *after 10 years. The initial population was 1000 tubers in the field centre*

Dispersal mechanism		Prediction after 10 years	
Adherence	Mixing	Tuber number (thousands)	Infested area (m^2)
−	−	135	253
+	−	339	455
−	+	385	602
+	+	422	693

Source: After Schippers *et al.* (1993).

variance of the distribution of horizontal tuber spread to be 0.47 m^2. Table 7.1 shows the predicted effects of the individual and combined influences of soil adherence and soil mixing. Soil mixing alone caused a three-fold increase in the number of tubers in the grid after ten years over that in undisturbed soil. The underlying reason for this increase was the reduction in intraspecific competition in dense shoot populations that arose due to tuber movement and spread. The area occupied by the population increased more slowly than total tuber number. Tuber movement through soil adherence had slightly less effect in increasing population growth. This difference is to be expected because only a small fraction of the tubers in the upper 5 cm of the soil were carried by soil adhesion whereas soil mixing involved all tubers in the top 25 cm of the soil profile where the bulk of the population resided.

In another simulation, the effects of a herbicide able to give 95% control of tubers were examined. Where control measures were initiated early in the development of the population, eradication of the weed had almost been achieved after 20 years. However, a delay of only one year in starting the control program meant that it took twice as long to achieve a similar result. Two other notable observations were made. Firstly, the number of tubers decreased faster than the size of the infested area, and secondly, in one instance the size of the area was still increasing while the total tuber number was declining.

The simulations for *Datura* and *Cyperus* illustrate the methods by which the relative importance of different dispersal agents to spatial dynamics can be assessed. In both cases it would be possible to impose constraints so as to take into account the direction of machinery movement at the field margins.

Interesting extensions would be to examine the rate of spread when the population is confronted by barriers of unsuitable habitat of different widths, or by mosaics of suitable and unsuitable habitats in different proportions.

Conclusions

This chapter has sought to expose the essential ecological questions that need to be answered if we are to begin to understand the dynamics of weed spread at a detailed level. There is a clear theoretical basis on which we can work but a paucity of data. Careful experimentation measuring the effects of different dispersal agents is needed and, when such data are combined with models of population density, simulation models may be used to provide valuable insights. Without these approaches, inferences about weed patch dynamics, particularly from single surveys, are very limited.

Many questions, however, remain unanswered. How rapidly will a particular weed (or indeed any weed) spread across a field? Is rate of spread related to its life history? How long will it take to occupy fully all available areas of suitable habitat within a field? What are the local rates of population increase within each habitat type within a field? Once the species has occupied all suitable habitats, how stable are the boundaries of the weed patches? Where there are stable boundaries, is this because the surrounding area is unsuitable, because of poor dispersal or because birth and death rates do not respond quickly enough to be able to track habitat changes? What is the limiting size of an area of suitable habitat which will allow the persistence of a species within it? How do these considerations determine how we might control common weeds or conserve rare ones? Clearly this is a fruitful area for future research.

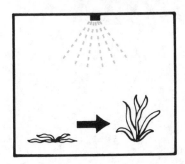

8

The evolution of
herbicide resistance

It has been estimated that in 1830 it required 58 hours of labour to cultivate and harvest one acre of a cereal crop; 150 years later this was achieved in just two (Kirby, 1980). Of the many improvements in agriculture in the last 50 years, the introduction of selective chemicals to control weeds has had one of the greatest effects on the magnitudes of yields, their stability and the human input requirements for intensive cropping. The early 1940s saw the first commercial introductions of synthesised herbicides including 2,4, dichloro-phenoxyacetic acid (2,4-D) and methyl-chloro-phenoxyacetic acid (MCPA) and there are now in excess of 200 phytotoxic chemicals that are commercially available world-wide in herbicide formulations (Hance & Holly, 1990). In intensive farming systems it is now common for farmers to make two herbicide applications in managing every cereal crop (Tottman & Wilson, 1990).

The persistent soil applied herbicide simazine (2-chloro-4,6-bis{ethyl-amino}-s-triazine) was introduced into commercial use in 1956 and in 1968 a biotype of *Senecio vulgaris* was found to be resistant to the recommended rate of application of simazine in Washington state in the USA (Ryan, 1970). This was the first confirmed instance of a herbicide resistant weed. LeBaron & McFarland (1990b) listed more than 50 weed species showing resistance to herbicides, whilst Darmency & Gasquez (1990) concluded that 19 grasses and 48 broad-leaved weed species exhibited biotypes with resistance to herbicides. In comparison with other xenobiotics the emergence of herbicide resistance is relatively recent, yet the pattern of appearance of cases appears similar (Fig. 8.1).

There has been much debate about the use of the terms resistance and tolerance. We will define 'herbicide resistance' as *evolved* tolerance in a weed population in response to selection through the application of herbicide(s). Tolerance is a response where there is survival of a plant,

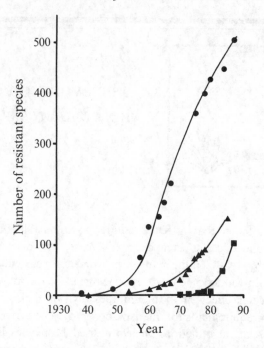

Fig. 8.1. Increase in the number of organisms resistant to pesticides (after Holt & LeBaron, 1990): arthropods (●); plant pathogens (▲); weeds (■).

though perhaps with some reduction in growth, at a herbicide concentration that would normally give complete mortality in a susceptible population (Putwain, 1990). Thus, plants may vary along a scale of sensitivity to herbicides, ranging from non-tolerant (as susceptible as individuals in unselected populations) through to highly tolerant ≈ resistant (all individuals perform as well on exposure to herbicide as they would in the absence of herbicide).

Over a wide range of herbicide doses applied, the response of a weed population (either in terms of number surviving or mean plant biomass) to the logarithm of dose is typically sigmoidal in form (Fig. 8.2). The evolution of resistance will result in a change of this response curve in the resistant biotype, at its simplest a horizontal shift in the position of the curve. In the cases of resistance recorded to date, the shift has been gradual or rapid (from as little as four years up to at least two decades) and it has been either small or large (from less than an order of magnitude to three orders of magnitude in terms of dose).

From a practical point of view, a useful definition of resistance should

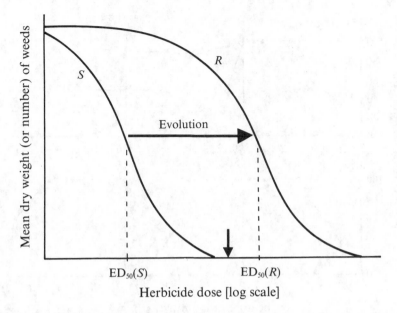

Fig. 8.2. Diagrammatic representation of the outcome of the evolution of herbicide resistance. As evolution develops, the population dose response curve moves from left to right, from susceptible (S) to resistant (R). The herbicide application rate required to decrease plant biomass by 50% (ED_{50}) is shown for both R and S. The ratio $ED_{50}(R)/ED_{50}(S)$ can be used as a measure of the degree of resistance which has evolved. The commercial application rate (pre-resistance) is shown by a vertical arrow.

relate to field agricultural doses of herbicides: LeBaron & Gressel (1982) argued that a working definition of a resistant weed species is one that survives and grows normally at the usually effective dose of a herbicide. Whilst farmers may readily identify with this definition, taken at face value it does not necessarily imply evolutionary change and selection of resistant individuals through herbicide spraying. Moreover, as we will show, resistant populations may vary in their ecological fitness (growth and survival). Fig. 8.3 shows some typical responses to herbicide dose for susceptible and resistant populations of a range of weeds. In these instances, performance is measured as a percentage of an unsprayed control and it is clear that in some cases resistance is almost absolute (*Eleusine indica*), whereas in others it is not; in addition, the level of resistance may vary amongst populations (*Solanum nigrum*). In this chapter we will discuss the factors which govern the rate of evolution of resistance, our current limited knowledge about the dynamics of its development and the design of

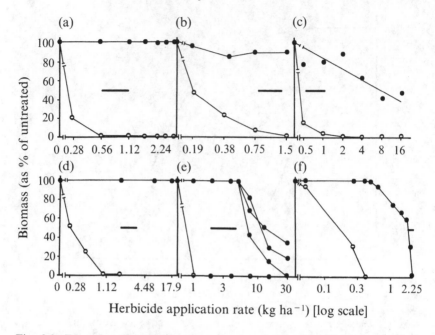

Fig. 8.3. Examples of herbicide dose response curves for non-selected (○ – susceptible) and selected (● – resistant) populations of weeds (redrawn after Gressel, 1986). Note different scales. (a) *Eleusine indica* – trifluralin; (b) *Lolium rigidum* – diclofop methyl; (c) *Erigeron philadelphicus* – paraquat; (d) *Senecio vulgaris* – atrazine; (e) *Solanum nigrum* (several populations) – atrazine; (f) *Amaranthus blitoides* – atrazine. Recommended application rate of the herbicide is shown as a horizontal bar; herbicide doses are on a logarithmic scale.

strategies of weed management to minimise the chance of resistance occurring.

Theoretical background

The development of herbicide resistance in a weed population is an example of evolution by natural selection, where the selective agent (the herbicide) is a specific component of weed management. By killing the more susceptible phenotypes[1] in a population, or by reducing their reproductive potential, and by having a lesser effect on the more tolerant phenotypes, herbicides have the potential to select for phenotypes which will persist when sprayed with herbicide. Over generations experiencing repeated herbicide application, genes conferring resistance in the phenotype will become more abundant in the population as resistant plants reproduce and susceptible plants are eliminated.

Typically, studies of herbicide resistance emanate from the practical concerns arising after the discovery of resistant populations. In consequence, the underlying factors determining the evolutionary process up to the point of discovery of a resistant phenotype have to be inferred retrospectively. It would therefore be useful to consider some population genetics as a baseline for discussing the dynamics of herbicide resistance.

There are two pre-requisites for the evolution of herbicide resistance in a weed population: the occurrence of heritable variation for resistance, and selection for increased resistance by herbicide application. Blackman (1950) pointed to the underlying similarities between the processes of mass selection of a character in crops and the selection of weeds after continuous herbicide treatment, and indeed Harper (1956) predicted the evolution of herbicide resistance some 12 years before its occurrence.

Starting from a zygote (seed), a typical diploid bisexual plant matures to produce gametes (pollen and ovules at flowering) and exchanges genes to varying extent with neighbouring plants to give zygotes in new gametic combinations. Consider two alleles (G and G') at a single locus and define p, the gene frequency of G, as a fraction of the total number of alleles at that locus. If the locus is autosomal, the frequency of GG homozygotes is a, of GG' heterozygotes is b and that of $G'G'$ is c, then at the zygotic stage, $p = a + b/2$. Then q, the corresponding frequency of the allele G' is $1 - p = c + b/2$. If we assume that the weed population is infinite in size, mating amongst plants is at random and there is no gene mutation, migration or selection, the three genotypes will be present one generation later and thereafter in the frequencies p^2, $2pq$ and q^2. This is known as the Hardy–Weinberg equilibrium (Crow & Kimura, 1970).

We can now follow the work of May & Dobson (1986), who have developed a general analysis of the evolution of pesticide resistance by considering the change in allele frequency over generations with repeated application of pesticide each generation. We begin by denoting the original susceptible allele as S and the resistant dominant allele as R and defining the genotype RR as resistant to a herbicide, the heterozygote RS as intermediate in performance, whilst genotype SS is susceptible. Let us assume that S and R alleles occur at frequencies s_t and r_t in generation t respectively ($s_t + r_t = 1$), and under a given dosage of pesticide, the fitnesses ('relative

---

[1] The *genotype* of an individual may be defined as the hereditary 'blueprint' of the individual, namely the complement of alleles (genes) present in the genome of the plant. This complement gives rise to the *phenotype*, the functional expression of gene products that is the living plant. Whilst plants surviving a normally lethal dose of herbicide will be of a resistant phenotype; they may or may not have the same alleles conferring the resistant trait. It is common to use the term *biotype* to describe a plant with a particular attribute (such as resistance) but where the genotype (e.g. homozygote or heterozygote) is not specified.

performance') of the three genotypes are w_{RR}, w_{RS} and w_{SS}, where $w_{RR} \geq w_{RS} \geq w_{SS}$.

The gene frequency of R in the next generation r_{t+1} will be

$$r_{t+1} = \frac{w_{RR}r_t^2 + w_{RS}r_t s_t}{w_{RR}r_t^2 + 2w_{RS}r_t s_t + w_{SS}s_t^2} \tag{8.1}$$

Before the application of recurrent selection by a herbicide, the frequency of the resistant allele r_t will be very low (e.g. 10^{-10}) and the frequency of the susceptible allele will be close to 1.0. The ratio r_t/s_t will be much smaller than both w_{RS}/w_{RR} and w_{SS}/w_{RS} and equation 8.1 may be simplified to

$$r_{t+1} = \left(\frac{w_{RS}}{w_{SS}}\right)r_t \tag{8.2}$$

By compounding equation 8.2 with an initial R frequency of r_0, we can estimate that after n generations, the frequency of the resistance allele, r_f, will be

$$r_f = r_0 \left(\frac{w_{RS}}{w_{SS}}\right)^n \tag{8.3}$$

If T_R is defined to be the absolute time taken for a significant degree of resistance ($r_f = 0.5$) to appear and T_g is the time taken for a generation of population growth, then $n = T_R/T_g$ and substituting into equation 8.3 we have an equation that provides the *approximate* time by which a specified resistance level r_f is reached. Thus

$$T_R = T_g \log_e\left(\frac{r_f}{r_0}\right) \Big/ \log_e\left(\frac{w_{RS}}{w_{SS}}\right) \tag{8.4}$$

In this approach, the evolution of resistance is considered to be an example of directional selection under recurrent application of pesticide, making the simple assumption that resistance is conferred by a single semi-dominant gene. As such, it leads to four general conclusions, namely that the rate of evolution depends on

1. the species generation time, T_g
2. the initial frequency of the resistance allele, r_0
3. the choice of threshold at which significant resistance is recognised, r_f
4. the strength of selection w_{RS}/w_{SS}, which is determined by the herbicide-dose applied and the degree of dominance of R.

As a general conclusion May & Dobson pointed out that even if r_0 varies in the range 10^{-5} to 10^{-16}, w_{RS}/w_{SS} in the range 10^{-1} to 10^{-4}, and T_g is 1 year (i.e. an annual weed), T_r will lie in the relatively narrow range of 10 to 100 years assuming recurrent selection.

Two components contribute to the selection 'pressure' exerted by herbicides – the *intensity* of selection and its *duration* (Maxwell & Mortimer, 1994). Selection intensity in response to herbicide application is a measure of the relative mortality exerted on a genotype and/or the relative reduction in the seed production of survivors and will be related, in some manner, to herbicide dose. Selection duration is a measure of the period of time over which phytotoxicity is imposed by a herbicide. Both intensity and duration will interact to give seasonal variation in the process of selection, which will in turn depend upon the phenology and growth of a weed species. For instance, with pre-emergence herbicide control of weeds that show germination over a protracted period, the intensity of selection may be much higher on weed seedlings recruited early in the life of a crop in comparison with seedlings emerging latterly. Formally, a coefficient of selection, s, may be defined most simply as the proportional reduction in the contribution of a particular genotype to future generations compared with a standard genotype (usually the most favoured), whose contribution is usually taken to be unity. Thus, if $s = 0.1$, for every 100 reproductively mature offspring produced by the most favoured genotype, only 90 are produced by the genotype against which selection acts. Since herbicides may cause both mortality and reduction in seed production of survivors, it is possible to calculate the coefficient of selection as $(1 - S_m/R_m \times S_f/R_f)$ where S_m and R_m are the respective numbers of resistant and susceptible genotypes surviving herbicide treatment and S_f/R_f is the ratio of their respective reproductive outputs. However, the selection 'pressure' exerted by herbicides has been defined in various ways in the literature, in part to suit the purposes of the particular author (see Table 8.1).

The importance of a bank of seeds in evolutionary terms is that it represents a 'memory' of past selection events which may buffer evolutionary processes (Templeton & Levins, 1979) and serve to delay the onset of resistance. May & Dobson (1986) modified equation 8.4 to incorporate both a seed bank and the fact that susceptible and resistant phenotypes may differ in their reproductive output. Their revision is shown in Table 8.1. In this form, their model is mathematically analogous to an alternative model, which was derived independently by Gressel & Segel (1978, 1990). Both models (Table 8.1) treat genotypic differences in a straightforward manner and selection is perceived to act against the susceptible phenotype in favour of the resistant ones.

Table 8.1. *Two models describing the rate of emergence of herbicide resistance*

Estimation of the approximate time (T_r) by which a specified resistance level r_f/r_o is reached within a weed population (May & Dobson, 1986)

$$T_r = T_g \log_e (r_f/r_o)/\log_e(1 + (w_{RS}/w_{SS})/(f_{RS}/f_{SS})(1/T_{soil})) \qquad (8.5)$$

where T_g is the species generation time (years);

r_o is the initial frequency of the resistance allele;

r_f is the choice of threshold at which resistance is recognised;

(w_{RS}/w_{SS}) is the strength of selection, where w_i denotes genotypic fitness (RS and RR, resistant; SS, susceptible);

$(f_{RS})/f_{SS})$ is the relative reproductive success of resistant and susceptible genotypes;

T_{soil} is the persistence of seeds in the seed bank (years).

Estimation of the proportion of resistant individuals in a weed population (Gressel & Segel, 1978; Gressel, 1991)

$$R_n = R_o \left(1 + \frac{F\alpha}{b}\right)^n \qquad (8.6)$$

where R_n is the fraction of resistant plants (or seeds) per unit area present in the population after n years of recurrent selection;

R_o is the initial frequency of resistant plants;

b is the average number of years a seed remains viable in the seed bank;

α is the ratio of resistant to susceptible individuals that survive after a generation of selection;

F is a factor that describes the relative fecundity of surviving plants.

Gressel & Segel (1978) use the term 'selection pressure' for α and $F\alpha$ is relative fitness $(1 - 1/F\alpha$ is the coefficient of selection defined earlier) over a whole generation of growth of the weed species.

So far, we have considered the evolution of resistance in general terms with respect to frequency of resistance alleles, selection and fitness. Table 8.2 lists the major biological factors that may interact in determining the evolution of herbicide resistance and at first sight any analysis of their relative importance may seem complex. For this reason, several researchers have turned to the use of simulation models as exploratory tools. In such an approach, difference equations (see Chapter 5) are used to describe the dynamics of competing populations of susceptible and resistant genotypes of annual species with discrete generations. Genotypes contribute alleles to a gene pool which are then recombined according to specific genetic models

Table 8.2. *Factors which determine the likelihood and speed of evolution of herbicide resistance*

- The number of alleles involved in the expression of functional resistance
- The frequency of resistance alleles in natural (unselected) populations of the weed
- The mode of inheritance of the resistance allele(s)
- The reproductive and breeding characteristics of the species
- The longevity of seeds in the soil
- The intensity of selection which differentiates resistant genotypes from susceptible ones
- The relative fitness of resistance and susceptible genotypes

---

that determine the genotypic structure in the seed pool of the succeeding generation. Fig. 8.4 illustrates the approach (Maxwell *et al.*, 1990), in which additionally the migration of genes both in time (from seeds in a persistent seed bank) and in space (by pollen flow from neighbouring areas) are incorporated (see below for a discussion of these).

With this type of model we can explore the generalised relationships between the key components in the evolution of herbicide resistance. Fig. 8.5 examines the relationship between the time (in generations) required for a significant (20%) level of resistance to be present in relation to selection in a *notional* weed species in which resistance is inherited as a single dominant nuclear encoded allele. Selection is assumed to act by killing plants at the seedling stage. These simulations demonstrate five key points:

1. If the resistance allele is rare (1 in 10^{16}) and selection pressures are low (*c.* 80% mortality) then at least 20 generations of recurrent selection are required before the evolution of resistance and this is only likely to arise in a species with a very high finite rate of increase ($\lambda = 1000$ fold) and no seed bank. In a species with a persistent seed bank and a much lower (more realistic) rate of increase (10 fold) then over 50 generations of selection causing 90% mortality are required.
2. Evolutionary rate will be proportional to finite rate of increase of the weed population, all other factors being equal.
3. In species with considerable seed longevity and persistent seed banks, the frequency of resistant/susceptible alleles in the seed bank will be different from the frequency in the growing plant population. The possession of a persistent seed bank by a species may therefore delay the appearance of herbicide resistance. This occurs because there is gene flow from the seed bank in the form of alleles for susceptibility derived

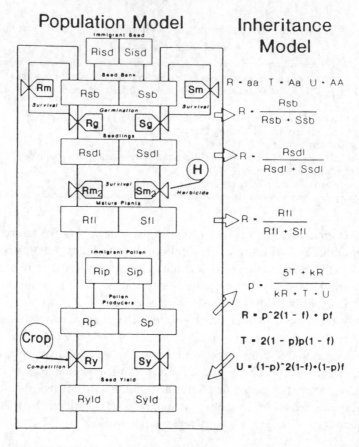

Fig. 8.4. Flow chart of a model of the dynamics of a population containing resistant and susceptible alleles (from Maxwell *et al.*, 1990). Open arrows indicate flow of information between a population sub-model and an inheritance sub-model. State variables are shown in each rectangular box; labels for processes are shown in italics.

from plants of past generations in which susceptible genotypes were more common.

4. The response to selection is slowest in species with low finite rates of increase and persistent seed banks.

5. If the resistance allele is relatively common (1 in 10^6), then the speed of evolution is substantially increased.

The conclusions drawn above are specific in that they relate to a dominant allele for resistance. If resistance is inherited maternally (as in the case of triazine resistance – see below) then the rate of evolution will only be

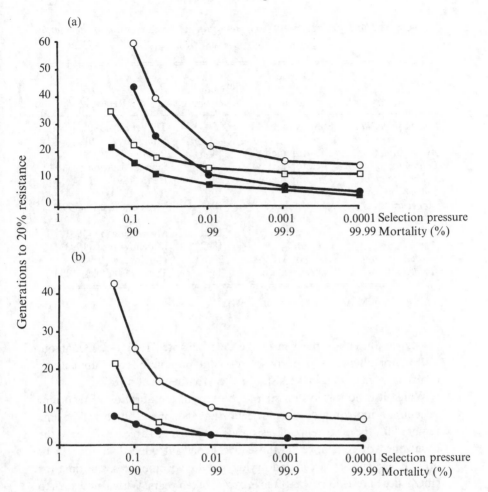

Fig. 8.5. The relationship between the rate of evolution and selection pressure for a hypothetical weed species with differing finite rates of increase and buried seed persistence (Mortimer *et al.*, 1992). Circles represent a low finite rate of increase, squares are for a higher rate of increase; solid symbols are for no persistent seed bank, open symbols are for a persistent seed bank. Data are derived by simulation with an initial allele frequency of (a) 1 in 10^{16} and (b) 1 in 10^{6} for herbicide resistance, inherited as a dominant nuclear encoded gene. In (b) ● and ■ are indistinguishable.

marginally faster than for a dominant allele (Macnair, 1981). Where resistance is inherited as a recessive allele then heterozygotes constitute susceptible phenotypes. For resistant homozygotes to increase significantly in the population the initial frequency of the allele in unselected populations is an important determinant of the likelihood of evolution, since

Table 8.3. *The estimated number of years for natural selection of herbicide resistance in some weed species*

Species	Chemical selection agent	Years for resistance to be recognised	Reference
Kochia scoparia	Sulfonylureas	3–5	Thill *et al.* (1990) McKinley (1990)
Avena fatua	Diclofop methyl	4–6	Piper (1990)
Lolium multiflorum	Diclofop methyl	7	Stanger & Appleby (1989)
Lolium rigidum	Diclofop methyl	4	Heap (1991)
Senecio vulgaris	Simazine	10	Ryan (1970)
Alopecurus myosuroides	Chlorotoluron	10	Moss & Cussans (1991)
Setaria viridis	Trifluralin	15	Morrison *et al.* (1991)
Avena fatua	Triallate	18–20	Malchow *et al.* (1993)
Carduus nutans	2,4-D or MCPA	20	Harrington (1990)
Hordeum leporinum	Paraquat/diquat	25	Tucker & Powles (1988)

Source: From Maxwell & Mortimer (1994).

resistance alleles are masked in the heterozygote (Taylor & Geoghiou, 1978). For all practical purposes resistant populations are unlikely to evolve if there is an initial resistance allele frequency of $< 10^{-6}$.

Whilst it is not always possible to be sure of the duration of recurrent selection by individual chemical selection agents, some estimates of the time required for the occurrence of resistance are given in Table 8.3. Prior to the occurrence of resistance to those herbicides introduced in the 1980s, notably the sulfonylureas and aryloxyphenoxypropionates, the time for emergence of resistance was often in excess of ten years. We will now turn to consider the genetics of resistance in more detail before reviewing some individual case studies so as to assess the relative importance of factors considered in Table 8.2.

The genetic basis of herbicide resistance: single and multigene systems

Any mechanism which interrupts the passage of a herbicide to its biochemical site(s) of action, reduces the sensitivity of the target site, detoxifies the chemical or enhances repair can potentially confer resistance. Specific known mechanisms include: sequestration of the herbicide in the apoplast; modification of cell membrane function and structure; change in the sensitivity of the key target enzyme; enhanced production of the herbicide target; enhanced metabolic breakdown and conjugation of the herbicide;

Table 8.4. *Inheritance of herbicide resistance in weed biotypes*

(a) Mendelian Inheritance in selected weed populations

Herbicide	Weed	Number of genes	Reference
Atrazine	*Abutilon theophrasti*	1 semi-dominant	Andersen & Gronwald (1987)
Chlorotoluron	*Alopecurus myosuroides*	2 additive	Chauvel (1991)
Diclofop	*Lolium multiflorum*	1 semi-dominant	Betts *et al.* (1992)
Fenoxoprop	*Avena sterilis*	1 semi-dominant	Barr *et al.* (1992)
Fluazifop	*Avena sterilis*	1 semi-dominant	Barr *et al.* (1992)
Haloxyfop	*Lolium rigidum*	1 semi-dominant	Tardif & Powles (1993)
Metsulfuron	*Lactuca serriola*	1 semi-dominant	Mallory-Smith *et al.* (1990)
Paraquat	*Arctotheca calendula*	1 semi-dominant	Purba *et al.* (1993)
Paraquat	*Conyza bonariensis*	1 dominant	Shaaltiel *et al.* (1988)
Paraquat	*Conyza philadelphicus*	1 dominant	Itoh & Miyahara (1984)
Paraquat	*Erigeron canadensis*	1 dominant	Yamasue *et al.* (1992)
Paraquat	*Hordeum glaucum*	1 semi-dominant	Islam & Powles (1988)
Paraquat	*Hordeum leporinum*	1 semi-dominant	Purba *et al.* (1993)
Trifluralin	*Setaria viridis*	1 recessive	Jasieniuk *et al.* (1993)

(b) Quantitative inheritance in weed and wild populations

Herbicide	Weed	Heritability	Reference
Barban	*Avena fatua*	0 to 0.63	Price *et al.* (1985)
Glyphosate	*Convolvulus arvensis*	Additive	Duncan & Weller (1987)
Simazine	*Senecio vulgaris*	0.22	Holliday & Putwain (1980)

Source: After Darmency (1994).

and enhanced degradation of herbicide generated toxic products (Dodge, 1992). It is therefore conceivable that many genetic loci may be involved in the expression of resistance, particularly if mechanisms result from reduced uptake or movement of herbicide. Conversely, resistance due to altered sensitivity of a target enzyme can arise as the result of a single allelic change. There is insufficient knowledge at present to appraise the underlying genetics of all of these mechanisms, but it is clear that both cytoplasmic and nuclear encoded single and multigene systems can be involved (Table 8.4).

Studies of triazine resistance have shown that in most cases resistance to these herbicides is determined by the chloroplast genome and hence maternally inherited, although other mechanisms including paternal plastid transmission through pollen may be involved in some cases (Darmency & Gasquez, 1981). An exception occurs in a biotype of *Abutilon theophrasti* in which atrazine resistance was found to be controlled by a single partially dominant allele (Andersen & Gronwald, 1987).

To date, nuclear encoded alleles for herbicide resistance that are inherited in a Mendelian manner have almost exclusively been found to be dominant or semi-dominant, the exception being recessive trifluralin resistance in *Setaria viridis* (Jasieniuk *et al.*, 1993). Semi or partial dominance of an allele results in resistance in the heterozygote being lower than in the resistant homozygote; the heterozygous phenotype may result from a single semi-dominant gene or a dominant gene whose effects are modified by pleiotropy or other factors. In many instances of herbicide resistance, genetic analyses have not been completed to identify fully the inheritance mechanisms and some apparent cases of semi-dominance may yet be shown to involve more than one gene, rather than simply gene dosage effects. Nevertheless, resistance acquired by altered target site of action has been shown to be controlled by a single allele in ACCase (acetyl coenzyme A carboxylase) and ALS (acetolactate synthase) inhibitor herbicides; respective examples are fluazifop resistance in *Avena sterilis* ssp. *ludoviciana* (Barr *et al.*, 1992) and metsulfuron resistance in *Lactuca serriola* (Mallory-Smith *et al.*, 1990).

Multigene or polygenic mechanisms of inheritance are suggested when plants in a population show continuous variation in response to the herbicide (e.g. Moss & Cussans, 1991, for *Alopecurus myosuroides* resistance to chlorotoluron) or in progeny of crosses amongst parent plants. Biometrical techniques may be used to quantify the component of this variation that contributes to evolutionary change and an increase in the mean population level of resistance (Lawrence, 1984). The *heritability* of a polygenic trait (which relates directly to additive genetic variation according to the breeding programme used) indicates, on a scale from 0 to 1, the potential for response to selection. Values in excess of 0.3 are considered to lead to significant response to selection (Falconer, 1981). Price *et al.* (1985) screened populations of *A. fatua* and *A. barbata* for tolerance to barban (4-chloro-2-butynyl *m*-chlorocarbanilate) and bromoxynil (3,5-dibromo-4-hydroxybenzonitrile) with sub-lethal doses of herbicide. Calculated 'broad sense' heritabilities from differing populations ranged from 0 to 0.64. Whilst these calculations do not precisely estimate heritable genetic variation, they lead us to two important conclusions. The first is that populations may differ in their genotypic structure and the alleles required for herbicide resistance may be absent. Hence, some populations will not respond to selection. Secondly, and conversely, other populations may contain varying amounts of genetic variation for resistance. Moreover, Price *et al.* concluded that the amount of genetic variation in response to herbicide was higher than to be expected on the basis of random mutation alone, since unrealistically high mutation rates of 1 in 10^2 would be required to account for the observed heritabilities.

The observation that variation in response to paraquat existed amongst commercial cultivars of perennial ryegrass, *Lolium perenne*, prompted Johnston & Faulkner (1991) to undertake a recurrent selection programme to evaluate herbicide resistance in this species for forage pasture use. A breeding programme confirmed that about 70% of the variation in response was due to additive genetic variation (Faulkner, 1974). Clones were selected from diverse sources, allowed to intercross by open pollination and seedling progenies sprayed with paraquat at a low application rate. Mass screening in this manner over successive generations increased the ED_{50} to paraquat of tolerant cultivars 20- to 30-fold in comparison with susceptible cultivars. This clearly illustrates that there was considerable natural variation within this species, which had not previously been exposed to selection by paraquat.

We can thus see that there are two ways in which the emergence of resistance may occur. Maternal inheritance or semi-dominant single alleles that confer target site resistance often lead to relatively high levels of resistance in the phenotype and resistant genotypes will be immediately selected. Alternatively, where inheritance is polygenic, the increase in the level of resistance (at the population level) under recurrent selection will be slower, since many loci are involved, and will also depend on the intensity of selection. High doses of herbicide may cause mortality of phenotypes intermediate in resistance, whereas lower doses select for enhanced resistance in the population through a gradual process of segregation and recombination. In either case the existence of alleles conferring resistance in previously unselected populations is a necessary precursor for evolution.

The probability that an allele for resistance will occur in a weed population that has hitherto been unexposed to selection by herbicides will depend on three factors: the mutation rate (μ), the size of the population (N) and the selective disadvantage of the allele in the absence of selection. Fig. 8.6 indicates the relationship between the size of population necessary to detect a resistant phenotype in a diploid plant and the mutation rate. As mutation rate declines, the number of individuals needing to be exposed to

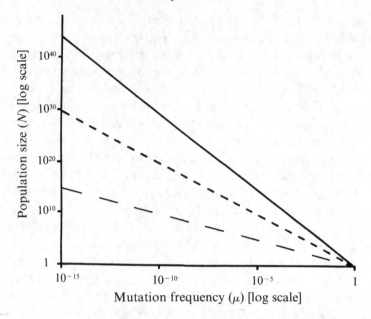

Fig. 8.6. Simple expectations of the size of population that must be examined in order to detect one resistant phenotype containing at least one dominant allele conferring resistance (after Maxwell & Mortimer, 1994). Estimates are calculated as $N = 1/[p^2 + (2pq)]^n$ where N is the sample size required to detect at least one genotype showing resistance, assuming dominant resistance, p is the natural frequency of a resistance allele, n is the number of loci and $q = (1-p)$. Mutation frequency is assumed to be the same for all loci and forward and backward mutation rates to be equal. ——— 3 loci; – – – – 2 loci; — — — 1 locus.

selection increases in a logarithmic manner. At its simplest these data suggest that at least $N > 1/\mu$ individuals will need to undergo selection before a mutant may emerge. In a non-selecting environment such mutants may be at a selective disadvantage and in consequence they may occur at even lower frequency. Consider the case of a weed species with a mutation rate of 10^{-7} at a single locus conferring resistance. If there were 10^6 individuals ha^{-1}, then we might expect 10 mutant plants in a 100 ha field. If the mutation rate is constant over generations and there is no selective disadvantage to the allele in the absence of herbicide, it will take 1000 generations for the frequency of mutants to achieve 1 m^{-2}. However, under herbicide selection the proportional increase may be very rapid.

Herbicide resistant weeds are only recognised as such by the practitioner in a retrospective manner, when there is failure of chemical control which cannot be explained by faults in application procedure or by weather. In consequence and not surprisingly there has been little opportunity to observe the process of selection in action. In one, perhaps unique, case it is known that there can be two steps in the evolutionary process. Gasquez (1991) and Darmency & Gasquez (1990) have investigated the occurrence of triazine resistance in *Chenopodium album* in France. This species has become one of the world's most widespread resistant weeds and occurs over at least 200000 ha in maize fields and vineyards in France. Darmency & Gasquez screened seeds from plants of *C. album*, from populations which had previously been unexposed to triazines, with a low dose of atrazine (500 g ha^{-1}). In the progeny of each plant from which seeds were collected a small (up to 12%) proportion of survivors was found. These survivors were shown (a) to be killed by normal (4.5 kg ha^{-1}) application rates (and termed intermediate '*I*' types), and (b) to be less fecund than both the susceptible phenotype and fully resistant (to 4.5 kg atrazine ha^{-1}) phenotypes (termed '*R*' types) that had appeared in maize fields. Gasquez *et al.* (1984) confirmed that the resistance of the *I* characteristic was maternally inherited and molecular analysis (Bettini *et al.*, 1987) showed that the same *psbA* gene in the chloroplast DNA was involved. A third and surprising finding was that the seed progeny of *I* types was fully resistant after exposure to a low dose of herbicide, surviving doses up to 40 kg atrazine ha^{-1} (Gasquez *et al.*, 1985). The underlying mechanism of this rather unusual case of resistance, which would not be predicted from any of the simple genetics described previously in this chapter, remains unknown. It serves to illustrate that the genetics of herbicide resistance is still poorly understood. An additional important finding from the study was that not all populations possessed plants that gave rise to the *I* phenotype.

The dynamics of selection

Using the same temporal and spatial themes developed elsewhere in this book, we can consider the dynamics of the evolution of resistance at two scales. On the one hand the dynamics can be viewed on a geographic scale as the emergence of new resistant weed populations, and on the other in terms of the rate of replacement of the susceptible phenotype by the resistant one within a single population. Each of these will be discussed in turn.

The emergence of resistant populations: evidence of independent evolutionary events in three species

On a regional/geographical scale resistance may be acquired by a population through (1) immigration of a resistance allele (gene flow) from a neighbouring population; or (2) as a consequence of independent evolutionary paths occurring in discrete populations. Distinguishing between these two as the causal process, on the discovery of a resistant population, is often difficult since it requires rigorous exclusion of the possibility of gene flow. Gene migration may occur naturally by seed movement or by pollen flow. Characteristically, the bulk of viable pollen from a point source is dispersed only over a short distance (*c.* 2 m), declining exponentially with distance in wind pollinated species (Levin & Kerster, 1984), although much further distances are possible in insect pollinated species. Conversely, and as discussed earlier, seed movement can potentially occur over considerable distances, particularly if animals or grain harvesting equipment move between farms (Porterfield, 1988)

Separate evolutionary paths may be inferred if (a) resistance emerges in a set of isolated locations either simultaneously or over a short time period; (b) differing levels of heritable variation are present in populations undergoing selection; and (c) if different mechanisms of heritable resistance are demonstrated between populations. Resistance to herbicides exhibited by three weed species supports all three aspects of this general hypothesis.

Resistance to sulfonylurea herbicides was first discovered in 1987 when failure of chlorsulfuron to control *Lactuca serriola* and *Kochia scoparia* in winter wheat was reported. Two years later resistance in *K. scoparia* had been found at locations in six different states of the USA (Colorado, Kansas, Montana, North Dakota, South Dakota and Texas) and in one province in Canada (Saskatchewan) (Thill *et al.*, 1991). Populations varied in the dose required for 90% growth reduction by chlorsulfuron when applied post-emergence – up to a 31-fold increase in dose being required in comparison with a susceptible standard. Resistance to chlorsulfuron in *L. serriola* is inherited as a dominant or semi-dominant nuclear gene that confers alteration in the target site of the ALS enzyme. It is likely that a similar pattern of inheritance occurs in *K. scoparia*, but this awaits confirmation.

All sites had a past history of recurrent chlorsulfuron use (15 g active ingredient ha^{-1} year^{-1}) in no-till continuous cropping, varying in duration from three to five years. Whilst *K. scoparia* may produce over 14 000 seeds per plant and as a tumbleweed can disperse seed at least 50 m from a source plant, evolution of resistance in such widely distant locations over a relatively short time span is strongly suggestive of separate evolutionary events. It also implies a relatively high frequency of resistance alleles in natural populations.

Resistance to the substituted urea group of herbicides, particularly chlorotoluron (3-(3-chloro-*p*-tolyl)-1,1-dimethylurea) and isoproturon (3-

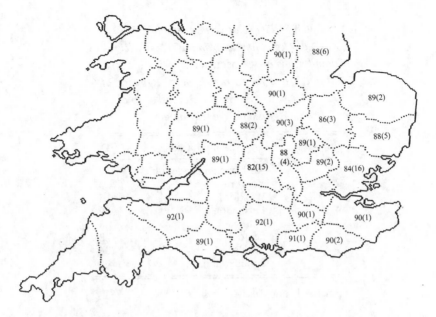

Fig. 8.7. Discovery of resistance of *Alopecurus myosuroides* to chlorotoluron in England (based on S. R. Moss, pers. commun.). The year of first record is shown in each county along with (in brackets) the number of confirmed cases up to 1992.

(4-isopropylphenyl)-1,1-dimethylurea) has emerged in *Alopecurus myosuroides* in the 1980s in the UK (Moss & Orson 1988), in Germany (Niemann & Pestemer, 1984) and in Spain (Jorrin *et al.*, 1992). Cases of resistant *A. myosuroides* have been associated mostly with continuous winter cereal cropping and recurrent use of the same herbicides (on average 1.6 applications per year for at least 10 years) and no-plough cultivation systems (Moss & Cussans, 1991). Fig. 8.7 shows the recorded pattern of discovery of resistant populations since the first confirmed case in 1982. Initially, a relatively low level of resistance was detected, but in 1984 a population was discovered that had an ED_{50} 16 times greater than that of a known susceptible population. Since then, populations resistant to substituted ureas have been confirmed on 71 farms in 22 counties in the UK (Moss & Clarke, 1994). The degree of resistance varies noticeably amongst populations and it is thought that the main mechanism of resistance is due to an enhanced ability of resistant plants to metabolise and detoxify substituted ureas (Caseley *et al.*, 1990), although target site resistance may also be present in some populations (L. M. Hall, pers. commun.).

As a grass, *A. myosuroides* is able to disperse by natural means only over a

Table 8.5. *The relative performance (* ED_{80} *)*
values (kg chlorotoluron ha $^{-1}$ *) of selected*
populations of Alopecurus myosuroides *from*
the UK and the former Federal Republic of
Germany. Plants were assessed in a
glasshouse environment

Population	Sampling site	ED_{80}
1	Rothamsted (susceptible)	0.2
2	Yoxford	0.3
3	Wickford	0.4
4	Faringdon	0.9
5	Veltheim	1.6
6	Peldon	3.2
7	Schulp	3.6
8	Sommerland	4.8

Source: From Moss & Cussans (1985).

very limited area. Seed dehisces before harvest and is therefore unlikely to be carried between farms as a contaminant of grain. Three other observations suggest that evolution of resistance to substituted ureas is occurring independently at different locations in *A. myosuroides*, the last of which we shall consider in conjunction with *Lolium rigidum*.

The first is that a comparative examination of levels of resistance in different populations has shown that resistance varies quantitatively both among and within populations. Table 8.5 shows a 24-fold range in mean chlorotoluron dose required to reduce fresh weight of plant populations by 80% in selected populations. Of equal interest is the level of variation within populations. Ulf-Hansen (1989) examined within-population variation by raising the progeny of individual plants in nutrient solution containing chlorotoluron at a concentration equivalent to the ED_{50} dose for a susceptible population. Root growth was assessed in plants at the three tiller stage (Fig. 8.8). It is noticeable that there was considerable intra-population variation especially in the population (*R*1) which showed the greatest level of resistance. Whilst all three populations included individuals as sensitive as some plants in the susceptible reference population (*S*), the populations previously exposed to chlorotoluron (*R*1 and *R*2) included segregants that exceeded the upper level of resistance in *S*.

The inference from these data is that populations are pursuing separate evolutionary paths for any of the reasons outlined in Table 8.2, such as differing past coefficients of selection or initial frequency of resistance

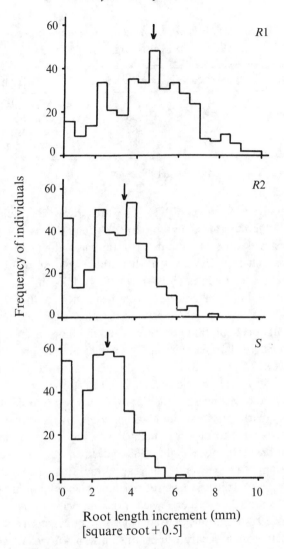

Fig. 8.8. Frequency distributions of root lengths of plants of *Alopecurus myosuroides* grown from tillers in nutrient solution containing chlorotoluron (redrawn from Ulf-Hansen, 1989). Seeds were collected from Peldon (*R*1), Brickhouse (*R*2) and Rothamsted (*S*). Arrows indicate the means.

alleles. Genetical studies provide a second inference. In *A. myosuroides*, open crossing of resistant and susceptible plants has yielded progeny segregating quantitative variation in resistance (Moss & Cussans, 1991). Thus, resistance by metabolism and detoxification processes may be argued to be inherited polygenically, although a relatively limited number of loci

Table 8.6. *History of herbicide use and the development of resistance in a*
population of Lolium rigidum *(SLR 31)*

Herbicide applied	First used	Resistance evident	Resistant to
Trifluralin	1970	1977	Trifluralin
Diclofop-methyl	1977	1980	Aryloxyphenoxypropionates
Chlorsulfuron	1983	1984	Sulfonylureas/imidazolinones
Sethoxydim	1986	1988	Cyclohexanediones

Source: From Powles & Matthews (1992).

may be involved (Chauvel, 1991). In which case, if populations have evolved at different rates, the amount of genetic variation retained by populations will differ in relation to the past intensity of selection. Populations that have low levels of heritable variation will be unable to respond further to selection (Bradshaw, 1984). Ulf-Hansen (1989) calculated the heritability of chlorotoluron resistance in three populations varying in relative resistance. Seeds were collected in the field from 25 mature plants; seedling families were raised in chlorotoluron hydroponic culture as described above. Heritability values of root growth increment varied between 0.21 and 0.34 amongst populations, indicating that populations may differ in their capacity to respond to further selection.

The third and strongest evidence to date for multiple independent evolutionary paths in weed populations comes from studies of *Lolium rigidum*. Resistance to diclofop-methyl in *L. rigidum* was discovered in Australia in 1981 (Heap & Knight, 1982) on a farm where the herbicide had been applied previously over a four year period. In a subsequent intensive survey (Heap, 1991) the frequency of resistance in populations occurring on farms throughout the southern wheat growing belt was assessed. Of 179 populations, 39 populations were considered to be resistant and came from geographically widely separate areas of southern Australia.

Table 8.6 shows the history of the development of resistance in one population which has been extensively analysed. This population has been found to exhibit multiple resistance, since more than one resistance mechanism is present, endowing the ability to withstand herbicides from different chemical classes.

Resistance to diclofop-methyl can be due to a slightly enhanced ability to metabolise the herbicide, but also because cell membranes have polarisation properties that allow recovery of function after herbicide exposure (Holtum & Powles, 1991; Häusler *et al.*, 1991). ACC-ase and ALS target

Table 8.7. *Herbicides to which cross-resistance has been found within a single population of diclofop-methyl resistant* Lolium rigidum. *Chemical groups are shown in italics*

Cross-resistance	Moderate cross-resistance	No cross-resistance
Aryloxyphenoxypropionate	*Substituted ureas*	*Carbamates*
Fluazifop-butyl	Isoproturon	Carbetamide
Haloxyfop-methyl		Asulam
Chlorazifop-propynil	*Miscellaneous*	Propham
Quizalofop-ethyl	Propyzamide	
Propaquizafop		*P-nitro diphenyl-ether*
		Oxyfluorfen
Cyclohexanedione		
Alloxydim-sodium		*Triazine*
Sethoxydim		Simazine
Sulfonylurea		*Bipyridyl*
Chlorsulfuron		Paraquat
Metsulfuron-methyl		
Triasulfuron		*N-phospho-methyl glycine ester*
		Glyphosate
Dinitroaniline		
Trifluralin		
Triazinone		
Metribuzin		

Source: From Heap (1991).

site insensitivity are also present, in addition to the ability to detoxify sulfonylurea herbicides (ALS inhibitors) by metabolic means (Christopher *et al.*, 1992). Screening of *L. rigidum* collected from different locations has shown that populations do not display consistent patterns of multiple resistance and that at least four mechanisms can be found at different frequencies within populations (Hall *et al.*, 1994).

Of equal significance to the phenomenon of multiple herbicide resistance is the occurrence of 'cross-resistance'. This is defined as resistance which has evolved to herbicides chemically unrelated to the original selecting herbicide and to which the population has not been previously exposed. For example, Gill (1993) found that 42% of *L. rigidum* populations resistant to diclofop-methyl were cross-resistant to triasulfuron. The pattern of cross-resistance amongst populations may be highly complex (Table 8.7), including resistance to several unrelated herbicide groups (Heap & Knight, 1990), but sometimes excluding some chemicals from within those groups (Kemp *et al.*, 1990). In Heap's survey described earlier, 11 of the diclofop-methyl resistant populations were assessed for cross-resistance to other herbicides and whilst they all exhibited some levels of resistance, there was noticeable variation in the patterns of cross-resistance. This variation could not be explained by the past history of herbicide usage at the farms from which populations were isolated. In a similar vein, complex patterns of cross-resistance have also been recorded in *Alopecurus myosuroides* (Moss, 1990b) and one population has been shown to exhibit cross-resistance to 18 different herbicides.

The genetical mechanisms underlying the phenomena of multiple- and cross-resistance in both *L. rigidum* and *A. myosuroides* have yet to be investigated. Whilst it may be that a single locus may be shown to confer resistance to a particular chemical class and give rise to a component of cross-resistance, the existence of multiple-resistance and the varied patterns of cross-resistance point strongest to different evolutionary events occurring within populations.

The evolution of resistance within a population

Earlier, we summarised the factors that determine the rate of evolution within a population, pointing out the broad differences to be expected as a result of differing modes of inheritance and in relation to relative fitness and selection. The process of replacement of susceptible phenotypes by resistant ones under recurrent selection has been called 'enrichment of resistance' by Gressel & Segel (1978). They used equation 8.6 (Table 8.1) to predict the number of seasons of repeated application required for the emergence of a significant proportion of resistant phenotypes. Fig. 8.9 illustrates some representative conclusions emphasizing the points made earlier, namely that four factors – persistence of individuals in the seed bank, the initial gene frequency, the relative fitness of resistant and susceptible phenotypes, and the intensity of selection – interact to determine the rate of evolution of resistance. Although these relationships have been known for some time and indeed more sophisticated modelling approaches have been applied (Maxwell *et al.*, 1990), the predictions remain largely untested. Of particular practical interest to the management of resistance is the persistence of susceptible phenotypes within the population, since their presence will allow for the opportunity for 'back-selection' towards susceptibility should resistance to a particular chemical group arise. In this area, however, there remains a considerable paucity of experimental data.

Before continuing further, it is necessary to define the term fitness more

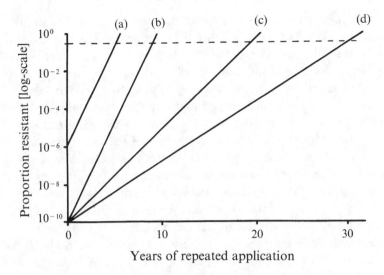

Fig. 8.9. Predicted rates of evolution of herbicide resistance in relation to the parameters of the Gressel & Segel (1978) model (see Table 8.1 for details): (a) and (b) $F=1$, $\alpha=0.1$, $b=1$ (or $\alpha=0.01$, $b=5$); (c) $F=1$, $\alpha=0.5$, $b=1$ (or $\alpha=0.1$, $b=5$); (d) $F=0.5$, $\alpha=0.5$, $b=1$ (or $\alpha=0.1$, $b=5$). The dashed line shows 30% resistance, a level which would be observable in the field.

rigorously. The fitness of a gene may be defined in terms of its effect on the survival and reproduction of the genotype in a given environment. Thus, the relative fitness of two genotypes differing by a single allele can be calculated as the ratio of their contributions in terms of progeny to future generations. In turn, their individual contributions will be the product of the probability of survival to reproduction and the number of offspring produced. If we are to talk correctly about relative fitness in susceptible and resistant weed populations then implicitly it is assumed that genotypes differ only with respect to the alleles conferring that resistance and in all other respects the genotypes are isogenic. Clearly, any measure of relative fitness is dependent upon the environment in which genotypes are compared.

It should be noted, however, that often in published studies the fitness of one or more resistant populations is compared with a single susceptible population from a different location. Firstly, it must be pointed out that within a 'resistant' population there will usually be more than one genotype present. Even in a population in which resistance has been prevalent for some time, there may be a proportion of heterozygotes and susceptible homozygotes. The average fitness of the population will therefore include all genotypes and may not give a true estimate of the 'cost' of a resistant allele to a plant. Secondly, it must be remembered that populations of weeds can differ considerably in their growth characteristics. A single susceptible population may not therefore be typical of the species as a whole. To determine the fitness differential conferred by resistance, we need to be able to assume that the resistant population was identical to the susceptible one before evolution of resistance occurred. Since it is not usually possible to make separate collections of resistant and susceptible individuals from within a population, the only option is to ensure that a range of susceptible populations is used against which to compare the resistant population. Only if the resistant population has a fitness measure outside of the range of the susceptible populations can we be reasonably sure that the difference is due to the resistance trait.

Caution is also required with the experimental conditions in which fitness is measured. We noted above that growing conditions affect the fitness of a plant. The growing conditions in a pot, which may contain an artificial soil and may be placed in a glasshouse, will be very different from a field. The presence of a crop will also affect the growth of a weed (see Chapter 4): for example, a slight difference in performance between resistant and susceptible plants may become magnified when they are grown with a crop. It is therefore very important to conduct fitness experiments in the environment in which the plants would normally occur. Rather than estimate fitness at a single weed density, it would be more relevant to grow biotypes at a range of densities and frequencies and to describe their relative seed production response curves (see p.95).

We will return later to consider the extent of our knowledge of the individual factors which affect the rate of evolution of resistant populations. Before doing so, we will describe a case study of the population dynamics of herbicide resistance. Almost all of our current knowledge of the dynamics of herbicide resistance is inferred, either from biological parameters possessed by the resistant phenotypes or from mathematical models. One of the few direct studies of selection for resistance is for triazine resistant *Senecio vulgaris*.

A case study: triazine resistant Senecio vulgaris

The triazine group of herbicides came into intensive world-wide use in the 1960s and by the mid 1980s triazine resistant weed species were common in many countries, particularly in North America and in western and eastern Europe. Multiple- and cross-resistance to other herbicides is also evident in some populations (Fuerst *et al.*, 1986). For most weeds, triazine resistance is inherited maternally through plastid genes (van Oorschot, 1991).

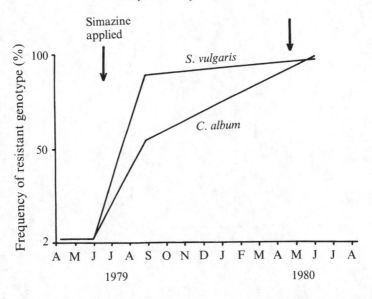

Fig. 8.10. Change in the proportion of resistant genotypes in populations of *Senecio vulgaris* and *Chenopodium album* (Bradshaw, 1984). Each population was initially seeded with a mixture containing 2% of a resistant genotype. The horizontal arrows show the approximate dates of simazine application at a rate of 2 kg ha^{-1}. This herbicide has residual soil activity and can last for several months.

The evolution of resistance within a population may be very rapid. Fig. 8.10 shows the overall change in the frequency of resistant genotypes in experimentally managed populations of *Senecio vulgaris* and *Chenopodium album*. Populations of each species containing 2% of a resistant genotype were introduced (as achenes) into blackcurrant plantations in spring. Simazine was applied at 2 kg ha^{-1} when plants were established (in June), and in April the following year. The frequency of resistant genotypes in the progeny of survivors in the following September exceeded 50%; after a further generation, in the following year, the majority of the population comprised the resistant genotype. The small fraction of susceptible genotypes remaining after two cycles of selection had probably escaped the effects of the herbicide.

The combination of maternal inheritance and recurrent selection may result in rapid increase of resistance within a population once the resistance gene is present, but the speed of response will depend upon the relative fitness of genotypes and the generation time and phenology of the weed. Putwain and co-workers examined the dynamics of *S. vulgaris* genotypes on a monthly basis, enabling analysis of the survivorship and fecundity of plants classified into cohorts. Genotypic identification of progeny was

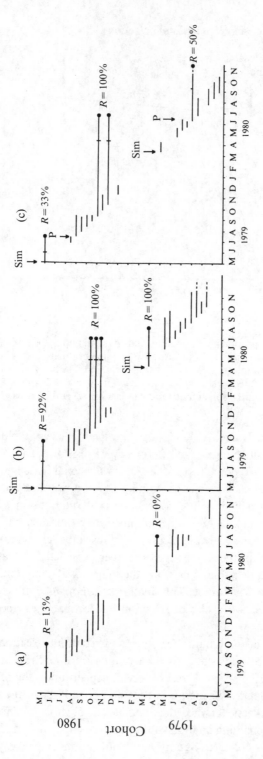

Fig. 8.11. Fates of cohorts within mixed populations of simazine resistant (R) and susceptible (S) genotypes of *Senecio vulgaris* (adapted from Mortimer, 1983): (a) no herbicides applied; (b) simazine (Sim) applied at $2.24\,kg\,ha^{-1}$; (c) simazine at $2.24\,kg\,ha^{-1}$, plus paraquat (P) spot treatment. Application dates are indicated by arrows. Each horizontal line shows the lifespan of a cohort. A vertical mark indicates the onset of seed production and ● indicates the death of the last flowering plant in the cohort. R: indicates the percentage of the R genotype in the progeny.

straightforward, since maternal inheritance was associated with a morphological marker (Scott & Putwain, 1981) which allowed the screening of samples of progeny for resistance. Experimental material was originally chosen from a common location in which resistance had emerged but where both genotypes were present. Typically, the life expectancy of those *S. vulgaris* plants that survived to flower was of the order of four months.

In an initial study, plots in a soft fruit plantation were either sprayed in spring (April/May) with a single application of simazine or left unsprayed. The herbicide has residual activity and can kill susceptible individuals for several weeks after application. Resistant (R) genotypes were able to survive and to reproduce, plants occurring as both summer annuals and winter annuals. As the spring application of simazine declined in efficacy, susceptible plants were able to establish in the treated plots by the following August and to overwinter to produce seed in spring of the following year. Thus, the susceptible (S) genotype behaved as a winter annual in the treated plots (Putwain *et al.*, 1982). Temporal escape from selection by the S genotype was possible by a switch in seasonal phenology, illustrating one means by which simazine sensitive individuals may persist on treated areas.

In a subsequent study of the effects of differing management practices (Mortimer, 1983), it was found in unsprayed plots that although cohorts of plants were recruited throughout the period May to December, it was only those individuals that emerged early in the year (April/May) that survived to disperse progeny (Fig. 8.11a). From an initial sowing in open ground, more achenes were produced by susceptible individuals than resistant ones, but in the following year, during which natural regeneration of a weed flora occurred, only susceptible genotypes successfully produced seeds. In contrast, in the spring simazine treated plots, the summer and winter annual phenology of R genotypes was again apparent (with partially overlapping generations), but in this study the S individuals failed to set seed in the following year (May/June 1980) (Fig. 8.11b).

In a third management regime, simazine was applied as before, together with spot applications of paraquat to control established *S. vulgaris* populations. Plants in those cohorts arising after spot treatment experienced alternative fates. In 1980 (Fig. 8.11c), seedlings in cohorts emerging in May and June, predominantly R genotypes, were killed by paraquat application at the beginning of July and the subsequent cohort dispersed seed in equal genotypic proportions.

From these field observations over just two seasons of study, it is difficult to infer unequivocally the individual components of population regulation. They do serve, however, to raise a number of points of ecological relevance in relation to the dynamics of herbicide resistance. The first is that whilst there was continuing recruitment to the adult plant population in all plots, it was only specific cohorts from which seeds were dispersed. Fitness (*sensu* reproductive success) was cohort (time) specific. Secondly, comparing the performance of *S* genotypes between the two studies in simazine treated plots, it is clear that winter climatic hazards varied between years and that the forced switch to a winter annual phenology by this genotype did not necessarily ensure reproductive success. Conversely, the fitness of the *R* genotype was zero in unsprayed plots which developed a natural weed flora in which the *S* genotype reproduced achenes. This third point suggests that there may be significant innate differences in survivorship and fecundity between genotypes. A fourth point, however, is that a chemical management regime that does not discriminate between the genotypes may allow both to reproduce successfully.

Factors determining the rate of evolution of resistance

Having examined closely the complexities which may be involved in the dynamics of a single case of resistance, we now need to consider more generally the extent of our knowledge of the factors governing the rate of evolution. We will discuss, in turn, our current knowledge of initial resistance gene frequency, seed bank persistence, relative fitness and selection intensity.

Throughout the development of herbicide resistance models, the gene frequency in unselected populations (r_0) has had to be assumed. Gressel (1991), for example, speculated that for triazine resistance the proportion in a population may be between 10^{-20} and 10^{-10}. In earlier sections we discussed the implications of whether the frequency was 10^{-16} or 10^{-6}. We would expect that the larger the number of genes required for resistance, the lower will be the initial frequency of resistant plants or seeds. In some populations and for some herbicides r_0 may be zero and the possibility of evolution of resistance then depends on the frequency of mutation events – again, commonly assumed to be 10^{-6} or less for a single locus (e.g. Maxwell *et al.*, 1990).

In fact, we have very few experimentally determined estimates of r_0, largely because of the difficulties and the scale of effort involved. Very large (known) numbers of plants need to be screened, perhaps as many as 10^7 individuals to determine accurately a frequency in the region of 10^{-6} (and 10^{11} individuals if r_0 is of the order of 10^{-10}). While this may be possible for a crop species (e.g. Mackenzie *et al.*, 1993) the difficulties posed by dormancy in many weeds poses considerable technical problems.

In the studies of the *I* biotype for triazine resistance in *Chenopodium album*, Gasquez *et al.* (1985) detected levels of 10^{-4} to 10^{-3} resistant phenotypes in garden populations. Matthews & Powles (1992) reported values of 2×10^{-2} and 0 respectively in farm and non-farm populations of *Lolium rigidum* not previously exposed to herbicides. In both of these studies the frequencies were surprisingly high. Perhaps by chance the researchers had chosen extreme populations, or perhaps again by chance they detected a few resistant individuals in a fairly small sample. An alternative explanation would be that the same mutation which confers herbicide resistance had already become established in the populations because of other benefits to plants in the absence of herbicides. In another study, however, Heap (1988) found 47 survivors from 10^6 seedlings of *L. rigidum* treated with diclofop methyl. For the present, it would seem from the paucity of data that we must continue to make assumptions about the value of r_o.

Seed persistence in the soil varies considerably amongst species which have developed resistance (see also Chapter 4). *Alopecurus myosuroides* and *Avena* spp., both of which have evolved resistance, have half-lives in the region of 6 months and the majority of seeds from a year's production will have disappeared within 2–4 years (e.g. Moss, 1985). The buffering effect of a susceptible seed bank will therefore diminish rapidly. *Senecio vulgaris* seeds may survive somewhat longer: Roberts & Feast (1972) recorded 13% remaining in undisturbed soil after six years. *Setaria viridis* and *Amaranthus retroflexus* can survive in the soil for at least ten years (Burnside *et al.*, 1981) and *Chenopodium album* can last more than 20 years (Lewis, 1973), although their half-lives may be closer to three to four years.

Amongst those species which have evolved resistance to sulfonylurea herbicides in only three to five years, *Kochia scoparia*, *Salsola iberica*, *Lactuca serriola* and *Lolium rigidum* all have seed bank half-lives of between a few months and 1.5 years (Thill *et al.*, 1991; Gramshaw, 1972); however, *Stellaria media* seeds can be extremely long lived. The data therefore do not support a contention that species with longer lived seeds necessarily take longer to evolve resistance. Presumably, other factors have a greater influence.

Although the seed bank longevity of resistant and susceptible pheno-types has not often been compared, there is some evidence that differences can occur. Watson *et al.* (1987) examined the survival of *Senecio vulgaris* achenes buried at differing depths in the soil. Achenes were buried in mesh packets and destructively harvested at monthly intervals over the following 24 months. Viability of recovered dormant seeds was then assessed by a

Table 8.8. *The survival of buried achene populations of* Senecio vulgaris *in a blackcurrant plantation under differing management regimes (see text for details). Data are mean rates of decline expressed as a half-life in days for simazine resistant and susceptible biotypes; in parentheses are the percentage losses per annum*

| | Depth of achene burial (cm) | | | |
| | 1 | | 7 | |
	Resistant	Susceptible	Resistant	Susceptible
Soil rotovation	329 (53.7)	454 (42.7)	884 (24.9)	696 (30.5)
Spring simazine	245 (64.4)	273 (60.4)	986 (22.6)	527 (38.2)
Control	501 (39.6)	433 (44.2)	1633 (14.4)	1189 (19.2)

Source: From Watson *et al.* (1987).

combination of germination and chemical viability tests. Seeds placed in undisturbed soil declined at a slower rate at 7 cm depth than at 1 cm (Table 8.8) and the rate of loss of the resistant genotype was significantly less than that of the susceptible. This difference was also evident at depth in simazine treated plots and in plots undergoing soil rotovation (in the latter treatment seed packets were removed from the soil before treatment application in darkness and returned under similar conditions). Conversely, the decline of achenes near the surface was greater in the R genotype where permanent vegetation cover did not persist (soil rotovation and spring simazine treatments). Although the exact mechanisms governing the differential decline are unknown, these findings suggest that subtle differences existed between the genotypes and that a management programme involving deep burial of achenes will cause more rapid depletion of the susceptible genotype than of the resistant. If cultivation subsequently returned resistant achenes to the surface, the use of triazine herbicides would lead to accelerated evolution of resistance.

Knowledge of the relative fitness of resistant and susceptible genotypes is important since lowered fitness of the resistant biotype in the absence of selection by herbicides may allow alternative control strategies to exploit this weakness. Differences in 'fitness' between susceptible and resistant biotypes in unselecting (herbicide free) environments have been reported for several weed species in respect of triazine resistance (e.g. Weaver & Warwick, 1983). In many cases there appears to be a 'cost' associated with the resistant trait such that growth and vigour of the S genotype exceeds

Table 8.9. *A comparison of life history parameters of triazine susceptible and resistant genotypes of* Senecio vulgaris. *Sample sizes were all in excess of 400 plants per genotype*

	Susceptible	Resistant
Time to first inflorescence bud production (weeks ± SE)	6.42 ± 0.05	6.73 ± 0.07
Time to first capitula production (weeks ± SE)	10.30 ± 0.05	10.55 ± 0.06
Mean life expectancy (weeks ± SE)	16.61 ± 0.3	16.52 ± 0.19
Mean number of seeds produced per surviving plant	3074	2198
Intrinsic rate of increase (per year)	0.0218	0.0208

Sources: From Watson (1987). Intrinsic rate of increase was calculated following Leverich & Levin (1979).

that of the *R* genotype. Gressel (1991) states that when in competition with a 'wild type' (susceptible) genotype, the fitness of triazine-resistant individuals may be 10–50% of that of the wild type. It is usually argued that a less efficient electron transport system, arising from alteration of the herbicide binding site on the thylakoid membrane in the acquisition of resistance, lowers photosynthetic potential and thus reduces both vigour and overall ecological fitness (see Holt & Thill, 1994, for a review).

In many cases, experiments to determine fitness differentials between genotypes simply measure either biomass or seed number at a single final harvest. It is unusual for the developmental patterns leading to those differences to be studied (although some researchers have examined differences in rates of germination and emergence – e.g. Dyer *et al.*, 1993). Watson (1987) contrasted the *S* and *R* genotypes of *Senecio vulgaris* discussed earlier, by raising plants individually in pots in the glasshouse and measuring developmental rates, life expectancy and age-specific seed production. As Table 8.9 and Fig. 8.12 illustrate, there were both subtle and major differences between the genotypes that culminated in different rates of increase. Susceptible plants were faster to achieve capitulum production, displayed a subtly different pattern of survivorship but similar mean life expectancy to resistant plants and a differing age-specific seed production schedule. These differences in life history traits, contributing to an overall fitness difference, were subsequently confirmed in comparisons of competing genotypes.

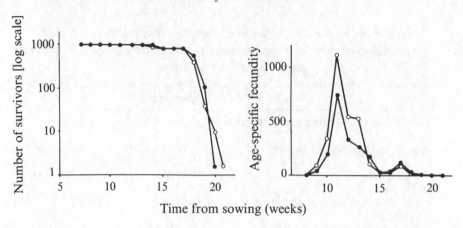

Fig. 8.12. Survival (a) and age-specific seed production (b) of simazine-resistant (●) and susceptible (○) genotypes of *Senecio vulgaris*, grown in a glasshouse (Watson, 1987). Age-specific seed production is defined as the number of seeds produced within a time period by the surviving plants. Both have been standardised by assuming that the population initially consisted of 1000 seedlings.

Reductions in fitness are not universal, even amongst triazine resistant weeds: certain biotypes of *Chenopodium album* and *Phalaris paradoxa* are exceptions where fitness appears unaffected (Rubin, 1991). In several recent reports of resistance to other herbicides, fitness of resistant populations does not appear to be reduced (e.g. Moss & Cussans, 1991). Other studies have been equivocal. In one case, a sulfonylurea resistant biotype of *Lactuca serriola* grew more rapidly in monoculture than a susceptible biotype, but was equally competitive when they were grown in mixture (Mallory-Smith *et al.*, 1992). In the same study, collections of a sulfony-lurea resistant and a susceptible biotype of *Kochia scoparia* from Kansas grew at similar rates in monoculture but the resistant biotype was more competitive in mixture; a resistant biotype from North Dakota grew faster than a susceptible biotype from the same state. Three factors contribute to the difficulty in interpretation of these results: there is no assurance that the resistant and susceptible biotypes did not differ in growth even before resistance evolved; relativities may change depending on the experimental design and presence of competitors; and biomass accumulation is only one component of fitness and may not reflect the true (overall) fitness differential.

Large reductions in fitness may, in fact, turn out to be unusual and may be confined to a few, very specific types of herbicide such as triazines. At

Table 8.10. *A comparative evaluation of selection coefficients on the performance of herbicide resistant and susceptible biotypes of* Alopecurus myosuroides. *Coefficients of selection are calculated as (1–[score of given biotype / score of best-performing biotype]). Scores were measured as either proportional survival of adult plants or seed production m $^{-2}$ per generation. The direction of selection is indicated by sign: negative values indicate selection against the susceptible biotype. Statistical comparison of biotype performance was based on G-tests of survival proportions or t-tests on seed numbers. Statistically significant values are indicated by: *, $P \leq 0.05$; **, $P \leq 0.01$; all other values are non-significant, $P > 0.05$*

Herbicide		Sowing density (seeds m $^{-2}$)			
		≤ 10	30	100	300
Control	Plant survival	0.05	0.00	0.01	0.01
(No herbicide)	Seed production	0.15	0.52	0.38	−0.36
Half rate	Plant survival	−0.47**	−0.66**	−0.56**	−0.58**
Chlorotoluron	Seed production	−0.97**	−0.86**	−0.89*	−0.94*
Full rate	Plant survival	−0.52**	−0.89**	−0.82**	−0.83**
Chlorotoluron	Seed production	−0.82	−0.96*	−0.98**	−0.98**
Full rate	Plant survival	−1.00**	−0.92**	−0.78**	−0.91**
Isoproturon	Seed production	−1.00	−0.99	−0.87	−0.96

Source: From Ulf-Hausen (1989).

present, however, there are too few studies to draw firm conclusions. More studies are clearly required, particularly under field conditions and in which a crop is present. For modelling the dynamics of resistance caused by a semi-dominant gene, estimates of the fitness of both homozygotes and heterozygotes with and without herbicide are needed. It is only in rigorous experimentation varying the density and frequency of genotypes, ideally under differing selection intensities (s, see p.249), that fitness differentials can be measured precisely.

Few attempts have been made experimentally to quantify selection intensities in the field. Table 8.10 gives a summary of estimated selection intensities imposed by chlorotoluron and isoproturon on *Alopecurus myosuroides* growing in field plots in winter wheat. Selection coefficients are based upon both measurements of seed production per unit area over a single season and mortality. This study illustrates three important features:

1. that selection coefficients measured over the whole growing season

incorporating both mortality and fecundity are generally higher than those based solely on mortality;

2. the intensity of selection against susceptible biotypes of *A. myosuroides* under herbicide application varied in the range 0.82–1.00 and there were no consistent responses to density of weed infestation;

3. selection against the resistant biotype in the absence of herbicide was negligible (0.0–0.03) and of a much lower magnitude than that previously recorded for triazine resistant biotypes.

The management of herbicide resistance

The design of management strategies specifically to delay and/or prevent the occurrence of herbicide resistance has been discussed by a number of authors (e.g. Gressel & Segel, 1990; Powles & Howat, 1990; Roush *et al.*, 1990). The varying options that have been proposed have a basis resting on only two biological processes – alteration of selection pressure and/or back-selection. The duration of recurrent selection has already been argued to be a major determinant in the evolution of herbicide resistance, whilst back-selection refers to the reversal of the direction of selection to favour susceptible alleles. The other factors that contribute to the control of evolutionary rate, namely the initial frequency of resistance genes, mode of inheritance and differential fitness of genotypes, are clearly not amenable to managerial control.

Alteration of selection pressure

Major gene systems

When resistance is determined by major genes, a lowering of selection pressure may delay the onset of resistance. This may be achieved by direct reduction of the dosage of the selecting herbicide, provided that the reduced dose results in a lower level of mortality. Alternatively, it may be achieved by switching over entirely to the use of herbicides with chemistries that are known to have a fundamentally different mode of action and hence alter the existing process of selection. Using mixtures of herbicides is a third way of reducing selection for resistance. Co-evolution of resistance to two different herbicide chemistries will necessarily be slower than resistance to a single chemical because the frequency of dual-resistant plants will be the compounded frequency of resistance to each individual herbicide (Gressel, 1986). A final method is by rotation of herbicides with chemicals of different modes of action over cropping seasons.

Any of these methods of lowering selection pressure may delay the evolution of resistance for the following reasons. Susceptible plants may be more likely to escape mortality. They may then contribute susceptible progeny to the next generation and hence lower the frequency of resistant alleles in the total population. Such an approach, however, may well incur yield penalties or force a change in cropping programme.

If the species has a persistent seed bank, seeds from susceptible plants will be added to the dormant population in the soil. Since only a fraction of the total (plant + seed) population will be exposed to selection in each cropping season, these susceptible progeny will act to buffer the plant population over several years by dilution with susceptible genes.

Plants that escape mortality may be recipients of immigrant susceptible pollen (from external sources), enabling an influx of susceptible alleles and hence may lead to lowering of resistance gene frequency. Maxwell *et al.* (1990) have argued that management tactics should include strips of unsprayed weeds in close proximity to resistant populations to act in this manner. The effectiveness of this gene flow will be determined by the mode of inheritance of resistance and the frequency of resistance alleles at the time when gene flow occurs. The influence of such immigration may be noticeable if resistance is conferred as a recessive allele that is at low frequency in the population; where resistance is controlled by a dominant gene the effect is likely to be small, and if maternally inherited, zero. Implementation of such a strategy would depend on a clear knowledge of pollen spread in field environments and a deliberate management programme that encouraged susceptible genotypes. At present, this remains a purely theoretical option.

The reverse situation, in which resistance alleles emigrate into surrounding populations of susceptible genotypes, is significant for the management of existing herbicide resistant populations. The movement of pollen may result in the spread of resistance to other parts of a farm. Where selection is relaxed due to a change in management, the frequency of resistance alleles in the neighbouring susceptible populations will reduce more slowly than in the absence of pollen flow. However, Heap (1991) found no evidence that genes were spreading from a resistant population across a field boundary and into the next field. It is probable that gene flow by pollen movement (at least in grasses) will only occur over relatively short distances (Copeland & Hardin, 1970). However, recent studies of the hybridisation between transgenic crops and wild relatives suggest that there is considerable variation amongst species and populations and that generalisations are risky (Kohn & Casper, 1992; Raybould & Gray, 1993).

Polygenic systems

In contrast to major genes, a reduction of selection on a weed species exhibiting polygenic resistance is likely to have diametrically the reverse effect and potentially encourage the evolution of resistance. The response to selection based on polygenes depends on genetic recombination causing several or many genes (each contributing in a minor way) to 'coalesce' in a single genotype. Relatively rapid response to selection will occur if low selection pressures are applied, since this will strongly select for genotypes showing elevated resistance as individual genes become combined within a genotype. Application of an increasingly strong dose of herbicide will intensify the response to selection. In consequence, if the selection pressure is high initially then those genotypes with a small enhancement of resistance will be lost from the population and the frequency of recombinations of polygenes or multiple gene amplifications will be greatly reduced.

Back-selection

In the absence of selection by herbicide, a reversal in the direction of selection favouring susceptible alleles may occur for two reasons. Firstly, with relaxed selection, fitness differentials between genotypes may change significantly in comparison with their performance in the presence of the herbicide. Secondly, the relative fitness of resistant genotypes may be reduced further in competition with susceptible ones by the action of other environmental factors under the control of farm managers. Response to tillage and increased susceptibility to other herbicides by resistant biotypes ('negative cross resistance' – see Gressel, 1991) has been reported in a range of triazine resistant weeds. The success of management strategies that invoke back-selection depends, however, upon noticeable differences in fitness between resistant and susceptible phenotypes in different environments. As has been discussed, such fitness differentials may be limited to certain cases of resistance.

Whilst either of the above mechanisms may form the basis of a management strategy, in order to determine accurately the efficacy of any proposal for the management of resistance a full understanding of the dynamics of individual genotypes (homozygous resistant/susceptible and heterozygotes) is required. Moreover, the entire life-cycle of the plant needs to be considered since selection may act to differing extents and possibly in differing directions at individual stages in the life of a plant. In perennial weeds reproducing by vegetative means, measuring the changes in selection pressure may be much more complex.

Strategies for the management of herbicide resistance have been the focus of considerable speculation and simulation modelling. Gressel & Segel (1990) have considered the effect of rotational strategies in the use of herbicides; Maxwell *et al.* (1990) have explored in depth the role of breeding systems and gene flow; and Mortimer *et al.* (1992) have examined the importance of density-dependent fitness differentials between biotypes.

A major conclusion from the modelling work of Gressel (1991) is that reduction in selection pressure due to herbicide rotations will only be of significance in cases where there is a considerable difference in the relative fitness of resistant and susceptible individuals when selection is relaxed. Roush *et al.* (1990) have argued that model predictions point to the potential for manipulating gene flow as a key component to the management of resistance. However, such predictions remain to be tested.

What does all this mean in terms of developing recommendations for the farmer? In the first instance it is clear that a knowledge of the genetics and mechanism(s) of resistance in a weed species is important in defining a long term strategy of resistance management. As Gressel (in press) points out, the situation may become particularly difficult if more than one mechanism is present within a weed species, one controlled by polygenes and one by a major gene. Clearly, recommendations for lowering selection pressure to discourage evolution of major gene resistance will favour evolution of polygenic resistance. Equally, the sole use of a high rate of herbicide will favour major gene resistance in the attempt to manage polygenic resistance. Moreover, if evolution is occurring by differing mechanisms in separate populations a simple 'blanket' recommendation will have little value. Such a scenario appears to have been realised in *Lolium rigidum* in Australia. The existence of multiple mechanisms of resistance and cross-resistance, inter-population variation, the confirmation of a single semi-dominant gene giving target site resistance (Tardif & Powles, 1993) and the possibility of polygenes controlling metabolic detoxification led Powles & Matthews (1992) and Matthews (1994) to argue strongly for truly integrated weed management practices. It is inevitable that non-chemical methods of weed control will be essential components of such programmes. This in turn demands a clear understanding of the population ecology of both resistant and susceptible biotypes in the absence of chemical selection. Whilst such knowledge is slow to be gained, the cessation of selection is the necessary first step!

Conclusions

Resistance to herbicides in weeds was largely confined to the triazine group until the early 1980s. Moreover, the existence of alternative chemical and cultural methods of control for triazine resistant weeds had a strong tendency to reduce herbicide resistance to the level of academic and scientific curiosity. The rapid emergence of resistance to sulfonylurea herbicides, coupled with the increasing incidence of multiple and cross resistance in major grass weeds in the late 1980s, has changed this perception.

In this chapter we have reviewed the underlying factors governing the evolution of resistance. The ability to predict both the likelihood and the rate of evolution of resistance to a particular chemical in advance of its occurrence in the field is a highly desirable goal. But how realistic is this? As we have seen, the evolution of resistance will depend on several factors, and whilst some of these can be studied in advance of the resistance occurring, others cannot. Whilst it is possible experimentally to measure the selection intensity imposed by a herbicide, estimation of the frequency of resistance alleles in unselected populations is much more difficult. Genetic analysis can, of course, only be completed on finding resistant biotypes and the approaches need to be as rigorous as for any conventional crop breeding programme. Similarly, rigorous fitness analysis can only be completed with isogenic lines after resistance has evolved. In order to predict which species in a weed flora will become resistant, we would need to screen every species to every chemical used in the cropping system – a daunting task indeed! Since we know very little about the breeding systems of most weeds, these would also need to be the subject of a major comparative study.

As a practical consequence, an analysis of herbicide resistance in a weed species can only be retrospective. In turn, therefore, the management of resistance has to be inherently proactive so that weed control is practised in a manner which does not ensure the build up of resistance alleles. The rotation of herbicide chemistries and the incorporation of cultural practices in a fully integrated and scientifically understood programme of weed management is the solution to such increasingly pressing problems (Powles & Matthews, 1992).

9

Weed population dynamics: synthesis and prognosis

The farmer or land manager has to make decisions about land use and habitat manipulation by predicting, either qualitatively or quantitatively, the consequences of the available weed control options. By studying the theoretical basis of weed management, the intention of weed scientists is to help these practitioners to make decisions on the basis of understanding rather than just on 'gut-feeling'. '*A knowledge of the seed cycle aids the understanding of the factors affecting populations and hence assists in the development of control strategies*' (Moss, 1990a). Statements such as this have been made repeatedly in the literature over the past 10–15 years. They are perhaps more ambitions than statements of fact. How often have farming practices been changed as the direct result of such ecological work? To date, probably very rarely.

Is the study of weed population dynamics therefore simply an esoteric subject within weed science? Why have its findings been applied so seldom? Are weed scientists not innovative enough? Are there methodological problems still to be solved? Or are other events dictating the changes within farming systems, such that weed science is unable to have a guiding influence?

In this final chapter, we will review the state of the subject at the present time and then speculate on the directions which further development should take. As a framework, we pose four pertinent questions:

1. Is the current data base on weed ecology extensive enough?
2. To what extent can changes in weed populations be predicted?
3. Is research trying to answer the right sorts of questions?
4. Can studies of weed biology be expected to drive changes in farm or land management?

Is the data base extensive enough?

The reference list of this book is very lengthy. Even so, we have only given sufficient examples to illustrate our discussion, rather than attempting an exhaustive review. There is much published work which has not been discussed. Initially then, the data base appears considerable. But it is small when compared with the data available on herbicides and their use (Norris, 1992). Weed scientists have concentrated in the past on maximising the effectiveness of the array of chemical tools at their disposal. It is only relatively recently, with concerns about pollution, legislation in some countries to reduce pesticide use and the emergence of herbicide resistance, that more attention has been diverted to integrated weed management and the biological data needed to support its development.

The only species to have been mentioned in every chapter of this book is *Avena fatua*. Although this may reflect our own familiarity with the species, it may also illustrate the fact that the biology of only a small number of species is known in any depth, whereas the majority of species have been studied superficially or not at all. Even for the most studied species, there is little or no information on dispersal, the influence of soil type, weather and sub-lethal doses of herbicides on their biology or population dynamics and the causes of seed mortality (Chapters 3 and 4). There are few species for which there are data for all parts of the life-cycle and therefore for which multi-stage population models can be produced without guessing at values for at least some parameters. Hence, although understanding of the subject of population biology in general is very good, detailed knowledge of most species is poor.

Most arable fields contain at least six weed species, while some fields have 25 or more. For most of these species there are limited ecological data, either collected locally or elsewhere in the world. Cousens & Medd (1994) surveyed the extent of Australian data on weeds in Australian crops and pastures. It was reported that while there were data on the persistence of seeds in the soil for 30 species (perhaps less than 30% of the total number in the continent), data relating seed production to plant density were only available for six species in crops, ten species in annual pastures and six species in perennial pastures. It is unlikely that this situation would be much different in any other country. Hence, although it might be possible to predict the dynamics of one or two species in a field, there is little likelihood of predicting the dynamics of an entire weed community. As an example, even though cultivation has been studied more than any other management factor, it is still not possible to say which broad-leaved species are likely to respond to each tillage method (Chapter 6).

Is it worth trying to obtain data experimentally on all of the weed species in a plant community? Starting from scratch, the effort required even for a simple community would be considerable and would be difficult to fund. For example, the present data base for *Alopecurus myosuroides*, which includes both average values for life-cycle transition probabilities and an assessment of their variability (Moss, 1990a), has taken at least 30 researcher-years to generate. A fixed value research grant will be more efficiently spent, from both pragmatic and academic viewpoints, by study-ing one species in depth rather than many species superficially. Most future studies of weeds are therefore likely to continue to be of single species which *have* become problems, rather than of multi-species communities and the prediction of which species *will* become problems. It is likely that few weed communities will be analysed intensively to understand all of the interac-tions that determine species abundance.

In summary, we must accept that our understanding of the ecology and biology of all but the major species is likely to remain incomplete. Nevertheless, at the community level attempts to define the 'assembly rules' for weed species are important. Rather than call for 'more of the same' types of data, covering new species or habitats, it would be worthwhile turning attention to those *subject areas* and lines of approach that focus on the ecology of weeds in agro-ecosystems. Perhaps an area in need of most attention is the spatial dynamics of weed populations within fields (Chapter 7). Little is known about rates of spread of any species at this scale; although seed dispersal has been studied, especially in members of the Asteraceae, there is little knowledge of how this translates into the spread of populations. In species with multiple methods of dispersal, which agencies contribute most to spread? What makes some communities apparently able to resist invasion or limit spread by certain species but not by others? These are just some facets which need to be addressed.

Can changes in weed populations be predicted?

It was argued in Chapter 5 that a central reason for studying weed population dynamics is to generate predictive ability. But, having reviewed the methodology for doing this, how good is our current ability to predict changes in weed populations? If population trends of a species were to be thoroughly studied under a particular management regime it would be possible, without the need for models, to say that if a farmer introduces that type of management a particular trend will probably occur. It might only be possible to point to the direction of a trend, or if the data were consistent, to predict the average rate of increase of the weed species. As we have already argued, with the exception of grass weeds and tillage (Chapter 6), there are few cases where such predictions currently could be made. For most weeds in most habitats it is not possible to predict the dynamics of the populations.

In examining the question of why population changes cannot be predicted, there are a number of issues which need to be addressed: in particular, are the types of model being used capable of accurate predictions; are there adequate data to assign values to their parameters; and, for these reasons, can accurate predictions be expected?

There has been much attention by weed ecologists to the development of simple mathematical models describing the life-cycles of species (Chapter 5), in the anticipation that they will have predictive power. Indeed, such models have proliferated. In life-cycle models, experimental data from different management regimes are used to derive parameter values (usually one value per management regime for each stage in the cycle); the models are then used to make simulations.

A simulation using one set of data merely allows prediction from short term experiments of what would theoretically happen if those conditions were invariant over many generations. They allow estimation of the rate of increase, the equilibrium density and the type of trajectory which population density would follow. If data are available for more than one management regime, then simulations can examine the implications of rotations of management conditions (Cussans & Moss, 1982). The sensitivity of the predictions to events in particular parts of the life-cycle can also be examined: how much will the dynamics be affected by the cessation of stubble burning, or by use of a herbicide giving slightly poorer control? Life-cycle models have also been used effectively in examining the economic effect of different decision rules for herbicide use (e.g. Cousens *et al.*, 1986). However, these models cannot predict what will actually happen: they are merely a *best guess* made under the assumption that all parameters are known precisely, do not vary and that the model is an adequate description. It is not possible to say, for example, that the rate of increase of *Avena fatua* in southern England in early sown winter wheat where the soil has been mouldboard ploughed and where no herbicides are used will be exactly 2.7 per annum rather than 2.2 or even 5.6; but it is possible to be confident that the population will increase. Similarly, although it has been calculated that the maximum long term profit will be gained from spraying *A. fatua* when populations exceed 3.2 plants m^{-2}, it is safer to conclude only that the optimum threshold density will be considerably below the value derived from single year calculations (Chapter 6). Hence, although *quantitative* models may be used, they may have no more than *qualitative* predictive power, due to variability, the availability of data and simplistic model structure.

The most enlightening use of a model is for prediction *outside* of the experiences for which the data were generated. Can 'life-cycle' models be used to predict the consequences of future changes in management? Unless those new management regimes have already been studied, then the answer is 'no'. In order to calculate the changes in weed populations that may ensue, values of parameters under the new regimes must be estimated. The difficulty with this is that populations are described at a 'phenomenological' level, based on outcomes, rather than at a 'mechanistic' level based on processes. For example, if only a mean value is used for the annual proportion of weed seeds germinating in an early-sown crop, it is not possible to predict what would happen if crop sowing is delayed. It might be possible to hypothesise what would happen to seedling numbers, based on experience, but it is over-generous to call this a rigorous scientific prediction. If, however, seedling establishment in the model were based on relationships between germination, temperature and moisture, typical patterns of meteorological data could be used to make informed predictions as to what might be expected to occur. A useful future development would therefore be the construction of such 'mechanistic' population models. Mechanistic simulation models for plant growth are already being used to gain understanding of competition between weeds and crops (Kropff, 1988).

Even given a suitable model structure, capable of informative predictions, the problem dealt with in the previous section must be faced: the paucity of data. Data on some aspects of weed ecology required for models are difficult or time-consuming to gather. A particular example is the study of the fates of seeds in the soil. Parameter values in many cases may often therefore be gleaned from the published literature. As a result, some models include data collected in totally different environments from those in which simulations are being made. For example, there are at least nine life-cycle models of *Avena fatua* populations world-wide, produced in North America, Europe and Australia (Rauber & Koch, 1975; Sagar & Mortimer, 1976; Manlove *et al.*, 1982; Murdoch & Roberts, 1982; Wilson *et al.*, 1984; Taylor & Burt, 1984; Spitters, 1986; Medd & Ridings, 1989; Martin, 1992). All but one of them draw on at least some data from either the work of J. D. Banting in Canada or from various researchers at the former Weed Research Organization in England. Since plant and seed behaviour are expected to vary with environment, the use of data from another climate must add to the uncertainty of the predictions.

Another reason that present population models may have low quantitative predictive ability is that the parameter values are averages from experiments made perhaps over a number of sites and years. If there is significant site-to-site and year-to-year variation, it is unrealistic to expect to predict the precise population trajectory in a particular site or year. Models cannot, therefore, be used with sufficient confidence to advise a farmer on what will happen in a particular field in the following cropping season. Errors in predictions must also be expected to be compounded through time; even after only a few generations confidence intervals are likely to exceed the prediction. If predictions are made over several generations, it must be assumed that the environment remains constant; but this is known not to be the case.

What use, then, are long term predictions of patterns of change assuming constant conditions? They must surely be considered as theoretical predictions only, helping to understand the ways in which populations *might* change and how different factors may lead to different types of dynamics. For example, models of competing weeds and crops can tell us what magnitudes of population reduction we *might* expect in the long term from changes in crop density or from crop varieties with greater vigour; they will not tell us precisely what *will* happen. Retrospectively, models can also help to explain events which have occurred and enable similar situations to be avoided in the future. Population models are therefore perhaps best considered as powerful educational tools.

Are the right questions being answered?

In any subject area, an attempt to comment on attitudes and approaches without a formal survey will risk being damned as purely subjective and most probably biased. Generalisations within a diverse subject such as weed science are likely to be shown to have so many exceptions that the very concept of generality will be stretched to the limit. This being so, we will still present our impressions based on discussions at numerous conferences throughout the world and over a number of years. We may have gained false impressions, but at least they will lead to an open discussion of objectives.

It is a trivial but important statement that if the wrong question is asked, an inappropriate or incomplete answer to the real question will be obtained. No matter how well the experiments are conducted, we will have progressed little. So are the right questions being asked? Here we will identify the kinds of questions currently being answered, and then propose an alternative.

Questions involving only outcomes

A great many studies of weed population dynamics aim simply to *monitor* population density: what will happen to numbers if management regime A is followed? How will the numbers differ from those under regime B? They are descriptive only and inform us of outcomes, not of processes. Hence, there are numerous papers world-wide showing that under minimum tillage some annual broad-leaved species sometimes increase, sometimes decrease and often show no change at all (Chapter 6). But they give little indication as to why this may be so. By stopping at the *observation* and not proceeding to an *investigation* poor insight is achieved. Faced with apparent unpredictability, it is important to ask why this should be the case. *Why* does species X increase more rapidly than species Y? *Why* does species Z not respond at all? In particular, an understanding is required of the ecological factors and processes which are likely to determine the responses to particular management regimes. If events are merely described, it is not possible to extrapolate to other conditions.

Questions involving solutions

Few non-herbicidal studies have as their aim the need to find a *solution* to a specific weed management problem. Perhaps this is the most straightforward explanation of why weed population biology has seldom led to changes in farming practices. But how would one go about finding a solution? It is often held that by obtaining information about survival and seed production at every stage of a species' life-cycle, it might be possible to identify the 'Achilles' Heel' of the species (see Chapter 6). Perhaps the data could be put into a model and sensitivity analysis used to identify those parameters having the greatest influence on dynamics, and hence in which small changes might elicit large effects. Although this is clearly achievable, we know of no detailed examples where it has been done for weeds.

Rather than ask which parameters have the greatest effect when changed by a small amount, one could ask which parameters can be changed by a large amount most easily. An excellent example of this is the work on controlling herbicide resistant *Lolium rigidum* in Australia (Chapter 8). Managers now have a more restricted range of chemicals which work and in order to keep weed population levels down they need to achieve mortality by alternative means. Traditionally, before the advent of herbicides, *L. rigidum* was controlled by sowing later in combination with further cultivations. Many seedlings were allowed to emerge but were then killed while still young. However, because of concerns over soil erosion from tillage and reduced yield potential from later sowings, a return to such a regime has little appeal. It has been observed that, in *L. rigidum*, a large proportion of seeds can be taken up into the combine harvester. If these could be retained rather than being returned to the field with the chaff and straw, significant extra losses of seed could be induced. Both locally developed and imported seed-catching machinery has been found to be effective, removing up to 80% of the seeds produced (G. S. Gill, pers. commun.). The collected seeds can be fed to stock or burnt. If such results can be obtained consistently, this goes a considerable way towards replacing the 90 + % control formerly given by herbicides. Another technique being considered is the use of non-selective herbicides close to crop maturity in order to reduce weed seed production.

Other examples of the search for a solution to a specified problem have occurred in countries where governments have legislated to reduce pesticide use. For example, in Denmark the requirement to use less herbicide has led to projects to develop mechanical weed control techniques to minimise weed populations (e.g. Rasmussen, 1992), although these are usually only examined within a generation.

Questions involving the search for optima

A manager will need to know not just the effects of specific weed control options, but the best practices for their use in order to minimise the rate of weed population increase. Again, studies leading to such information are uncommon outside of herbicide research. As we have shown, population models can be used to derive the minimum frequency of ploughing in a direct drilling system and the economic optimum threshold density for spraying (Chapter 6). Field studies of mechanical weed control techniques have been used to identify optimum timings and frequencies of operation. Another question requiring such an approach would be the determination of the period of time by which to delay sowing in order to achieve better weed control. Although it is well known that by delaying sowing until after seedling emergence more weeds can be controlled by tillage, there is little information on the relationship between seedling number and sowing/cultivation date for any species. In addition, to make the relevant economic calculations, data would be needed on the relationship between sowing date and yield loss–weed density curves and on weed-free yield in relation to sowing date.

Questions involving long term dynamics

The emphasis of many ecologists working in natural communities has been the prediction of long term patterns in behaviour (Chapter 5). By studying plant populations over a period of perhaps one or two generations, models can be formulated to predict changes over many generations. Particular interest has been shown in those species and conditions in which complex population behaviour is predicted and in deciding whether or not species will co-exist. Although seldom stated explicitly, it might be thought that in cases of complex dynamics, such as chaotic behaviour, weed management will be made more difficult because of the unpredictability that is intrinsic to the regulatory processes within populations. In an applied context, Mortimer *et al.* (1989) examined the types of trajectory likely to occur from different seed production–weed density curves. Bazzaz *et al.* (1992) predicted, on the basis of models parameterised under artificial conditions, that chaotic population behaviour would be more likely under an elevated CO_2 environment.

Variation in the environment will inevitably act to cause substantial modification of population dynamics. Both intrinsically determined complex population behaviour and complex behaviour arising from unpredictable extrinsic factors will appear the same to a manager and will have similar implications. Hence, long term predictions of dynamics can only help us to understand weed population behaviour in a general sense; they will be of doubtful practical significance for planning weed control programmes.

A different type of question

An *implicit* assumption of most weed population studies is that knowledge of long term behaviour is important. After all, if the aim is to develop sustainable systems, there is an explicit long term goal. However, we have argued above that it is difficult, if not impossible, to know what will happen in the long term. Rather than plan long term scenarios and rotations, why not plan weed management over the shorter term? The manager needs to minimise the rate of population increase (λ); whether in the short or the long term, λ needs to be driven downwards (Chapter 6). The amount which the farmer is prepared to spend to do this will depend more on the gross margins and variable costs in that year than on long term economic calculations. Many farm management decisions are made in this way. For example, although farmers have an overall long term crop rotation strategy, they will not enforce it rigidly: they make repeated tactical decisions about which crops to grow and how to manage them. This is also the way in which they have always used herbicides: although they may have a general plan of pre- and post-emergence control, their choice of specific chemicals will change according to weather conditions, time constraints and year to year changes in density and species dominance. For management of herbicide resistance, although there may be an overall desire to rotate chemicals with different modes of action (Chapter 8), the chemical used will depend on how cross-resistance changes from year to year, the density of the weed infestation and the prices of different products.

Perhaps, therefore, the search for ways in which weed population biology can be used by managers should focus on the current population size (N_t) and prediction of the *immediate* trajectory (λ) (Fig. 9.1). Together, these two factors determine the size of the seed bank in the following generation (Chapter 5). Each aspect of habitat management should then be examined according to how much influence it will have on λ (this is in contrast to most weed control studies which only examine mortality, i.e. within a generation). This can be done as described previously through sensitivity studies of multi-stage population models. It is somewhat surprising, however, that it has so seldom been studied experimentally. For example, although we have some idea of the influence of straw burning or tillage on λ in some species, there is little or no information on the effects of fertiliser application (see Lintell-Smith *et al.*, 1991), crop seeding rate, crop species selection, fallow grazing or even most herbicides (Chapter 6).

Can studies of weed biology drive changes in farm or land management?

Most changes in farming systems are dictated by the prices of commodities, access to markets and the availability of suitable crop cultivars. For example, farmers in southern Canada began to grow maize when short season varieties became available, because it made economic sense to do so. The world-wide increase in rapeseed (canola) production would not have occurred unless a market had developed. Weed control methods have always followed such profit-driven changes and aided crop production

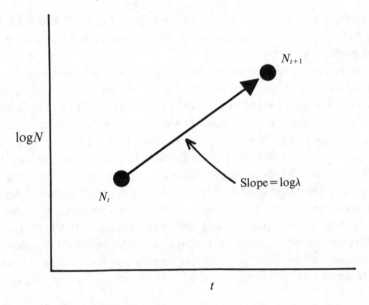

Fig. 9.1. Illustration of two parameters, the current population size (N_t) and the rate of increase (λ), which together determine the population size one generation later (N_{t+1}). Future research needs to focus on the measurement of λ, rather than just on estimating within-generation mortality from weed control practices.

rather than driving the changes. Unless a crop is so weedy and control methods so ineffective that the crop cannot be grown successfully, it is unlikely that weed control will drive future choices of crops or sequences of rotations.

More subtle adjustments of crop management practices, however, could be driven by weed control issues. For example, yield increases from earlier cereal sowings have only been possible because of the availability of effective herbicides. If resistance to those herbicides becomes prevalent, then a move back towards later sowings may occur. Powles & Matthews (1992) have illustrated the imperative need for changes in agricultural practice where multiple mechanisms of herbicide resistance are present in a weed species. However, it is unlikely that farmers will wait for researchers to provide data before they make such a change. An accepted protocol in scientific research is that sufficient data are gathered in order to establish generality of findings before communication. By the time research had identified the benefits of occasionally ploughing in a direct drilling system in Britain, farmers had already implemented the method. Innovative farmers act more quickly than the research process and develop their techniques iteratively by trial and error. Hence, much weed research is conducted to confirm or refute the benefits of an innovation, rather than making the innovation in the first place.

Many changes in farming come through technological advances made by engineers and chemists. Although there are exceptions in genetical engineering, biologists cannot be expected a priori to invent new techniques and to solve problems. If novel solutions are required, weed scientists need to be working more closely with engineers, rather than waiting to see what the engineers come up with next and then react. The example of *Lolium rigidum* seed capture at harvest in Australia has already been mentioned; weed scientists are working with those developing the machinery to help them identify the most effective systems and optimal timings. Studies of the dynamics of weed patches and the development of 'patch sprayers' by engineers is another example. Perhaps this is the most valuable role of the weed population biologist – to *test* innovations and to identify the best combinations – rather than to expect to initiate changes in farming systems.

Conclusions

It is difficult to come to the end of a wide-ranging book and to make take-away statements which will be both succinct and stimulatory. Much has already been said in the conclusions to individual chapters. It would be easy to end here by simply adding that there should be more research in most areas, more comparative studies, more attempts to relate dynamics to biology (rather than to demography), and more studies of mechanisms that drive population processes. However, it is easier to generate ideas of topics for study than it is to propose radical new directions. We have tried in this book to establish a framework for the study of weed populations, based almost entirely on existing work. It is difficult to claim that we have said anything new and it is clear that much of the conceptual understanding of the subject has already been developed.

It has been argued that the origins of the academic study of plant population ecology lie in part in investigations of weed species (Mortimer, 1984). Weed science encompasses the application of plant population ecology and as such relies on the tools provided by academic study. It is our belief that these tools need to be applied in the testing and fine-tuning of innovations suggested by farmers and technologists. We would hope too that the agricultural field is a proving ground for the development of new scientific tools.

References

Abrahamson, W. G. (1980). Demography and vegetative reproduction. In *Demography and Evolution in Plant Populations*, ed. O. T. Solbrig, pp. 89–106. Oxford: Blackwell Scientific.

Adkins, S. W., Loewen, M. & Symons, S. J. (1987). Variation within pure lines of wild oats (*Avena fatua*) in relation to temperature of development. *Weed Science*, **35**, 169–72.

Ahrens, C., Cramer, H. H., Mogk, M. & Peschel, H. (1981). Economic impact of crop losses. *Proceedings of the 10th International Congress of Plant Protection*, pp. 65–73. Croyden: BCPC.

Amor, R. L. (1984). Weed flora of Victorian cereal crops. *Australian Weeds*, **3**, 70–3.

Amor, R. L. & Harris, R. V. (1975). Seedling establishment and vegetative spread of *Cirsium arvense* (L.) Scop. in Victoria, Australia. *Weed Research*, **15**, 407–11.

Ampong-Nyarko, K. & De Datta, S. K. (1993). Effects of nitrogen application on growth, nitrogen use efficiency and rice–weed interaction. *Weed Research*, **33**, 269–76.

Andersen, R. N. & Gronwald, J. W. (1987). Non-cytoplasmic inheritance of atrazine tolerance in velvetleaf (*Abutilon theophrasti*). *Weed Science*, **35**, 496–8.

Andujar, J. L. G., Fernandez, A. & Fernandez-Quintanilla, C. (1986). Modelizacion y simulacion de la dinamica de poblaciones de plantas mediante el modelo matricial de Leslie. *Investigaciones Agraria: Produccion y Proteccion Vegetales (Madrid)*, **1**, 209–18.

Auld, B. A. (1988). Dynamics of pasture invasion by three weeds, *Avena fatua* L., *Carduus tenuiflorus* Curt. and *Onopordum acanthium* L. *Australian Journal of Agricultural Research*, **39**, 589–96.

Auld, B. A. & Coote, B. G. (1980). A model of a spreading plant population. *Oikos*, **34**, 287–92.

Auld, B. A. & Coote, B. G. (1981) Prediction of pasture invasion by *Nassella trichotoma* (Gramineae) in south east Australia. *Protection Ecology*, **3**, 271–7.

Auld, B. A., Hosking, J. & McFadyen, R. E. (1982). Analysis of the spread of tiger pear and parthenium weed in Australia. *Australian Weeds*, **2**, 56–60.

Auld, B. A. & Medd, R. W. (1987). *Weeds: An Illustrated Botanical Guide to the Weeds of Australia*. Melbourne: Inkata Press.

295

Auld, B. A., Menz, K. M. & Monaghan, N. M. (1978/9). Dynamics of weed spread: implications for policies of public control. *Protection Ecology*, **1**, 141–8.

Auld, B. A., Menz, K. M. & Tisdell, C. A. (1987). *Weed Control Economics*. London: Academic Press.

Austin, M. P., Groves, R. H., Fresco, L. M. F. & Kaye, P. E. (1985). Relative growth of six thistle species along a nutrient gradient with multispecies competition. *Journal of Ecology*, **73**, 667–84.

Babiker, A. G. T., Ibrahim, N. E. & Edwards, W. G. (1988). Persistence of GR7 and *Striga* germination stimulants from *Euphorbia aegyptiaca* Boiss, in soil and in solution. *Weed Research*, **28**, 1–6.

Bachthaler, G. (1974). The development of the weed flora after several years direct drilling in cereal rotations on different soils. *Proceedings of the 12th British Weed Control Conference*, pp. 1063–71.

Baker, H. G. (1947). Infection of species of *Melandrium* by *Ustilago violacea* (Pers.) Fuckel and the transmission of the resultant disease. *Annals of Botany, N.S.*, **11**, 333–48.

Baker, H. G. (1962). Weeds – native and introduced. *Journal of the California Horticultural Society*, **23**, 97–104.

Baker, H. G. (1965). Characteristics and modes of origin of weeds. In *The Genetics of Colonizing Species*, ed. H. G. Baker & G. L. Stebbins, pp. 147–69. New York: Academic Press.

Ball, D. A. & Miller, S. D. (1990). Weed seed population response to tillage and herbicide use in three irrigated cropping sequences. *Weed Science*, **38**, 511–17.

Ballaré, C. L., Scopel, A. L., Ghersa, C. M. & Sanchez, R. A. (1987a). The demography of *Datura ferox* (L.) in soybean crops. *Weed Research*, **27**, 91–102.

Ballaré, C. L., Scopel, A. L., Ghersa, C. M. & Sanchez, R. A. (1987b). The population ecology of *Datura ferox* in soybean crops. A simulation approach incorporating seed dispersal. *Agriculture, Ecosystems and Environment*, **19**, 177–88.

Banting, J. D. (1974). Growth habit and control of wild oats. *Publication 1531, Agriculture Canada*.

Barr, A. R., Mansooji, A. L., Holtum, J. A. M. & Powles, S. B. (1992). The inheritance of herbicide resistance in *Avena sterilis* ssp. *ludoviciana*, biotype SAS 1. *Proceedings of the First International Weed Control Congress*, pp. 70–2.

Baskin, J. M. & Baskin, C. C. (1985). The annual dormancy cycle in buried weed seeds: a continuum. *BioScience*, **35**, 492–8.

Bastow Wilson, B. & Lee, W. G. (1989). Infiltration invasion. *Functional Ecology*, **3**, 379–80.

Bazzaz, F. A., Ackerly, D. D., Woodward, F. I. & Rochefort, L. (1992). CO_2 enrichment and dependence of reproduction on density in an annual plant and a simulation of its population dynamics. *Journal of Ecology*, **80**, 643–51.

Begon, M. & Mortimer, M. (1986). *Population Ecology: A Unified Study of Animals and Plants*, 2nd edn. Oxford: Blackwell Scientific.

Benech Arnold, R. L., Ghersa, C. M., Sanchez, R. A. & Insausti, P. (1990). A mathematical model to predict *Sorghum halepense* (L.) Pers. seedling emergence in relation to soil temperature. *Weed Research*, **30**, 91–9.

Benoit, D. L., Derksen, D. A. & Panneton, B. (1992). Innovative approaches to seedbank studies. *Weed Science*, **40**, 660–9.

Berg, M. A. van den (1977). Natural enemies of certain *Acacias* in Australia. *Proceedings of the 2nd National Weeds Conference of South Africa*, pp. 75–82.

Bettini, P., McNally, S., Sevignac, M., Darmency, H., Gasquez, J. & Dron, M. (1987). Atrazine resistance in *Chenopodium album*: low and high levels of resistance to the herbicide are related to the same chloroplast psbA gene mutation. *Plant Physiology*, **84**, 1442–6.

Betts, K. J., Ehlke, N. J., Wyse, D. L., Gronwald, J. W. & Somers, D. A. (1992). Mechanism of inheritance of diclofop resistance in Italian ryegrass (*Lolium multiflorum*). *Weed Science*, **40**, 184–9.

Bewley, J. D. & Black, M. (1985). *Seeds: Physiology of Development and Germination.* New York: Plenum Press.

Blackman, G. E. (1950). Selective toxicity and the development of selective weedkillers. *Journal of the Royal Society of Arts*, **98**, 499–517.

Bough, M., Colosi, J. C. & Cavers, P. B. (1986). The major weedy biotypes of proso millet (*Panicum miliaceum*) in Canada. *Canadian Journal of Botany*, **64**, 1188–98.

Boydston, R. A. (1990). Time of emergence and seed production of longspine sandbur (*Cenchrus longispinus*) and puncturevine (*Tribulus terrestris*). *Weed Science*, **38**, 16–21.

Bradbeer, J. W. (1988). *Seed Dormancy and Germination.* Glasgow: Blackie.

Bradshaw, A. D. (1984). The importance of evolutionary ideas in ecology – and vice versa. In *Evolutionary Ecology*, ed. B. Shorrocks, pp. 1–25. Oxford: Blackwell Scientific.

Brain, P. & Cousens, R. (1990). The effect of weed distribution on predictions of yield loss. *Journal of Applied Ecology*, **27**, 735–42.

Brenchley, W. E. (1918). Buried weed seeds. *Journal of Agricultural Science*, **9**, 1–31.

Brenchley, W. E. & Warrington, K. (1936). The weed population of arable soil. III. The re-establishment of weed species after reduction by fallowing. *Journal of Ecology*, **24**, 479–501.

Bridges, D. C. (1994). Impact of weeds on human endeavors. *Weed Technology*, **8**, 392–5.

Buchanan, F. S., Swanton, C. J. & Gillespie, T. J. (1990). Postemergence control of weeds in winter rapeseed, *Brassica napus*, with DPX-A7881. *Weed Science*, **38**, 389–95.

Burdon, J. J., Groves, R. H. & Cullen, J. M. (1981). The impact of biological control on the distribution and abundance of *Chondrilla juncea* in south-eastern Australia. *Journal of Applied Ecology*, **18**, 957–66.

Burnside, O. C., Fenster, C. R., Evetts, L. L. & Mumm, R. F. (1981). Germination of exhumed weed seeds in Nebraska. *Weed Science*, **29**, 577–86.

Burry, J. N. & Kloot, P. M. (1982). The spread of Composite (Compositae) weeds in Australia. *Contact Dermatitis*, **8**, 410–3.

Caseley, J. C., Kueh, J., Jones, O. T. G., Hedden, P. & Cross, A. R. (1990). Mechanism of chlorotoluron resistance in *Alopecurus myosuroides*. *Abstracts of the 7th International Congress on Pesticide Chemistry*, p. 417.

Caswell, H. (1989). *Matrix Population Models.* Sunderland, MA: Sinauer.

Cavers, P. B. & Benoit, D. L. (1989). Seed banks in arable land. In *Ecology of Soil Seed Banks*, ed. M. A. Leck, V. T. Parker & R. L. Simpson, pp. 309–28. San Diego: Academic Press.

Challaiah, Burnside, O. C., Wicks, G. A. & Johnson, V. A. (1986). Competition

References

between winter wheat (*Triticum aestivum*) cultivars and downy brome
  (*Bromus tectorum*). *Weed Science*, **34**, 689–93.
Chancellor, R. J. (1964). The depth of weed seed germination in the field.
  Proceedings of the 7th British Weed Control Conference, pp. 607–13.
Chancellor, R. J. (1970). Biological background to the control of three perennial
  broad-leaved weeds. *Proceedings of the 10th British Weed Control
  Conference*, pp. 1114–20.
Chancellor, R. J. (1985a). Changes in the weed flora of an arable field cultivated
  for 20 years. *Journal of Applied Ecology*, **22**, 491–501.
Chancellor, R. J. (1985b). Maps of the changes in the weeds of Boddington Barn
  field over twenty years (1961–1981). *Technical Report of the Agricultural and
  Food Research Council Weed Research Organization*, **84**, 1–38.
Chancellor, R. J. (1986) Decline of arable weed seeds during 20 years in soil
  under grass and the periodicity of seedling emergence after cultivation.
  Journal of Applied Ecology, **23**, 631–7.
Chauvel, B. (1991). Polymorphism genetique et selection de la resistance aux
  urees substituees chez *Alopecurus myosuroides* Huds. PhD thesis, University
  of Paris-Orsay.
Chauvel, B., Gasquez, J. & Darmency, H. (1989). Changes of weed seed bank
  parameters according to species, time and environment. *Weed Research*, **29**,
  213–20.
Cheam, A. H. (1986). Seed production and seed dormancy in wild radish
  (*Raphanus raphanistrum* L.) and some possibilities for improving control.
  Weed Research, **26**, 405–13.
Cheam, A. H. & Lee, S. I. (1991). Management of brome grass in relation to its
  population dynamics. *Proceedings of the Thirteenth Asian-Pacific Weed
  Science Conference*, pp. 229–36.
Chepil, W. S. (1946). Germination of weed seeds. I. Longevity, periodicity of
  germination, and vitality of seeds in cultivated soil. *Scientific Agriculture*, **26**,
  307–46.
Christopher, J. T., Powles, S. B. & Holtum, J. A. M. (1992). Resistance to
  acetolactate synthase-inhibiting herbicides in annual ryegrass (*Lolium
  rigidum*) involves at least two mechanisms. *Plant Physiology*, **100**, 1909–13.
Coble, H. D. & Ritter, R. L. (1978). Pennsylvania smartweed (*Polygonum
  pensylvanicum*) interference in soybeans (*Glycine max*). *Weed Science*, **26**,
  556–9.
Cofrancesco, A. F., Stewart, R. M. & Sanders, D. R. (1984). The impact of
  Neochetina eichhorniae (Coleoptera: Curculionidae) on water hyacinth in
  Louisiana. *Proceedings of the VI International Symposium on Biological
  Control of Weeds, 19–25 August 1984, Vancouver, Canada*, pp. 525–35.
Combellack, J. H. (1989). Resource allocations for future weed control activities.
  Proceedings of the 42nd New Zealand Weed and Pest Control Conference, pp.
  15–31.
Cooper, M. M. (1983). British agriculture. In *Fream's Agriculture*, 16th edn, ed.
  C. R. W. Spedding, pp. 61–77. London: John Murray.
Copeland, L. O. & Hardin, E. E. (1970). Outcrossing in ryegrass (*Lolium* spp.)
  as determined by fluorescene tests. *Crop Science*, **20**, 254–7.
Cousens, R. (1985a). A simple model relating yield loss to weed density. *Annals
  of Applied Biology*, **107**, 239–52.
Cousens, R. (1985b). A comparison of empirical models relating crop yield to
  weed and crop density. *Journal of Agricultural Science*, **105**, 513–21.
Cousens, R. (1987). Theory and reality of weed control thresholds. *Plant

Protection Quarterly, **2**, 13–20.

Cousens, R. (1988). Misinterpretation of results in weed research through inappropriate use of statistics. *Weed Research*, **28**, 281–9.

Cousens, R. (1991). Aspects of the design, analysis and interpretation of competition (interference) experiments. *Weed Technology*, **5**, 664–73.

Cousens, R., Doyle, C. J., Wilson, B. J. & Cussans, G. W. (1986). Modelling the economics of controlling *Avena fatua* in winter wheat. *Pesticide Science*, **17**, 1–12.

Cousens, R., Firbank, L. G., Mortimer, A. M. & Smith, R. G. R. (1988a). Variability in the relationship between crop yield and weed density for winter wheat and *Bromus sterilis*. *Journal of Applied Ecology*, **25**, 1033–44.

Cousens, R., Marshall, E. J. P. & Arnold, G. M. (1988b). Problems in the interpretation of effects of herbicides on plant communities. *Proceedings of the BCPC Symposium, Field Methods for the Study of Environmental Effects of Pesticides*, ed. M. P. Greaves, B. D. Smith & P. W. Greig-Smith , pp. 275–82.

Cousens, R. D. & Medd, R. W. (1994). Discussion of the extent of Australian ecological and economic data on weeds. *Plant Protection Quarterly*, **9**, 69–72.

Cousens, R. & Moss, S. R. (1990). A model of the effects of cultivation on the vertical distribution of weed seeds within the soil. *Weed Research*, **30**, 61–70.

Cousens, R., Moss, S. R., Cussans, G. W. & Wilson, B. J. (1987). Modeling weed populations in cereals. *Reviews of Weed Science*, **3**, 93–112.

Cousens, R. & Peters, N.C.B. (1993). Predicting the emergence of weeds in the field. In *Pests of Pastures: Weed, Invertebrate and Disease Pests of Australian Sheep Pastures*, ed. E. S. Delfosse, pp. 133–8. CSIRO: Melbourne.

Cousens, R., Pollard, F. & Denner, R. A. P. (1985b). Competition between *Bromus sterilis* and winter cereals. *Aspects of Applied Biology 9, The Biology and Control of Weeds in Cereals*, pp. 67–74.

Cousens, R., Wilson, B. J. & Cussans, G. W. (1985a). To spray or not to spray: the theory behind the practice. *Proceedings of the British Crop Protection Conference – Weeds – 1985*, pp. 671–8.

Cramer, H. H. (1967). Plant protection and world crop production. *Pflanzenschutznachrichten*, **20**, 15–24.

Crawley, M. J. (1987). What makes a community invasible? In *Colonization, Succession and Stability*, ed. A. J. Gray, M. J. Crawley & P. J. Edwards, pp. 429–53. Oxford: Blackwell Scientific.

Cresswell, E. (1960). Ranging behaviour studies with Romney Marsh and Cheviot sheep in New Zealand. *Animal Behaviour*, **8**, 32–8.

Crow, J. F. & Kimura, M. (1970). *An Introduction to Population Genetics Theory*. New York: Harper & Row.

Cullen, J. M. & Delfosse, E. S. (1990). Progress and prospects in biological control of weeds. *Proceedings of the 9th Australian Weeds Conference*, pp. 452–76.

Curran, P. L. & MacNaeidhe, F. S. (1986). Weed invasion of milled-over bog. *Weed Research*, **26**, 45–50.

Cussans, G. W. & Moss, S. R. (1982). Population dynamics of annual grass weeds. *Proceedings of the 1982 British Crop Protection Symposium 'Decision Making in the Practice of Crop Protection'*, pp. 91–8.

Cussans, G. W. & Wilson, B. J. (1975). Some effects of crop row width and seed rate on competition between spring barley and wild oat, *Avena fatua* L., or common couch *Agropyron repens* (L.) Beauv. *Proceedings of the EWRS*

Symposium 'Status, Biology and Control of Grass Weeds in Europe', pp. 77–86.

Cuthbertson, E. G. (1967). Skeleton weed distribution and control. *New South Wales Department of Agriculture Bulletin*, **68**, 1–47.

Darmency, H. (1994). Genetics of herbicide resistance in weeds and crops. In *Herbicide Resistance in Plants: Biology and Biochemistry*, ed. S. B. Powles & J. A. M. Holtum, pp. 263–98. Boca Raton, FL: Lewis.

Darmency, H. & Gasquez, J. (1981). Inheritance of triazine resistance in *Poa annua*: consequences for population dynamics. *New Phytologist*, **89**, 487–93.

Darmency, H. & Gasquez, J. (1990). The fate of herbicide resistant genes in weeds. In *Managing Resistance to Agrochemicals: From Fundamental Research to Practical Strategies*, ed. M. B. Green, H. M. LeBaron & W. K. Moberg, pp. 231–46. Washington DC: American Chemical Society.

Davies, D. H. K. (1985). Patterns of emergence of grass weeds in winter cereals in south-east Scotland – a long-term study. *Aspects of Applied Biology 9, The Biology and Controls of Weeds in Cereals*, pp. 19–30.

Debaeke, P. (1988). Dynamique de quelques dicotyledones adventices en culture de cereale. II. Survie, floraison et fructification. *Weed Research*, **28**, 265–79.

Debaeke, P. & Barralis, G. (1988). Essai de modelisation de l'evolution du stock semencier: application a une dicotyledone adventice *Anagallis arvensis* L. sur 3 sites pedoclimatiques. *Annales de VIIIe Colloque International sur la Biologie, l'Ecologie et la Systematique des Mauvaises Herbes*, pp. 91–102.

Deevey, E. S. (1947). Life tables for natural populations of animals. *Quarterly Reviews of Biology*, **22**, 283–314.

Derksen, D. A., Swanton, C. J. & Thomas, A. G. (1991). Weed community changes over time in reduced tillage systems. *Abstracts of the Weed Science Society of America*, **31**, 40.

Dessaint, F., Chadoeuf, R. & Barralis, G. (1990). Etude de la dynamique d'une communante adventice: III. Influence a long terme des techniques culturales sur la composition specifique du stock semencier. *Weed Research*, **30**, 319–30.

Dewey, S. A. & Whitesides, R. E. (1990). Weed seed analysis from four decades of Utah small grain drillbox surveys. *Proceedings of the Western Society of Weed Science*, p. 69.

Dirzo, R. & Harper, J. L. (1980). Experimental studies on slug–plant interactions. II. The effect of grazing by slugs on high density monocultures of *Capsella bursa-pastoris* and *Poa annua*. *Journal of Ecology*, **68**, 999–1011.

Dodd, J. (1987). An analysis of spread of skeleton weed, *Chondrilla juncea* L. in Western Australia. *Proceedings of the Eighth Australian Weeds Conference*, pp. 286–9.

Dodge, A. D. (1992). Mechanisms of resistance to herbicides. In *Achievements and Developments in Combating Pesticide Resistance*, ed. I. Denholm, A. L. Devonshire & D. W. Hollomon, pp. 201–16. London: Elsevier.

Doley, D. (1977). Parthenium weed (*Parthenium hysterophorus* L.): gas exchange characteristics as a basis for prediction of its geographical distribution. *Australian Journal of Agricultural Research*, **28**, 449–60.

Dorph-Peterson, K. (1925). Examinations of the occurrence and vitality of various weed species under different conditions. *Report of the 4th International Seed-Testing Congress*, pp. 7–12.

Douglas, B. J., Thomas, A. G. & Derksen, D. A. (1990). Downy brome (*Bromus tectorum*) invasion into southwestern Saskatchewan. *Canadian Journal of Plant Science*, **70**, 1143–51.

Doyle, C. J., Cousens, R. & Moss, S. R. (1986). A model of the economics of controlling *Alopecurus myosuroides* Huds. in winter wheat. *Crop Protection*, **5**, 143–50.

Druce, G. C. (1886). *The Flora of Oxfordshire, being a topographical and historical account of the flowering plants and ferns found in the county.* Oxford: Parker.

Duncan, C. N. & Weller, S. C. (1987). Heritability of glyphosate susceptibility among biotypes of field bindweed. *Journal of Heredity*, **78**, 257–60.

Dyer, W. E., Chee, P. W. & Fay, P. K. (1993). Rapid germination of sulfonylurea-resistant *Kochia scoparia* L. accessions is associated with elevated seed levels of branched chain amino acids. *Weed Science*, **41**, 18–22

Egley, G. H. & Dale, J. E. (1970). Ethylene, 2–chloroethylphosphonic acid, and witchweed germination. *Weed Science*, **18**, 586–9.

Elliott, J. G. (1982). Weed control in cereals – strategy and tactics. In *Decision Making In the Practice of Crop Protection, Monograph 25*, ed. R. B. Austin, pp. 115–19. Croydon: BCPC Publications.

England, G. J. (1954). Observations on the grazing behaviour of different breeds of sheep at Pantyrhuad Farm, Carmarthenshire. *British Journal of Animal Behaviour*, **2**, 56–60.

Eplee, R. E. (1975). Ethylene: a witchweed seed germination stimulant. *Weed Science*, **23**, 433–6.

Eplee, R. E. (1983). Progress in control of *Striga asiatica* in the United States. *Proceedings of the Second International Workshop on Striga, 5–8 October 1981, IDRC/ICRISAT, Ouagadougou, Upper Volta*, pp. 99–102.

Ervio, L-R. & Salonen, J. (1986). Changes in the Finnish weed population. *Proceedings of the 27th Swedish Weed Conference*, pp. 230–5.

Falconer, D. S. (1981). *Introduction to Quantitative Genetics*, 2nd edn. London: Longman.

Faulkner, J. S. (1974). Heritability of paraquat tolerance in *Lolium perenne* L. *Euphytica*, **23**, 281–8.

Fay, P. K. & Olson, W. A. (1978). Technique for separating weed seed from soil. *Weed Science*, **26**, 530–3.

Fenner, M. (1985). *Seed Ecology.* London: Chapman & Hall.

Firbank, L. G. (1989). Forecasting weed infestations – the desirable and the possible. *Proceedings of the Brighton Crop Protection Conference – Weeds – 1989*, pp. 567–72.

Firbank, L. G., Mortimer, A. M. & Putwain, P. D. (1985). *Bromus sterilis* in winter wheat: a test of a predictive population model. *Aspects of Applied Biology 9, 1985, The Biology and Control of Weeds in Cereals*, pp. 59–66.

Firbank, L. G. & Watkinson, A. R. (1985). On the analysis of competition within two-species mixtures of plants. *Journal of Applied Ecology*, **22**, 503–17.

Firbank, L. G. & Watkinson, A. R. (1986). Modelling the population dynamics of an arable weed and its effects upon crop yield. *Journal of Applied Ecology*, **23**, 147–59.

Fogelfors, H. (1982). Collection of chaff, awns and straw when combining and its influence on the seed bank and the composition of the weed flora. *Proceedings of the 23rd Swedish Weed Conference*, pp. 339–45.

Forbes, J. C. (1977). Population flux and mortality in a ragwort (*Senecio jacobaea* L.) infestation. *Weed Research*, **17**, 387–91.

Forcella, F. (1985). Final distribution is related to rate of spread in alien weeds. *Weed Research*, **25**, 181–91.

Forcella, F. & Harvey, S. J. (1988). Patterns of weed migration in north western

USA. *Weed Science*, **36**, 194–201.

Forcella, F. & Lindstrom, M. J. (1988). Weed seed populations in ridge and conventional tillage. *Weed Science*, **36**, 500–3.

Forcella, F., Wilson, R. G., Renner, K. A., Dekker, J., Harvey, R. G., Alm, D. A., Buhler, D. D. & Cardina, J. (1992). Weed seedbanks of the U.S.A. Corn Belt: Magnitude, variation, emergence, and application. *Weed Science*, **40**, 636–44.

Forcella, F., Wood, J. T. & Dillon, S. P. (1986). Characteristics distinguishing invasive weeds within *Echium* (Bugloss). *Weed Research*, **26**, 351–64.

Froud-Williams, R. J. (1983). The influence of straw disposal and cultivation regime on the population dynamics of *Bromus sterilis*. *Annals of Applied Biology*, **103**, 139–48.

Froud-Williams, R. J., Chancellor, R. J. & Drennan, D. S. H. (1984). The effects of seed burial and soil disturbance on emergence and survival of arable weeds in relation to minimal cultivation. *Journal of Applied Ecology*, **21**, 629–41.

Froud-Williams, R. J. & Ferris-Kaan, R. (1991). Intraspecific variation among populations of cleavers (*Galium aparine*). *Proceedings of the Brighton Crop Protection Conference – Weeds – 1991*, pp. 1007–14.

Fryer, J. D. & Chancellor, R. J. (1970) Herbicides and our changing weeds. In *The Flora of a Changing Britain*. Botanical Society of the British Isles, Report No.11, pp. 105–18.

Fuerst, E. P., Arntzen, C. J. & Penner, D. (1986). Herbicide cross-resistance in triazine-resistant biotypes of four species. *Weed Science*, **34**, 344–53.

Gasquez J. (1991). Mutation for triazine resistance within susceptible populations of *Chenopodium album* L. In *Herbicide Resistance in Weeds and Crops: Proceedings of the 19th Long Ashton Symposium*, ed. J. C. Caseley, G. W. Cussans & R. K. Atkin, pp. 103–15. Oxford: Butterworth Heinemann.

Gasquez, J. & Darmency, H. (1989). Appearance and spread of triazine resistance in common lambsquarters, *Chenopodium album* L. *Abstracts of the Weed Science Society of America*, p. 133.

Gasquez, J., Al Mouemar, A. & Darmency, H. (1984). Quels gènes pour la résistance chloroplastique aux triazines chez *Chenopodium album*? *VIIe Colloque International sur l'Ecologie, la Biologie et la Systematique des Mauvaises Herbes*, pp. 281–6.

Gasquez, J., Al Mouemar, A. & Darmency, H. (1985). Triazine herbicide resistance in *Chenopodium album* L. Occurrence and characteristics of an intermediate biotype. *Pesticide Science*, **16**, 390–5.

Gill, G. S. (1993). Development of herbicide resistance in annual ryegrass in the cropping belt of Western Australia. *Proceedings of the 10th Australian and 14th Asian-Pacific Weed Conference*, pp. 282–5.

Goloff, A. A. & Bazzaz, F. A. (1975). A germination model for natural seed populations. *Journal of Theoretical Biology*, **52**, 259–83.

Gonzalez Ponce, R., Lamela, A. & Salas, M. L. (1988). Effets concurrentiels entre *Avena sterilis* L. et deux varietes de ble a des doses differentes de fertilisation azotee. *Annales VIIIe Colloque International sur la Biologie, L'ecologie et la Systematique des Mauvaises Herbes*, pp. 573–9.

Goodman, D. (1987). The demography of chance extinction. In *Viable Populations for Conservation*, ed. M. E. Soulé, pp. 11–34. Cambridge: Cambridge University Press.

Goodman, P. J., Braybrooks, E. M., Marchant, C. J. & Lambert, J. M. (1969). Biological flora of the British Isles. *Spartina × townsendii* H. & J. Groves *sensu lato. Journal of Ecology*, **57**, 285–313.

Gordon, A. J., Kluge, R. L. & Neser, S. (1984). Effect of the gall midge *Zeuxidiplosis giardi* (Diptera: Cecidomyiidae) on seedlings of St. John's Wort, *Hypericum perforatum. Proceedings of the VI International Symposium on Biological Control of Weeds, 19–25 August 1984, Vancouver, Canada*, pp. 743–8.

Goss-Custard, J. D. & Moser, M. E. (1988). Rates of change in numbers of Dunlin, *Calidris alpina*, wintering in British estuaries in relation to the spread of *Spartina anglica. Journal of Applied Ecology*, **25**, 95–109.

Gramshaw, D. (1972). Germination of annual ryegrass seeds (*Lolium rigidum* Gaud.) as influenced by temperature, light, storage environment, and age. *Australian Journal of Agricultural Research*, **23**, 779–87.

Gray, P. A. (1960). The water hyacinth in the Sudan. In *The Biology of Weeds*, ed. J. L. Harper, pp. 184–8. Oxford: Blackwell Scientific.

Gregory, P. H. (1973). *The Microbiology of the Atmosphere*. London: Leonard Hill.

Gressel, J. (1986). Modes and genetics of herbicide resistance in plants. In *Pesticide Resistance: Strategies and Tactics for Management*, pp. 54–73. Washington DC: National Academy Press.

Gressel, J. (1991). Why get resistance? It can be prevented or delayed. In *Herbicide Resistance in Weeds and Crops: Proceedings of the 19th Long Ashton Symposium, 1989*, ed. J. Caseley, G. W. Cussans & R. K. Atkin, pp. 1–25. Oxford: Butterworth Heinemann.

Gressel, J. (in press). Catch 22 – mutually exclusive strategies for delaying/ preventing polygenically vs. monogenically inherited resistances. In *Options 2000*, ed. N. Ragsdale. Washington DC: ACS.

Gressel, J. & Segel, L. A. (1978). The paucity of genetic adaptive resistance of plants to herbicides: possible biological reasons and implications. *Journal of Theoretical Biology*, **75**, 349–71.

Gressel, J. & Segel, L. A. (1990). Modelling the effectiveness of herbicide rotations and mixtures as strategies to delay or preclude resistance. *Weed Technology*, **4**, 186–98.

Groenendael, J. M. van & Habekotté, B. (1988). *Cyperus esculentus* L. – biology, population dynamics and possibilities to control this neophyte. *Zeitschrift fur Pflanzenkrankheit und Pflanzenschutz, Sonderheft*, **XI**, 61–9.

Groves, R. H. (1986). Invasion of mediterranean ecosystems by weeds. In *Resilience in Mediterranean-type Ecosystems*, ed. B. Dell, A. J. M. Hopkins & B. B. Lamont, pp. 129–45. Dordrecht: Junk.

Haas, H. & Streibig, J. C. (1982). Changing patterns of weed distribution as a result of herbicide use and other agronomic factors. In *Herbicide Resistance in Plants*, ed. H. M. LeBaron & J. Gressel, pp. 57–79. New York: Wiley.

Hall, L. M., Holtum, J. A. M. & Powles, S. B. (1994). Mechanisms responsible for cross resistance and multiple resistance. In *Herbicide Resistance in Plants: Biology and Biochemistry*, ed. S. B. Powles & J. A. M. Holtum, pp. 243–62. Boca Raton, FL: Lewis.

Hance, R. J, & Holly, K. (1990). The properties of herbicides. In *Weed Control Handbook: Principles*, 8th edn, ed. R. J. Hance & K. Holly, pp. 75–125. Oxford: Blackwell Scientific.

Hancock, J. (1953). Grazing behaviour of cattle. *Animal Breeding Abstracts*, **21**, 1–13.

Harper, J. L. (1956). The evolution of weeds in relation to the resistance to herbicides. *Proceedings of the 3rd British Weed Control Conference*, pp. 179–88.

Harper, J. L. (1969). The role of predation in vegetational diversity. *Brookhaven Symposium 'Diversity and Stability in Ecological Systems'*. New York: Brookhaven National Laboratory.

Harper, J. L. (1977). *Population Biology of Plants*. London: Academic Press.

Harper, J. L. & Gajic, D. (1961). Experimental studies of the mortality and plasticity of a weed. *Weed Research*, 1, 91–104.

Harradine, A. R. (1985). Dispersal and establishment of slender thistle (*Carduus pycnocephalus*) as affected by ground cover. *Australian Journal of Agricultural Research*, 36, 791–7.

Harrington, K. C. (1990). Spraying history and fitness of nodding thistle, *Carduus nutans*, populations resistant to MCPA and 2,4-D. *Proceedings of the 9th Australian Weeds Conference*, pp. 201–4.

Harris, P. (1973). The selection of effective agents for the biological control of weeds. *Proceedings of the 3rd International Symposium on the Biological Control of Weeds*, pp. 75–85.

Hartley, M. J., Lyttle, L. A. & Popay, A. I. (1984). Control of Californian thistle by grazing management. *Proceedings of the 37th New Zealand Weed and Pest Control Conference*, pp. 24–7.

Hassell, M. P. & Comins, H. N. (1976). Discrete time models for two-species competition. *Theoretical Population Biology*, 9, 202–21.

Häusler, R. E., Holtum, J. A. M. & Powles, S. B. (1991). Cross-resistance to herbicides in annual ryegrass (*Lolium rigidum*) IV. Correlation between membrane effect and resistance to graminicides. *Plant Physiology*, 97, 1035–43.

Heap, I. M. (1988). Resistance to herbicides in annual ryegrass (*Lolium rigidum*). PhD thesis, University of Adelaide.

Heap, I. M. (1991). Resistance to herbicides in annual ryegrass (*Lolium rigidum*) in Australia. In *Herbicide Resistance in Weeds and Crops: Proceedings of the 19th Long Ashton Symposium, 1989*, ed. J. Caseley, G. W. Cussans & R. K. Atkin, pp. 57–66. Oxford: Butterworth Heinemann.

Heap, I. M. & Knight, R. (1990). Variations in herbicide cross-resistance among populations of annual ryegrass (*Lolium rigidum*) resistant to diclofop-methyl. *Australian Journal of Agricultural Research*, 41, 121–8.

Heap, J. & Knight, R. (1982). A population of ryegrass tolerant to the herbicide diclofop-methyl. *Journal of the Australian Institute of Agricultural Science*, 48, 156–7.

Hengeveld, R. (1989). *Dynamics of Biological Invasions*. London: Chapman & Hall.

Hocking, P. J. & Liddle, M. J. (1986). The biology of Australian weeds: 15. *Xanthium occidentale* Bertol. complex and *Xanthium spinosum* L. *Journal of the Australian Institute of Agricultural Science*, 52, 191–221.

Holliday, R. J. & Putwain, P. D. (1980). Evolution of herbicide resistance in *Senecio vulgaris:* variation in susceptibility to simazine between and within populations. *Journal of Applied Ecology*, 17, 779–91.

Holm, L., Plunknett, D. L., Pancho, J. V. & Herberger, J. P. (1977). *The Worlds Worst Weeds: Distribution and Biology*. Honolulu: University Press of Hawaii.

Holmes, E. E. (1993). Are diffusion models too simple? A comparison with telegraph models of invasion. *American Naturalist*, 142, 403–19.

Holmes, E. E., Lewis, M. A., Banks, J. E. & Veit, R. R. (1994). Partial differential equations in ecology: spatial interactions and population dynamics. *Ecology*, **75**, 17–29.

Holroyd, J. (1964). The emergence and growth of *Avena fatua* from different depths in the soil. *Proceedings of the 7th British Weed Control Conference*, **2**, 621–7.

Holt, J.S. & LeBaron, H. M. (1990). Significance and distribution of herbicide resistance. *Weed Technology*, **4**, 141–9.

Holt, J. S. & Thill, D. C. (1994). Growth and productivity of resistant plants. In *Herbicide Resistance in Plants: Biology and Biochemistry*, ed. S. B. Powles & J. A. M. Holtum, pp. 299–316. Boca Raton, FL: Lewis.

Holtum, J. A. M. & Powles, S. B. (1991). Annual ryegrass: an abundance of resistance, a plethora of mechanisms. *Proceedings of the British Crop Protection Conference – Weeds – 1991*, pp. 1071–7.

Hoppensteadt, F. C. (1982). *Mathematical Methods in Population Biology*. Cambridge: Cambridge University Press.

Horowitz, M. (1973). Spatial growth of *Sorghum halepense* (L.) Pers. *Weed Research*, **13**, 200–8.

Howard, C. L. (1991). Comparative Ecology of Four Brome Grasses. PhD thesis, University of Liverpool.

Howard, C. L., Mortimer, A. M., Gould, P., Putwain, P. D., Cousens, R. & Cussans, G. W. (1991). The dispersal of weeds – seed movement in arable agriculture. *Proceedings of the Brighton Crop Protection Conference – Weeds*, pp. 664–73.

Hume, L. (1987). Long-term effects of 2,4-D application on plants. 1. Effects on the weed community in a wheat crop. *Canadian Journal of Botany*, **65**, 2530–6.

Hurtt, W. & Taylorson, R. B. (1986). Chemical manipulation of weed emergence. *Weed Research*, **26**, 259–67.

Islam, A. & Powles, S. (1988). Inheritance of resistance to paraquat in barley grass *Hordeum glaucum* Steud. *Weed Research*, **28**, 393–7.

Itoh, K. & Miyahara, M. (1984). Inheritance of paraquat resistance in *Erigeron philadelphicus* L. *Weed Research (Japan)*, **29**, 301–7.

Jackson, J. B. C., Buss, L. W. & Cook, R. E. (1985). *Population Biology and Evolution of Clonal Organisms*. New Haven: Yale University Press.

Jarvis, R. H. (1981). The integrated effect of herbicides and cultural methods on grass weed control at Boxworth E.H.F. *Proceedings of the Conference 'Grass Weeds in Cereals in the United Kingdom'*. Association of Applied Biologists, Warwick. pp. 367–76.

Jasieniuk, M., Brûlé-Babel, A. L. & Morrison, I. N. (1993). The genetics of trifluralin resistance in green foxtail (*Setaria viridis* (L.) Beauv.). *Weed Science Society of America Abstracts*, p. 61.

Jeffers, J. N. R. (1982). *Modelling*. Outline Studies in Ecology. London: Chapman & Hall.

Jetsum, A. R. (1988). Commercial application of biological control: status and prospects. *Philosophical Transactions of the Royal Society of London*, B **318**, 357–73.

Johnson, W. C., Sharpe, D. M., de Angelis, D. L., Fields, D. E. & Olson, R. J. (1981). Modelling seed dispersal and forest island dynamics. In *Forest Dynamics in Man-dominated Landscapes*, ed. R. L. Burgess & D. M. Sharpe, pp. 215–24. New York: Springer Verlag.

Johnston, D. T. & Faulkner, J. S. (1991). Herbicide resistance in the Graminaceae – a plant breeder's view. In *Herbicide Resistance in Weeds and*

Crops: Proceedings of the 19th Long Ashton Symposium, ed. J. C. Caseley, G. W. Cussans & R. K. Atkin, pp. 319–30. Oxford: Butterworth Heinemann.

Jordano, P. (1982). Migrant birds are the main seed dispersers of blackberries in southern Spain. *Oikos*, **38**, 183–93.

Jorrin, J., Menendez, J., Romera, E., Taberner, A., Tena, M. & De Prado, R. (1992). Chlorotoluron resistance in a black-grass *Alopecurus myosuroides* biotype is due to herbicide detoxification. *Mededelingen van de Faculteit Landbouwwetenschappen, Rijksuniversiteit Gent*, **57 (3b)**, pp. 1047–52.

Julien, M. H. (1992). *Biological Control of Weeds: A World Catalogue of Agents and their Target Weeds*, 3rd ed. Wallingford: CAB International.

Julien, M. H., Kerr, J. D. & Chan, R. R. (1984). Biological control of weeds: an evaluation. *Protection Ecology*, **7**, 3–25.

Karssen, C. M. & Bouwmeester, H. J. (1992). Annual dormancy patterns of weed seeds influence weed control. *Proceedings of the First International Weed Control Congress, Melbourne 1992*, **1**, 98–104.

Kays, S. & Harper, J. L. (1974). The regulation of plant and tiller density in a grass sward. *Journal of Ecology*, **62**, 97–105.

Keddy, P. A. (1989). *Competition*. London: Chapman & Hall.

Kelley, A. D. & Bruns, V. F. (1975). Dissemination of weed seeds by irrigation water. *Weed Science*, **23**, 486–93.

Kelly, D. & McCallum, K. (1990). Demography, seed biology and biological control of *Carduus nutans* in New Zealand. In *Biology and Control of Invasive Plants*, pp. 72–9.

Kemp, M. S., Moss, S. R. & Thomas, T. H. (1990). Herbicide resistance in *Alopecurus myosuroides*. In *Managing Resistance to Agrochemicals: From Fundamental Research to Practical Strategies*, ed. M. B. Green, H. M. LeBaron & W. K. Moberg, pp. 376–93. Washington DC: American Chemical Society.

Kendall, M. G. (1948). A form of wave propagation associated with the equation of heat conduction. *Proceedings of the Cambridge Philosophical Society*, **44**, 591–3.

Kiewnick, L. (1964). Untersuchungen uber den einfluss der samen- und bodenmikroflora auf die lebensdauer der spelzfruchte des flughafers (*Avena fatua* L.). II. Zum einfluss der mikroflora auf die lebensdauer der samen im boden. *Weed Research*, **4**, 31–43.

King, R. P., Lybecker, D. W., Schweizer, E. E. & Zimdahl, R. L. (1986). Bioeconomic modeling to simulate weed control strategies for continuous corn (*Zea mays*). *Weed Science*, **34**, 972–9.

Kirby, C. (1980). *The Hormone Weedkillers*. Croydon: BCPC Publications.

Kirkpatrick, B. L. & Bazzaz, F. A. (1979). Influence of certain fungi on seed germination and seedling survival of four colonizing annuals. *Journal of Applied Ecology*, **16**, 515–27.

Kloot, P. M. (1982). The naturalisation of *Echium plantagineum* L. in Australia. *Australian Weeds*, **1**, 29–31.

Kloot, P. M. (1986). The naturalised flora of South Australia. 3. Its origin, introduction, distribution, growth forms and significance. *Journal of the Adelaide Botanic Gardens*, **10**, 99–111.

Kloot, P. M. (1987). The naturalised flora of South Australia. 4. Its manner of introduction. *Journal of the Adelaide Botanic Gardens*, **10**, 223–40.

Kohn, J. R. & Casper, B. B. (1992). Pollen-mediated gene flow in *Cucurbita foetidissima* (Cucurbitaceae). *American Journal of Botany*, **79**, 57–62.

Krebs, C. J. (1972). *Ecology: The Experimental Analysis of Distribution and*

Abundance. New York: Harper & Row.

Kropff, M. J. (1988). Modelling the effects of weeds on crop production. *Weed Research*, **28**, 465–71.

Kudsk, P. & Kristensen, J. L. (1992). Effect of environmental factors on herbicide performance. *Proceedings of the First International Weed Control Congress, Melbourne*, pp. 173–86.

Kunin, W. E. (1993). Sex and the single mustard: population density and pollinator behaviour affects on seed-set. *Ecology*, **74**, 2145–60.

Lacey, J. R., Wallander, R. & Olson-Rutz, K. (1992). Recovery, germinability, and viability of leafy spurge (*Euphorbia esula*) seeds ingested by sheep and goats. *Weed Technology*, **6**, 599–602.

Lacey, W. S. (1957). A comparison of the spread of *Galinsoga parviflora* and in Britain. In *Progress in the Study of the British Flora*, ed. J. E. Lousley, pp. 109–15. London: Botanical Society of the British Isles.

Lapham, J. (1985). Unrestricted growth, tuber formation and spread of *Cyperus esculentus* L. in Zimbabwe. *Weed Research*, **25**, 323–9.

Lapham, J. (1987). Population dynamics and competitive effects of *Cyperus esculentus* (yellow nutsedge) – prediction of cost-effective control strategies. *Proceedings of the 1987 British Crop Protection Conference – Weeds*, pp. 1043–50.

Lapham, J., Drennan, D. S. H. & Francis, L. (1985). Population dynamics of *Cyperus esculentus* L. (yellow nutsedge) in Zimbabwe. *Proceedings of the 1985 British Crop Protection Conference – Weeds*, pp. 395–402.

Law, R. (1975). Colonisation and the evolution of life histories in *Poa annua*. PhD thesis, University of Liverpool.

Lawrence, M. J. (1984). The genetical analysis of ecological traits. In *Evolutionary Ecology*, ed. B. Shorrocks, pp. 27–63. Oxford: Blackwell Scientific.

LeBaron, H. M. & Gressel, J. (1982). Preface. In *Herbicide Resistance in Plants*, ed. H. M. LeBaron & J. Gressel, p. xv. New York: Wiley Interscience.

LeBaron, H. M. & McFarland, J. E. (1990a). Resistance to herbicides. *CHEMTECH*, **20**, 508–11.

LeBaron, H. M. & McFarland, J. E. (1990b). Herbicide resistance in weeds and crops: an overview and prognosis. In *Managing Resistance to Agrochemicals: From Fundamental Research to Practical Strategies*, ed. M. B. Green, H. M. LeBaron & W. K. Moberg, pp. 336–52. Washington DC: American Chemical Society.

Lefkovich, L. P. (1965). The study of population growth in organisms grouped by stages. *Biometrics*, **21**, 1–18.

Leslie, P. H. (1945). On the use of matrices in certain population mathematics. *Biometrika*, **33**, 183–212.

Leverich, W.J. & Levin, D.A. (1979). Age specific survivorship and reproduction on *Phlox drummondii*. *American Naturalist*, **113**, 881–903.

Levin, D. A. & Kerster, H. W. (1984). Gene flow in seed plants. *Evolutionary Biology*, **7**, 139–220.

Levins, R. (1970). Extinction. In *Some Mathematical Questions in Biology*, ed. J. D. Cowan, pp. 75–107. Providence, RI: American Mathematical Society.

Lewis, J. (1973). Longevity of crop and weed seeds: survival after 20 years in soil. *Weed Research*, **13**, 179–91.

Lintell-Smith, G., Watkinson, A. R. & Firbank, L. G. (1991). The effects of reduced nitrogen and weed-weed competition on the populations of three common cereal weeds. *Proceedings of the Brighton Crop Protection*

Conference – Weeds – 1991, pp. 135–40.

Long, S. P. & Mason, C. F. (1983). *Saltmarsh Ecology*. Glasgow: Blackies & Son.

Lonsdale, W. M. (1981). Studies on thinning in pure and mixed populations of plants. PhD thesis, University of East Anglia.

Lonsdale, W. M. (1990). The self-thinning rule: dead or alive? *Ecology*, **71**, 1373–88.

Lonsdale, W. M. (1993a). Rates of spread of an invading species – *Mimosa pigra* in northern Australia. *Journal of Ecology*, **81**, 513–21.

Lonsdale, W. M. (1993b). Losses from the seed bank of *Mimosa pigra*: soil micro-organisms vs. temperature fluctuations. *Journal of Applied Ecology*, **30**, 654–60.

Lonsdale, W. M. & Watkinson, A. R. (1983). Plant geometry and self-thinning. *Journal of Ecology*, **71**, 285–97.

Lotz, L. A. P., Groeneveld, R. M. W., Habekotté, B. & Oene, H. van (1991). Reduction of growth and reproduction of *Cyperus esculentus* by specific crops. *Weed Research*, **31**, 153–60.

Lovett Doust, L. (1981). Population dynamics and local specialization in a clonal perennial (*Ranunculus repens*). I. The dynamics of ramets in contrasting habitats. *Journal of Ecology*, **69**, 743–55.

Loyn, R. H. & French, K. (1991). Birds and environmental weeds in south-eastern Australia. *Plant Protection Quarterly*, **6**, 137–48.

Lueschen, W. E. & Andersen, R. N. (1980). Longevity of velvetleaf (*Abutilon theophrasti*) seeds in soil under agricultural practices. *Weed Science*, **28**, 341–6.

Mack, R. N. (1981). Invasion of *Bromus tectorum* L. into western North America: an ecological chronicle. *Agro-Ecosystems*, **7**, 145–65.

Mack, R. N. (1985). Invading plants: their potential contribution to population biology. In *Studies on Plant Demography: A Festscrift for John L.Harper*, ed. J.White, pp. 127–142. New York: Academic Press.

Mackenzie, R., Mortimer, A. M., Putwain, P. D., Bryan, I. B. & Hawkes, T. R. (1993). The evolution of herbicide resistance: deliberate selection for chlorsulfuron resistance in perennial ryegrass. *Proceedings of the Brighton Crop Protection Conference – Weeds- 1993*, pp. 645–6.

Macnair, M.R. (1981). Tolerance of higher plants to toxic materials. In *Genetic Consequences of Man-Made Changes*, ed. J. A. Bishop & L. M. Cook, pp. 177–207. London: Academic Press.

Mahn, E. G. & Helmecke, K. (1979). Effects of herbicide treatment on the structure and functioning of agro-ecosystems. II. Structural changes in the plant community after the application of herbicides over several years. *Agro-Ecosystems*, **5**, 159–79.

Malchow, W. E., Maxwell, B. D., Fay, P. K. & Dyer, W. E. (1993). Frequency of triallate resistance in Montana. *Proceedings of the Western Society of Weed Science*, **46**, 75.

Mallory-Smith, C. A., Thill, D. C., Dial, M. J. & Zemetra, R. S. (1990). Inheritance of sulfonylurea herbicide resistance in *Lactuca* sp. *Weed Technology*, **4**, 787–90.

Mallory-Smith, C. A., Thill, D. C., Alcocer-Ruthling, M. & Thompson, C. R. (1992). Growth comparisons of sulfonylurea resistant and susceptible biotypes. *Proceedings of the First International Weed Control Congress*, **2**, 301–3.

Manlove, R. J. (1985). On the population ecology of *Avena fatua* L. PhD thesis, University of Liverpool.

Manlove, R. J., Mortimer, A. M. & Putwain, P. D. (1982). Modelling wild oat populations and their control. *Proceedings 1982 British Crop Protection Conference – Weeds*, pp. 749–56.

Marchant, C. J. (1963). Corrected chromosome numbers for *S. × townsendii* and its parent species. *Nature (London)*, **199**, 929.

Marshall, E. J. P. (1985). Weed distributions associated with cereal field edges – some preliminary observations. *Aspects of Applied Biology 9, The Biology and Control of Weeds in Cereals*, pp. 49–58.

Marshall, E. J. P. (1988). Field-scale estimates of grass weed populations in arable land. *Weed Research*, **28**, 191–8.

Marshall, E. J. P. (1990). Interference between sown grasses and the growth of rhizome of *Elymus repens* (couch grass). *Agriculture, Ecosystems and Environment*, **33**, 11–22.

Marshall, E. J. P. & Butler, R. (1991). Seed rain patterns. *Institute of Arable Crops Research Report 1990*, p. 52.

Marshall, E. J. P. & Hopkins, A. (1990). Plant species composition and dispersal in agricultural land. In *Species Dispersal in Agricultural Habitats*, ed. R. G. H. Bunce & D. C. Howard, pp. 98–116. London: Belhaven Press.

Martin, R. J. (1992). Simulation of the effects of herbicide and crop rotation practices on the population dynamics of wild oats. *Proceedings of the Fourth International Oat Conference, Volume II, Wild Oats in World Agriculture*, pp. 88–90.

Martin, R. J. & Felton, W. L. (1990). Effect of crop rotation, tillage practice and herbicide use on the population dynamics of wild oats. *Proceedings of the 9th Australian Weeds Conference*, pp. 20–3.

Martin, R. J. & McMillan, M. G. (1984). Some results of a weed survey in northern New South Wales. *Australian Weeds*, **3**, 115–6.

Matthews, J. M. (1994). Management of herbicide resistant populations. In *Herbicide Resistance in Plants: Biology and Biochemistry*, ed. S. B. Powles & J. A. M. Holtum, pp. 317–27. Boca Raton, FL: Lewis.

Matthews, J. M. & Powles, S. B. (1992). Aspects of the population dynamics of selection for herbicide resistance in *Lolium rigidum* (Gaud). *Proceedings of the First International Weed Control Congress*, **2**, 318–20.

Maxwell, B. D. & Mortimer, A. M. (1994). Selection for herbicide resistance. In *Herbicide Resistance in Plants: Biology and Biochemistry*, ed. S. B. Powles & J. A. M. Holtum, pp. 1–26. Boca Raton, FL: Lewis.

Maxwell, B. D., Roush, M. L. & Radosevich, S. R. (1990). Predicting the evolution and dynamics of herbicide resistance in weed populations. *Weed Technology*, **4**, 2–13.

Maxwell, B. D., Wilson, M. V. & Radosevich, S. R. (1988). Population modeling approach for estimating leafy spurge (*Euphorbia esula*) development and control. *Weed Technology*, **2**, 132–8.

May, R. M. & Dobson, A. P. (1986). Population dynamics and the rate of evolution of pesticide resistance. In *Pesticide Resistance: Strategies and Tactics for Management*, pp. 170–93. Washington DC: National Academy Press.

McCanny, S. J. & Cavers, P. B. (1988). Spread of proso millet (*Panicum miliaceum* L.) in Ontario, Canada. II. Dispersal by combines. *Weed Research*, **28**, 67–72.

McCartney, H. A. (1990). Dispersal mechanisms through the air. In *Species Dispersal in Agricultural Habitats*, ed. R. G. H. Bunce & D. C. Howard, pp. 133–58. London: Belhaven Press.

McFadyen, P. J. (1985). Introduction of the gall fly *Rhopalomyia californica* from the USA into Australia for the control of the weed *Baccharis halimifolia*. *Proceedings of the VI International Symposium on Biological Control of Weeds*, pp. 779–87.

McIntyre, S., Finlayson, C. M., Ladiges, P. Y. & Mitchell, D. S. (1991). Weed community composition and rice husbandry practices in NSW, Australia. *Agriculture, Ecosystems and Environment*, **35**, 27–45.

McKinley, N. D. (1990). Sulfonylurea herbicide resistant weeds in cereals and non-crop areas in the US and Canada. *Proceedings of the 9th Australian Weeds Conference*, pp. 268–9.

McMahon, D. J. & Mortimer, A. M. (1980). The prediction of couch infestations – a modelling approach. *Proceedings of the 1980 British Crop Protection Conference – Weeds*, pp. 601–8.

McRill, M. & Sagar, G. R. (1973). Earthworms and seeds. *Nature (London)*, **243**, 482.

Medd, R. W. (1987a). Weed management on arable lands. In *Tillage: New Directions in Australian Agriculture*, ed. P. S. Cornish & J. E. Pratley, pp. 222–59. Melbourne: Inkata Press.

Medd, R. W. (1987b). Impact of legislative actions on the invasion of *Carduus nutans*. *Proceedings of the Eighth Australian Weeds Conference*, p. 292.

Medd, R. W., Auld, B. A., Kemp, D. R. & Murison, R. D. (1985). The influence of wheat density and spatial arrangement on annual ryegrass, *Lolium rigdum*, competition. *Australian Journal of Agricultural Research*, **36**, 361–71.

Medd, R. W. & Ridings, H. I. (1989). Relevance of seed kill for the control of annual grass weeds in crops. *Proceedings of the VIIth International Symposium on the Biological Control of Weeds, 1988, Rome, Italy*, pp. 645–50.

Medd, R. W. & Smith, R. C. G. (1978). Prediction of the potential distribution of *Carduus nutans* (nodding thistle) in Australia. *Journal of Applied Ecology*, **15**, 603–12.

Menz, K. M., Coote, B. G. & Auld, B. A. (1980/1). Spatial aspects of weed control. *Agricultural Systems*, **6**, 67–75.

Merrill, E. D. (1954). The botany of Cook's voyages. *Chronica Botanica*, **14**, 161–383.

Michael, P. W. (1981). Alien plants. In *Australian Vegetation*, ed. R. H. Groves, pp. 44–64. Cambridge: Cambridge University Press.

Michaux, B. (1989). Reproductive and vegetative biology of *Cirsium vulgare* (Savi) Ten. (Compositae: Cynareae). *New Zealand Journal of Botany*, **27**, 401–14.

Miller, I. L. (1988). Aspects of the biology and control of *Mimosa pigra* L. MScAgr thesis, University of Sydney.

Miller, S. D. & Nalewaja, J. D. (1990). Influence of burial depth on wild oats (*Avena fatua*) seed longevity. *Weed Technology*, **4**, 514–17.

Mohler, C. L. (1993). A model of the effects of tillage on emergence of weed seedlings. *Ecological Applications*, **3**, 53–73.

Mollison, D. (1977). Spatial contact models for ecological and epidemic spread. *Journal of the Royal Statistical Society*, **B 39**, 283–326.

Moody, M. E. & Mack, R. N. (1988). Controlling the spread of plant invasions: the importance of nascent foci. *Journal of Applied Ecology*, **25**, 1009–21.

Moolani, M. K., Knake, E. L. & Slife, F. W. (1964). Competition of smooth pigweed with corn and soybeans. *Weeds*, **12**, 126–8.

Morrison, I. N., Beckie, H. & Nawolsky, K. (1991). The occurrence of trifluralin resistant *Setaria viridis* (green foxtail) in western Canada. In *Herbicide Resistance in Weeds and Crops: Proceedings of the 19th Long Ashton Symposium, 1989*, ed. J. Caseley, G. W. Cussans & R. K. Atkin, pp. 67–75. Oxford: Butterworth Heinemann.

Mortimer, A. M. (1983). On weed demography. In *Recent Advances in Weed Research*, ed. W. W. Fletcher, pp. 3–40. Farnham Royal:Commonwealth Agricultural Bureau.

Mortimer, A. M. (1984). Population ecology and weed science. In *Perspectives on Plant Population Ecology*, ed. R. Dirzo & J. Sarukhan, pp. 363–88. Sunderland, MA: Sinauer.

Mortimer, A. M. (1985). Intractable weeds: a failure to appreciate ecological principles in weed control? *Proceedings of the 1985 British Crop Protection Conference – Weeds*, pp. 377–86.

Mortimer, A. M. (1987). The population ecology of weeds – implications for integrated weed management, forecasting and conservation. *Proceedings of the 1987 British Crop Protection Conference – Weeds*, pp. 935–44.

Mortimer, A. M. (1990). The biology of weeds. In *Weed Control Handbook: Principles*, 8th edn, ed. R. J. Hance & K. Holly, pp. 1–42. Oxford: Blackwell Scientific.

Mortimer, A. M., Gould, P. & Putwain, P. D. (1993). Difference equations in the description of weed population dynamics. *Proceedings of the European Weed Research Society Symposium 1993, 'Quantitative Approaches in Weed and Herbicide Research and Their Practical Application'*, pp. 615–22.

Mortimer, A. M., Putwain, P. D. & McMahon, D. J. (1978). A theoretical approach to the prediction of weed population sizes. *Proceedings of the 14th British Weed Control Conference*, pp. 467–74.

Mortimer, A. M., Sutton, J. J. & Gould, P. (1989). On robust weed population models. *Weed Research*, **29**, 229–38.

Mortimer, A. M., Sutton, J. J., Putwain, P. D. & Gould, P. (1990). The dynamics of mixtures of arable weed species. *Proceedings of the European Weed Research Society Symposium 1990, 'Integrated Weed Management in Cereals'*, pp. 19–26.

Mortimer, A. M., Ulf-Hansen, P. F. & Putwain, P. D. (1992). Modelling herbicide resistance: a study of ecological fitness. In *Achievements and developments in combating pesticide resistance*, ed. I. Denholm, A. L. Devonshire & D. W. Hollomon, pp. 148–64. London: Elsevier.

Moss, S. R. (1979). The influence of tillage and method of straw disposal on the survival and growth of black-grass, *Alopecurus myosuroides*, and its control by chlortoluron and isoproturon. *Annals of Applied Biology*, **91**, 91–100.

Moss, S. R. (1980a). Some effects of burning cereal straw on seed viability, seedling establishment and control of *Alopecurus myosuroides* Huds. *Weed Research*, **20**, 271–6.

Moss, S. R. (1980b). A study of black-grass (*Alopecurus myosuroides*) in winter wheat, as influenced by seed shed in the previous crop, cultivation system and straw disposal method. *Annals of Applied Biology*, **94**, 121–6.

Moss, S. R. (1983). The production and shedding of *Alopecurus myosuroides* Huds. seeds in winter cereal crops. *Weed Research*, **23**, 45–51.

Moss, S. R. (1985). The survival of *Alopecurus myosuroides* Huds. seeds in soil. *Weed Research*, **25**, 201–11.

Moss, S. R. (1987). Influence of tillage, straw disposal system and seed return on the population dynamics of *Alopecurus myosuroides* Huds. in winter wheat.

Weed Research, **27**, 313–20.

Moss, S. R. (1990a). The seed cycle of *Alopecurus myosuroides* in winter cereals: a quantitative analysis. *Proceedings of the EWRS Symposium 1990, 'Integrated Weed Management in Cereals*, pp. 27–35.

Moss, S. R. (1990b). Herbicide cross-resistance in slender foxtail (*Alopecurus myosuroides*). *Weed Science*, **38**, 492–6.

Moss, S. R. & Clarke, J. H. (1994). Guidelines for the prevention and control of herbicide-resistant black-grass (*Alopecurus myosuroides* Huds.). *Crop Protection*, **13**, 230–4.

Moss, S. R. & Cussans, G. W. (1985). Variability in the susceptibility of *Alopecurus myosuroides* (blackgrass) to chlorotoluron and isoproturon. *Aspects of Applied Biology 9, The Biology and Control of Weeds in Cereals*, pp. 91–8.

Moss, S. R. & Cussans, G. W. (1991). The development of herbicide resistant populations of *Alopecurus myosuroides* (blackgrass) in England. In *Herbicide Resistance in Weeds and Crops: Proceedings of the 19th Long Ashton Symposium, 1989*, ed. J. Caseley, G. W. Cussans & R. K. Atkin, pp. 34–48. Oxford: Butterworth Heinemann.

Moss, S. R. & Orson, J. (1988). The distribution of herbicide-resistant *Alopecurus myosuroides* (black-grass) in England. *Aspects of Applied Biology 18, Weed Control in Cereals and the Impact of Legislation on Pesticide Application*, pp. 177–85.

Muenscher, W. C. (1955) *Weeds*, 2nd edn. New York: MacMillan.

Mulligan, G. A. & Bailey, L. G. (1975). The biology of Canadian weeds. 8. *Sinapis arvensis* L. *Canadian Journal of Plant Science*, **55**, 171–83.

Murdoch, A. J. (1983). Environmental control of germination and emergence in *Avena fatua*. *Aspects of Applied Biology 4, 1983, Influence of Environmental Factors on Herbicide Performance and Crop and Weed Biology*, pp. 63–9.

Murdoch, A. J. (1988). Long-term profit from weed control. *Aspects of Applied Biology 18, 1988, Weed Control in Cereals and the Impact of Legislation on Pesticide Application*, pp. 91–8.

Murdoch, A. J. & Roberts, E. H. (1982). Biological and financial criteria of long-term control strategies for annual weeds. *Proceedings of the 1982 British Crop Protection Conference – Weeds*, pp. 741–8.

Myerscough, P. J. & Whitehead, F. H. (1966). Comparative biology of *Tussilago farfara* L., *Chamaenerion angustifolium* (L.) Scop., *Epilobium montanum* L. and *Epilobium adenocaulon* Hausskn. I. General biology and germination. *New Phytologist*, **65**, 192–210.

Nadeau, L. & King, J. R. (1991). Seed dispersal and seedling establishment of *Linaria vulgaris* Mill. *Canadian Journal of Plant Science*, **71**, 771–82.

Naylor, R. E. L. (1970). The prediction of blackgrass infestations. *Weed Research*, **10**, 296–9.

Naylor, R. E. L. (1972). Aspects of the population dynamics of the weed *Alopecurus myosuroides* Huds. in winter cereal crops. *Journal of Applied Ecology*, **9**, 127–39.

Niemann, P. & Pestemer, W. (1984). Resistance of blackgrass (*Alopecurus myosuroides*) from different sites to herbicides. *Nachrichtenblatt des Deutschen Pflanzenschutzdienstes*, **36**, 113–18.

Norris, R. F. (1981). Zero tolerance for weeds? *Proceedings of the 33rd Annual California Weed Conference*, pp. 46–8.

Norris, R. F. (1992). Have ecological and biological studies improved weed control strategies? *Proceedings of the First International Weed Control Congress*, pp. 7–33.

Norton, G. A. & Conway, G. R. (1977). The economic and social context of pest, disease and weed problems. In *Origin of Pest, Parasite, Disease and Weed Problems*, ed. J. M. Cherret & G. R. Sagar, pp. 205–26. Oxford: Blackwell Scientific.

Nussbaum, E. S., Wiese, A. F., Crutchfield, D. E., Chenault, E. W. & Lavake, D. (1985). The effects of temperature and rainfall on emergence and growth of eight weeds. *Weed Science*, **33**, 165–70.

O'Donovan, J. T., de St. Remy, E. A., O'Sullivan, P. A., Dew, D. A. & Sharma, A. K. (1985). Influence of the relative time of emergence of wild oat (*Avena fatua*) on yield loss of barley (*Hordeum vulgare*) and wheat (*Triticum aestivum*). *Weed Science*, **33**, 498–503.

O'Toole, J. J. & Cavers, P. B. (1983). Input to seed banks of proso millet (*Panicum miliaceum*) in southern Ontario. *Canadian Journal of Plant Science*, **63**, 1023–30.

Okubo, A. (1980). *Diffusion and Ecological Problems: Mathematical Models*. Berlin: Springer-Verlag.

Olivieri, I., Swan, M. & Gouyon, P-H. (1983). Reproductive system and colonizing strategy of two species of *Carduus* (Compositae). *Oecologia (Berlin)*, **60**, 114–17.

Oorschot, J. L. P. van (1991). Chloroplastic resistance of weeds to triazines in Europe. In *Herbicide Resistance in Weeds and Crops: Proceedings of the 19th Long Ashton Symposium*, ed. J. C. Caseley, G. W. Cussans & R. K. Atkin, pp. 87–102. Oxford: Butterworth Heinemann.

Opdam, P. (1990). Dispersal in fragmented populations: the key to survival. In *Species Dispersal in Agricultural Habitats*, ed. R. G. H. Bunce & D. C. Howard, pp. 3–17. London: Belhaven Press.

Pacala, S. W. & Silander, J. A. (1987). Neighbourhood interference among velvetleaf, *Abutilon theophrasti*, and pigweed, *Amaranthus retroflexus*. *Oikos*, **48**, 217–24.

Pandey, S. & Medd, R. W. (1990). Integration of seed and plant kill tactics for control of wild oats: an economic evaluation. *Agricultural Systems*, **34**, 65–76.

Panetta, F. D. (1988). Factors determining seed persistence of *Chondrilla juncea* L. (skeleton weed) in southern Western Australia. *Australian Journal of Ecology*, **13**, 211–24.

Panetta, F. D. & Dodd, J. (1987). Bioclimatic prediction of the potential distribution of skeleton weed *Chondrilla juncea* L. in Western Australia. *Journal of the Australian Institute of Agricultural Science*, **53**, 11–16.

Panetta, F. D. & Mitchell, N. D. (1991). Homoclime analysis and the prediction of weediness. *Weed Research*, **31**, 273–84.

Panetta, F. D., Ridsdill-Smith, T. J., Barbetti, M. J. & Jones, R. A. C. (1993) Ecology of weed, invertebrate and disease pests of Australian sheep pastures. In *Pests of Pastures: Weed, Invertebrate and Disease Pests of Australian Sheep Pastures*, ed. E. S. Delfosse, pp. 87–114. CSIRO: Melbourne.

Parker, C. (1983). Factors influencing *Striga* seed germination and host–parasite specificity. *Proceedings of the Second International Workshop on Striga, 5–8 October 1981, IDRC/ICRISAT, Ouagadougou, Upper Volta*, pp. 31–8.

Parsons, W. T. & Cuthbertson, E. G. (1992). *Noxious Weeds of Australia*. Melbourne: Inkata Press.

Paterson, J. G., Boyd, W. J. R. & Goodchild, N. A. (1976). Vernalization and photoperiod requirement of naturalized *Avena fatua* and *A. barbata* Pott ex Link in Western Australia. *Journal of Applied Ecology*, **13**, 265–72.

Patterson, D. T. (1990). Effects of day and night temperature on vegetative growth of Texas Panicum (*Panicum texanum*). *Weed Science*, **38**, 365–73.

Patterson, D. T., Meyer, C. R., Flint, E. P. & Quimby, P. C. (1979). Temperature responses and potential distribution of itchgrass (*Rottboellia exultata*) in the United States. *Weed Science*, **27**, 77–82.

Paul, N. D. & Ayres, P. G. (1987a). Effects of rust infection of *Senecio vulgaris* on competition with lettuce. *Weed Research*, **27**, 431–41.

Paul, N. D. & Ayres, P. G. (1987b). Survival, growth and reproduction of groundsel (*Senecio vulgaris*) infected by rust (*Puccinia lagenophorae*) in the field during summer. *Journal of Ecology*, **75**, 61–71.

Peters, N. C. B. (1978). Factors influencing the emergence and competition of *Avena fatua* L. with spring barley. PhD thesis, University of Reading.

Peters, N. C. B. (1991). Seed dormancy and seedling emergence studies in *Avena fatua* L. *Weed Research*, **31**, 107–16.

Petzold, K. (1956). Combine-harvesting and weeds. *Journal of Agricultural Engineering Research*, **1**, 178–81.

Pielou, E. C. (1977). *Mathematical Ecology*. New York: Wiley.

Piggin, C. M. (1978). Dispersal of *Echium plantagineum* L. by sheep. *Weed Research*, **18**, 155–60.

Pijl, L. van der (1969). *Principles of Dispersal in Higher Plants*. Berlin: Springer Verlag.

Piper, T. J. (1990). Field trials on diclofop-methyl tolerant wild oats (*Avena fatua*). *Proceedings of the 9th Australian Weeds Conference*, pp. 211–15.

Plummer, G. L. & Keever, C. (1963). Autumnal daylight weather and camphor-weed dispersal in the Georgia piedmont region. *Botanical Gazette*, **124**, 283–9.

Pollard, F. & Cussans, G. W. (1976). The influence of tillage on the weed flora of four sites sown to successive crops of spring barley. *Proceedings of the 1976 British Crop Protection Conference – Weeds*, pp. 1019–28.

Poole, A. L. & Cairns, D. (1940). Botanical aspects of ragwort (*Senecio jacobaea* L.) control. *Bulletin of the New Zealand Department of Scientific and Industrial Research*, **82**, 1–66.

Porterfield, J. W. (1988). *Harvesting Equipment Should Address Weed Seed Problems. Agricultural Engineering*, Jan/Feb, 11.

Powles, S. B. & Howat, P. D. (1990). Herbicide-resistant weeds in Australia. *Weed Technology*, **4**, 178–85.

Powles, S. B. & Matthews, J. M. (1992). Multiple herbicide resistance in annual ryegrass *(Lolium rigidum)*: a driving force for the adoption of integrated weed management. In *Achievements and Developments in Combating Pesticide Resistance*, ed. I. Denholm, A. L. Devonshire & D. W. Hollomon, pp. 75–87. London: Elsevier.

Price, S. C., Allard, R. W., Hill, J. E. & Naylor, J. (1985). Associations between discrete genetic loci and genetic variability for herbicide reaction in plant populations. *Weed Science*, **33**, 650–3.

Priestley, D. A. (1986) *Seed Aging: Implications for Seed Storage and Persistence in the Soil*. Ithaca, NY: Cornell University Press.

Proctor, V. W. (1968). Long distance dispersal of seeds by retention in digestive tract of birds. *Science*, **160**, 321–2.

Purba, E., Preston, C. & Powles, S. B. (1993). Inheritance of bipyridyl herbicide resistance in *Arctotheca calendula* and *Hordeum leporinum*. *Theoretical and Applied Genetics*, **87**, 598–602.

Putwain, P. D. (1990). The resistance of plants to herbicides. In *Weed Control*

Handbook: Principles, 8th edn, ed. R. J. Hance & K. Holly, pp. 217–42. Oxford: Blackwell Scientific.

Putwain, P. D., Scott, K. R. & Holliday, R. J. (1982). The nature of resistance of triazine herbicides: case histories of phenology and population studies. In *Herbicide Resistance in Plants*, ed. H. M. LeBaron & J. Gressel, pp. 99–116. New York: Wiley.

Quinlivan, B. J. (1971). Seed coat impermeability in legumes. *Journal of the Australian Institute of Agricultural Science*, **37**, 283–95.

Rademacher, B., Koch, W. & Hurle, K. (1970). Changes in the weed flora as a result of continuous cropping of cereals and the annual use of the same weed control measures since 1956. *Proceedings of the 10th British Weed Control Conference*, pp. 1–6.

Radosevich, S. R. & Holt, J. S. (1984). *Weed Ecology: Implications for Vegetation Management*. New York: Wiley.

Rai, J. P. N. & Tripathi, R. S. (1983). Population regulation of *Galinsoga ciliata* and *G. parviflora*. *Weed Research*, **23**, 151–63.

Ralphs, M. H., Turner, D. L., Mickelsen, L. V., Evans, J. O. & Dewey, S. A. (1990). Herbicides for control of tall larkspur (*Delphinium barbeyi*). *Weed Science*, **38**, 573–7.

Rasmussen, J. (1992). Testing harrows for mechanical control of annual weeds in agricultural crops. *Weed Research*, **32**, 267–74.

Rauber, R. & Koch, W. (1975). Zur populations dynamik des flughafers (*Avena fatua* L.) unter dem aspekt der langfristigen befallsprognose. *Proceedings of the European Weed Research Society Symposium 'Status and Control of Grass Weeds in Europe'*, pp. 113–23.

Raybould, A. F. & Gray, A. J. (1993). Genetically modified crops and hybridization with wild relatives: a UK perspective. *Journal of Applied Ecology*, **30**, 199–219.

Reader, R. J. (1993). Control of seedling emergence by ground cover and seed predation in relation to seed size for some old-field species. *Journal of Ecology*, **81**, 169–75.

Richardson, J. M., Gealy, D. R. & Morrow, L. A. (1989). Influence of moisture deficits on the reproductive ability of downy brome (*Bromus tectorum*). *Weed Science*, **37**, 525–30.

Ridley, H. N. (1930). *The Dispersal of Plants Throughout the World*. Ashford, UK: Reeve.

Roberts, H. A. (1962). Studies on the weeds of vegetable crops. II. Effect of six years of cropping on the weed seeds in the soil. *Journal of Ecology*, **50**, 803–13.

Roberts, H. A. (1986). Seed persistence in soil and seasonal emergence in plant species from different habitats. *Journal of Applied Ecology*, **23**, 639–56.

Roberts, H. A. & Feast, P. M. (1972). Fate of seeds of some annual weeds in different depths of cultivated and undisturbed soil. *Weed Research*, **12**, 316–24.

Roberts, H. A. & Feast, P. M. (1973). Emergence and longevity of seeds of annual weeds in cultivated and disturbed soil. *Journal of Applied Ecology*, **10**, 133–43.

Roberts, H. A. & Lockett, P. M. (1978). Seed dormancy and periodicity of seedling emergence in *Veronica hederifolia* L. *Weed Research*, **18**, 41–8.

Roush, M. L., Radosevich, S. R. & Maxwell, B. D. (1990). Future outlook for herbicide-resistance research. *Weed Technology*, **4**, 208–14.

Rubin, B. (1991). Herbicide resistance in weeds and crops, progress and prospects. In *Herbicide Resistance in Weeds and Crops: Proceedings of the 19th Long Ashton Symposium, 1989*, ed. J. Caseley, G. W. Cussans & R. K. Atkin, pp. 387–414. Oxford: Butterworth Heinemann.

Ryan, G. F. (1970). Resistance of common groundsel to simazine and atrazine. *Weed Science*, **18**, 614–16.

Ryle, G. J. A. (1966). Physiological aspects of seed yield in grasses. In *The Growth of Cereals and Grasses*, ed. F. L. Milthorpe & J. D. Ivins, pp. 106–20. London: Butterworths.

Sagar, G. R. & Mortimer, A. M. (1976). An approach to the study of the population dynamics of plants with special reference to weeds. *Applied Biology*, **1**, 1–47.

Salisbury, E. J. (1942). *The Reproductive Capacity of Plants*. London: G.Bell

Salisbury, E. J. (1961). *Weeds and Aliens*. London: Collins.

Sarukhan, J. (1970). A study of the population dynamics of three *Ranunculus* species. *Proceedings of the 10th British Weed Control Conference 1970*, pp. 20–5.

Sarukhan, J. (1974). Studies on plant demography: *Ranunculus repens* L., *R.bulbosus* L. and *R.acris* L. II. Reproductive strategies and seed population dynamics. *Journal of Ecology*, **62**, 151–77.

Sarukhan, J. & Gadgil, M. (1974). Studies on plant demography: *Ranunculus repens* L., *R.bulbosus* L. and *R.acris* L. III. A mathematical model incorporating multiple modes of reproduction. *Journal of Ecology*, **62**, 921–36.

Satorre, E. H., Ghersa, C. M. & Pataro, A. M. (1985). Prediction of *Sorghum halepense* (L.) Pers. rhizome sprout emergence in relation to air temperature. *Weed Research*, **25**, 103–9.

Schippers, P., Borg, S. J. ter, Groenendael, J. M. van & Habekotte, B. (1993). What makes *Cyperus esculentus* (yellow nutsedge) an invasive species? – a spatial model approach. *Proceedings of the Brighton Crop Protection Conference – Weeds*, pp. 495–504.

Schuler, B. (1986). Taxonomie, ökologie und verbreitung von *Avena sterilis* L. und anderen *Avena*-arten unter besonderer berücksichtigung des getreidebaus im westlichen mittelmeergebeit. *PLITS*, **4**, 1–266.

Scott, K. R. & Putwain, P. D. (1981). Maternal inheritance of simazine resistance in a population of *Senecio vulgaris*. *Weed Research*, **21**, 137–40.

Selman, M. (1970). The population dynamics of *Avena fatua* (wild oats) in continuous spring barley; desirable frequency of spraying with tri-allate. *Proceedings of the 10th British Weed Control Conference*, pp. 1176–88.

Shaaltiel, Y., Chua, N. H., Gepstein, S. & Gressel, J. (1988). Dominant pleitropy controls enzymes co-segregating with paraquat resistance in *Conyza bonariensis*. *Theoretical Applied Genetics*, **75**, 850–6.

Sharkey, M. J., Davis, I. F. & Kenney, P. A. (1964). The effect of rate of stocking with sheep on the botanical composition of an annual pasture in southern Victoria. *Australian Journal of Experimental Agriculture and Animal Husbandry*, **4**, 34–8.

Sheldon, J. C. & Burrows, F. M. (1973). The dispersal effectiveness of the achene–pappus units of selected compositae in steady winds with convection. *New Phytologist*, **72**, 665–75.

Sheppard, A. W., Cullen, J. M., Aeschlimann, J-P., Sagliocco, J-L. & Vitou, J. (1989). The importance of insect herbivores relative to other limiting factors on weed population dynamics: a case study of *Carduus nutans*. *Proceedings*

*of the VII International Symposium on the Biological Control of Weeds,
Rome*, pp. 211–19.

Sibbesen, E., Andersen, C. E., Andersen, S. & Flensted-Jensen, M. (1985). Soil
movement in long term field experiments as a result of cultivations. 1. A
model for approximating soil movement in one dimension by repeated
tillage. *Experimental Agriculture*, **21**, 101–7.

Silvertown, J., Franco, M., Pisanty, I. & Mendoza, A. (1993). Comparative plant
demography – relative importance of life-cycle components to the finite rate
of increase in woody and herbaceous perennials. *Journal of Ecology*, **81**,
465–76.

Sindel, B. M. (1991). A review of the ecology and control of thistles in Australia.
Weed Research, **31**, 189–201.

Skellam, J. G. (1951). Random dispersal in theoretical populations. *Biometrika*,
38, 196–218.

Smith, A. G. (1970). The influence of Mesolithic and Neolithic man on British
vegetation: a discussion. In *Studies in the Vegetational History of the British
Isles*, ed. D. Walker & R. G. West, pp. 81–96. Cambridge: Cambridge
University Press.

Smith, L. M. & Kok, L. T. (1984). Dispersal of musk thistle (*Carduus nutans*)
seeds. *Weed Science*, **32**, 120–5.

Smith, M. C., Holt, J. & Webb, M. (1993). Population model of the parasitic
weed *Striga hermonthica* (Scrophulariaceae) to investigate the potential of
Smicronyx umbrinus (Coleoptera: Curculionidae) for biological control.
Crop Protection, **12**, 470–6.

Snaydon, R. W. (1980). Plant demography in agricultural systems. In
Demography and Evolution in Plant Populations, Botanical Monographs 15,
ed. O. T. Solbrig, pp. 131–60. Oxford: Blackwell Scientific.

Spitters, C. J. T. (1986). Weeds: population dynamics, germination and
competition. In *Simulation and Systems Management in Crop Protection*.
Wageningen, The Netherlands: Pudoc.

St. John-Sweeting, R. S. & Morris, K. A. (1990). Seed transmission through the
digestive tract of the horse. *Proceedings of the Ninth Australian Weeds
Conference*, pp. 137–9.

Stamp, N. E. (1989). Seed dispersal of four sympatric grassland annual species of
Erodium. *Journal of Ecology*, **77**, 1005–20.

Stanger, C. & Appleby, A. (1989). Italian ryegrass (*Lolium multiflorum*)
accessions tolerant to diclofop. *Weed Science*, **37**, 350–2.

Staniforth, D. W. & Wiese, A. F. (1985). Weed biology and its relationship to
weed control in limited-tillage systems. In *Weed Control in Limited Tillage
Systems*, ed. A. F. Wiese, pp. 15–25. Champaign, IL: Weed Science Society
of America.

Staniforth, R. J. & Cavers, P. B. (1977). The importance of cottontail rabbits in
the dispersal of *Polygonum* spp. *Journal of Applied Ecology*, **14**, 261–7.

Stoller, E. W. (1973). Effect of minimum soil temperature on differential
distribution of *Cyperus rotundus* and *C. esculentus* in the United States.
Weed Research, **13**, 209–17.

Story, J. M. (1984). Status of biological weed control in Montana. *Proceedings of
the VI International Symposium on Biological Control of Weeds, 19–25
August 1984, Vancouver, Canada*, pp. 837–42.

Sutton, J. (1988). On the community dynamics of annual plants. PhD thesis,
University of Liverpool.

Symonides, E., Silvertown, J. & Andreasen, V. (1986). Population cycles caused by overcompensating density-dependence in an annual plant. *Oecologia* (Berlin), **71**, 156–8.

Tabor, P. (1952). Cogongrass in Mobile County, Alabama. *Agronomy Journal*, **44**, 50.

Tardif, F. J. & Powles, S. B. (1993). Target site-based resistance to herbicides inhibiting Acetyl-CoA carboxylase. *Proceedings of the Brighton Crop Protection Conference – Weeds – 1993*, pp. 533–40.

Taylor, C. E. & Georghiou, G. P. (1978) Suppression of insecticide resistance by alteration of gene dominance and migration. *Journal of Economic Entomology*, **72**, 105–9.

Taylor, C. R. & Burt, O. R. (1984). Near-optimal management strategies for controlling wild oats in spring wheat. *American Journal of Agricultural Economics*, **66**, 50–60.

Templeton, A. R. & Levins, D. A. (1979). Evolutionary consequences of seed pools. *American Naturalist*, **114**, 232–49.

Templeton, G. E., TeBeest, D. O. & Smith, R. J. (1979). Biological weed control with mycoherbicides. *Annual Reviews of Phytopathology*, **17**, 301–10.

Thill, D. C., Mallory-Smith, C. A., Alcocer-Ruthling, M. & Schumacher, W. J. (1990). Sulfonylurea herbicide resistant weeds in North America. *Proceedings of the 9th Australian Weeds Conference*, pp. 194–5.

Thill, D. C., Mallory-Smith, C. A., Saari, L. L., Cotterman, J. C., Primiani, M. M. & Saladini, J. L. (1991). Sulfonylurea herbicide resistant weeds: discovery, distribution, biology, mechanism and management. In *Herbicide Resistance in Weeds and Crops: Proceedings of the 19th Long Ashton Symposium*, ed. J. C. Caseley, G. W. Cussans & R. K. Atkin, pp. 115–28. Oxford: Butterworth Heinemann.

Thomas, A. G., Gill, A. M., Moore, P. H. R. & Forcella, F. (1984). Drought feeding and the dispersal of weeds. *Journal of the Australian Institute of Agricultural Science*, **50**, 103–7.

Thrall, P. H., Pacala, S. W. & Silander, J. A. (1989). Oscillatory dynamics in populations of an annual weed species *Abutilon theophrasti*. *Journal of Ecology*, **77**, 1135–49.

Thurston, J. M. (1964). Weed studies in winter wheat. *Proceedings of the 7th British Weed Control Conference*, pp. 592–8.

Thurston, J. M. (1966). Survival of seeds of wild oats (*Avena fatua* and *A.ludoviciana*) and charlock (*Sinapis arvensis*) in soil under leys. *Weed Research*, **6**, 67–80.

Tottman, D. R. & Wilson, B. J. (1990). Weed control in small grain cereals. In *Weed Control Handbook: Principles*, 8th ed, eds. R. J. Hance & K. Holly, pp. 301–28. Oxford: Blackwell Scientific.

Tribe, D. E. (1949). Some seasonal observations on the grazing habits of sheep. *Empire Journal of Experimental Agriculture*, **17**, 105–15.

Tripathi, R. S. (1985). Population dynamics of a few exotic weeds in north-east India. In *Studies on Plant Demography: A Festschrift for John L. Harper*, ed. J.White, pp. 157–70. London: Academic Press.

Truscott, A. (1984). Control of *Spartina anglica* on the amenity beaches of Southport. In Spartina anglica *in Great Britain. Focus on Nature Conservation No. 5*, ed. J. P. Doody, pp. 64–9. Cambridge: Nature Conservancy Council.

Tucker, E. & Powles, S. A. (1988). Occurrence and distribution in south-eastern Australia of barley grass (*Hordeum glaucum* Steud) resistant to paraquat.

Plant Protection Quarterly, **3**, 19–21.

Tumbleson, M. E. & Kommedahl, T. (1961). Reproductive potential of *Cyperus esculentus* by tubers. *Weeds*, **9**, 646–53.

Turner, J. (1970). Post-neolithic disturbance of British vegetation. In *Studies in the Vegetational History of the British Isles*, ed. D. Walker & R. G. West, pp. 97–116. Cambridge: Cambridge University Press.

Ulf-Hansen, P. F. (1989). The dynamics of natural selection for herbicide resistance in grass weeds. PhD thesis, University of Liverpool.

Usher, M. B. (1973). *Biological Management and Conservation*. London: Chapman & Hall.

Wall, G. G., McBryle, G. & Mock, T. (1979). Perception of weeds as an agricultural hazard. *Ontario Geography*, **14**, 5–19.

Warnes, D. D. & Andersen, R. N. (1984). Decline of wild mustard (*Brassica kaber*) seeds in soil under various cultural and chemical practices. *Weed Science*, **32**, 214–17.

Watkinson, A. R. (1980). Density-dependence in single-species populations of plants. *Journal of Theoretical Biology*, **83**, 345–57.

Watkinson, A. R. (1981). Interference in pure and mixed populations of *Agrostemma githago*. *Journal of Applied Ecology*, **18**, 967–76.

Watkinson, A. R. (1985). Plant responses to crowding. In *Studies on Plant Demography: A Festscrift for John L. Harper*, ed. J. White, pp. 275–89. London: Academic Press.

Watkinson, A. R. & White, J. (1985). Some life-history consequences of modular construction in plants. *Philosophical Transactions of the Royal Society,of London*, B **313**, 31–51.

Watson, A. K. (1989). Current advances in bioherbicide research. *Proceedings of the Brighton Crop Protection Conference – Weeds – 1989*, pp. 987–96.

Watson, D. (1987). Aspects of the population ecology of *Senecio vulgaris* L. PhD Thesis, University of Liverpool.

Watson, D., Mortimer, A. M. & Putwain, P. D. (1987). The seed bank dynamics of triazine resistant and susceptible biotypes of *Senecio vulgaris* – implications for control strategies. *Proceedings of the British Crop Protection Conference – Weeds, 1987*, pp. 917–24.

Weaver, S. E. (1985). Geographic spread of *Datura stramonium* in association with soybeans and maize in Ontario, Canada. *Proceedings of the 1985 British Crop Protection Conference – Weeds*, pp. 403–10.

Weaver, S. E. (1991). Size-dependent economic thresholds for three broadleaf weed species in soybeans. *Weed Technology*, **5**, 674–9.

Weaver, S. E., Tan, C. S. & Brain, P. (1988). Effect of temperature and soil moisture on time of emergence of tomatoes and four weed species. *Canadian Journal of Plant Science*, **68**, 877–86.

Weaver, S.E. & Warwick, S.I. (1983). Competitive relationships between atrazine resistant and susceptible populations of *Amaranthus retroflexus* and *A. powellii* from southern Ontario. *New Phytologist*, **92**, 131–9.

Wells, M. J. (1974). *Nassella trichotoma* (Nees) Hack. in South Africa. *Proceedings of the 1st National Weeds Conference of South Africa*, pp. 125–37.

Westbrooks, R. G. & Cross, G. (1993). Serrated tussock (*Nassella trichotoma*) in the United States. *Weed Technology*, **7**, 525–9.

White, J. (1980). Demographic factors in populations of plants. In *Demography and Evolution in Plant Populations*, ed. O. T. Solbrig, pp. 21–48. Oxford: Blackwell Scientific.

Whybrew, J. E. (1964). The survival of wild oats (*Avena fatua*) under continuous spring barley growing. *Proceedings of the 7th British Weed Control Conference*, pp. 614–20.

Wicks, G. A., Burnside, O. C. & Fenster, C. R. (1971). Influence of soil type and depth of planting on downy brome seed. *Weed Science*, **19**, 82–6.

Williams, J. D. & Groves, R. H. (1980). The influence of temperature and photoperiod on growth and development of *Parthenium hysterophorus* L. *Weed Research*, **20**, 47–52.

Wilson, B. J. (1981). A review of the population dynamics of *Avena fatua* L. in cereals with special reference to work at the Weed Research Organization. *Proceedings of the Association of Applied Biologists Conference 'Grass Weeds in Cereals in the United Kingdom'*, pp. 5–14.

Wilson, B. J. & Brain, P. (1990). Weed monitoring on a whole farm – patchiness and the stability of distribution of *Alopecurus myosuroides* over a ten year period. *Proceedings of the EWRS Symposium 1990, 'Integrated Weed Management in Cereals'*, pp. 45–52.

Wilson, B. J., Cousens, R. & Cussans, G. W. (1984). Exercises in modelling populations of *Avena fatua* to aid strategic planning for the long term control of this weed in cereals. *Proceedings of the 7th International Symposium on Weed Biology, Ecology and Systematics*, pp. 287–94.

Wilson, B. J. & Cussans, G. W. (1975). A study of the population dynamics of *Avena fatua* L. as influenced by straw burning, seed shedding and cultivations. *Weed Research*, **15**, 249–58.

Wilson, B. J. & Cussans, G. W. (1978). The effects of herbicides, applied alone and in sequence, on the control of wild-oats (*Avena fatua*) and broad-leaved weeds, and on yield of winter wheat. *Annals of Applied Biology*, **89**, 459–66.

Wilson, B. J. & Phipps, P. A. (1985). A long term experiment on tillage, rotation and herbicide use for the control of *A. fatua* in cereals. *Proceedings of the 1985 British Crop Protection Conference – Weeds*, pp. 693–700.

Wilson, R. G. (1980). Dissemination of weed seeds by surface irrigation water in western Nebraska. *Weed Science*, **28**, 87–92.

Wilson, R. G., Kerr, E. D. & Nelson, L. A. (1985). Potential for using weed seed content in the soil to predict future weed problems. *Weed Science*, **33**, 171–5.

Wilson-Jones, K. (1952). Three experiments in witchweed control. *Empire Journal of Experimental Agriculture*, **20**, 98–102.

Wood, G. M. (1987). Animals for biological brush control. *Agronomy Journal*, **79**, 319–21.

Wookey, B. (1987). *Rushall: The Story of an Organic Farm*. Oxford: Basil Blackwell.

Wright, A. D. & Stegeman, D. A. (1990). The weevil, *Neochetina bruchi*, could help control water hyacinth in Australia. *Proceedings of the 9th Australian Weeds Conference*, pp. 508–10.

Wright, K. J. (1993). Weed seed production as affected by crop density and nitrogen application. *Proceedings of the Brighton Crop Protection Conference – Weeds – 1993*, pp. 275–80.

Yamasue, Y., Kamiyama, K., Hanoika, Y. & Kusanagi, T. (1992). Paraquat resistance and its inheritance in seed germination of the foliar-resistant biotypes of *Erigeron canadensis* L. and *E. sumatrensis* Retz. *Pesticide Biochemistry and Physiology*, **44**, 21–7.

Yoda, K., Kira, T., Ogawa, H. & Hozumi, K. (1963). Self-thinning in overcrowded pure stands under cultivated and natural conditions (Intraspecific competition among higher plants XI). *Journal of Biology, Osaka City University*, **14**, 107–29.

Young, J. A., Evans, R. A. & Eckert, R. E. (1969). Population dynamics of downy brome. *Weed Science*, **17**, 20–6.

Zemanek, J. (1976). The influence of annual applications of herbicides on the changes of weed communities on ploughland. *Agrochemia*, **16**, 73–6.

Zollinger, R. K. & Kells, J. J. (1991). Effect of soil pH, soil water, light intensity, and temperature on perennial sowthistle (*Sonchus arvensis* L.). *Weed Science*, **39**, 376–84.

Zorner, P. S., Zimdahl, R. L. & Schweizer, E. E. (1984). Sources of viable seed loss in buried dormant and non-dormant populations of wild oat (*Avena fatua* L.) seed in Colorado. *Weed Research*, **24**, 143–50.

Zweep, W. van der (1982). Golden words and wisdom about weeds. In *Biology and Ecology of Weeds*, ed. W. Holzner & N. Numata, pp. 61–9. The Hague: Junk.

Organism index

Subject index